Sustainable Textiles: Production, Processing, Manufacturing & Chemistry

Series Editor

Subramanian Senthilkannan Muthu, Chief Sustainability Officer, Green Story Inc., Tung Chung, Hong Kong

This series aims to address all issues related to sustainability through the lifecycles of textiles from manufacturing to consumer behavior through sustainable disposal. Potential topics include but are not limited to: Environmental Footprints of Textile manufacturing; Environmental Life Cycle Assessment of Textile production; Environmental impact models of Textiles and Clothing Supply Chain; Clothing Supply Chain Sustainability; Carbon, energy and water footprints of textile products and in the clothing manufacturing chain; Functional life and reusability of textile products; Biodegradable textile products and the assessment of biodegradability; Waste management in textile industry; Pollution abatement in textile sector; Recycled textile materials and the evaluation of recycling; Consumer behavior in Sustainable Textiles; Eco-design in Clothing & Apparels; Sustainable polymers & fibers in Textiles; Sustainable waste water treatments in Textile manufacturing; Sustainable Textile Chemicals in Textile manufacturing. Innovative fibres, processes, methods and technologies for Sustainable textiles; Development of sustainable, eco-friendly textile products and processes; Environmental standards for textile industry; Modelling of environmental impacts of textile products; Green Chemistry, clean technology and their applications to textiles and clothing sector; Eco-production of Apparels, Energy and Water Efficient textiles. Sustainable Smart textiles & polymers, Sustainable Nano fibers and Textiles; Sustainable Innovations in Textile Chemistry & Manufacturing; Circular Economy, Advances in Sustainable Textiles Manufacturing; Sustainable Luxury & Craftsmanship; Zero Waste Textiles.

Subramanian Senthilkannan Muthu
Editor

Sustainable Coloration of Textiles

Springer

Editor
Subramanian Senthilkannan Muthu
Flat B 20th Floor
Green Story, Canada
Caribbean Coast, Hong Kong

ISSN 2662-7108 ISSN 2662-7116 (electronic)
Sustainable Textiles: Production, Processing, Manufacturing & Chemistry
ISBN 978-3-031-91216-0 ISBN 978-3-031-91217-7 (eBook)
https://doi.org/10.1007/978-3-031-91217-7

This Springer imprint is published by the registered company Springer Nature Switzerland AG
The registered company address is: Gewerbestrasse 11, 6330 Cham, Switzerland

This book is dedicated to:
The lotus feet of my beloved Lord
Pazhaniandavar
My beloved late Father
My beloved Mother
My beloved Wife Karpagam and Daughters
Anu and Karthika
My beloved Brother
Last but not least
To everyone working in various industrial
spaces to make the world "SUSTAINABLE"

Contents

Introduction to the Book

Subramanian Senthilkannan Muthu

Abstract The textile industry has long been a cornerstone of global economic development, contributing significantly to employment and trade. However, its environmental footprint, particularly in the dyeing and finishing processes, has raised substantial concerns in recent years. With dyeing processes accounting for some of the largest water and chemical usage in the industry, the shift towards sustainable practices is no longer optional—it is a necessity. This book explores a wide range of innovative, eco-friendly approaches to textile dyeing, from the utilization of natural pigments and algae-based dyes to revolutionary technologies like foam dyeing and waterless methods. By emphasizing sustainability, this compilation aims to provide researchers, industry professionals, and policymakers with actionable insights to reduce the environmental impact of textile production while maintaining product quality and meeting consumer demand.

Keywords Sustainable textile dyeing · Eco-friendly practices · Natural pigments · Environmental impact · Water conservation · Innovative technologies · Circular economy · Greenhouse gas reduction

The art and science of textile dyeing have evolved over centuries, transforming from rudimentary techniques using plant-based colourants to highly sophisticated chemical processes. However, the environmental cost of this evolution has become impossible to ignore. Traditional dyeing methods, reliant on synthetic dyes and excessive water usage, have led to severe ecological degradation, including water pollution, toxic waste generation, and greenhouse gas emissions. Today, as the world grapples with climate change and dwindling natural resources, the textile industry stands at crossroads. It must transition to sustainable practices, embracing innovations that align economic growth with environmental preservation. This book seeks to bridge the gap between tradition and technology, offering a holistic perspective on sustainable dyeing practices that are paving the way for a greener future. This chapter

S. S. Muthu (✉)
Green Story Inc., Toronto, ON, Canada

S. S. Muthu (ed.), *Sustainable Coloration of Textiles*, Sustainable Textiles: Production, Processing, Manufacturing & Chemistry,
https://doi.org/10.1007/978-3-031-91217-7_1

serves as an introduction, providing an overview of all the chapters and their content in detail.

The second chapter deals about the evaluation of post-consumer textile waste management systems. It details about the Case of Take-Back Systems in Mexico City. Mexico City discards approximately 66,000 tons of post-consumer textiles annually, producing toxic leachates and greenhouse gases in landfills, which exacerbate climate change and its associated economic and social impacts. The textile waste problem in Mexico City is compounded by the lack of effective waste management systems, insufficient processing capacity, and inadequate regulations. These issues contribute to environmental degradation and pose significant challenges to sustainability efforts. Selective textile collection systems have emerged as a circular solution to address this issue. However, the efficiency and sustainability of these systems are not well understood. To shed light on this, authors conducted a comparative Life Cycle Assessment (LCA) to evaluate the environmental impacts of various management alternatives for these wastes, including reuse, recycling, and landfill disposal. This study also explored the main social impacts of these systems using qualitative social interest indicators. Here, authors show that textile recycling and reuse can cause a rebound effect, increasing environmental impacts if the use of virgin fibre textiles is not entirely replaced. This increase is due to the higher use of materials, electricity, and fossil fuels, which elevate environmental impact in all categories except climate change. Notably, landfill disposal presents the highest environmental impact when recycled and reused products fail to replace new materials effectively. Our findings highlight several deficiencies within Mexico City's textile waste management framework. These include limited processing capacity, weak regulatory enforcement, and a lack of incentives to promote optimal waste management practices. Additionally, there is a critical need for increased public awareness about textile consumption and disposal. Socially, raising public awareness is crucial, as recycling and reuse could paradoxically promote hyper consumption, challenging sustainability. To mitigate these impacts, robust policy measures are essential. These should include the development of comprehensive regulations that mandate recycling and reuse while limiting landfill disposal. Incentives for businesses to adopt sustainable practices, such as tax breaks or subsidies, could further enhance the effectiveness of these measures. Public education campaigns aimed at promoting sustainable consumption and disposal behaviours are vital. These results emphasize the need for consolidating existing management systems to mitigate the environmental and social impacts of textile waste effectively. Moving towards a fair and sustainable circular economy requires robust, complementary systems that address both environmental and societal challenges. By implementing these policies, Mexico City can significantly reduce the environmental footprint of its textile waste and promote a more sustainable future.

The third chapter deals with the eco-friendly dyeing of cotton fabrics with natural pigments. Nature offers a rich colour palette, which is used extensively for colouring various textile materials through dyeing, printing, and other techniques. Generally, colours are linked to human emotions and qualities, and such hues are obtained from various natural resources ranging from plants, animals, and minerals.

There are various studies employed by researchers to identify new colouring resources that enhance the properties of textile materials rather than colouration. This study focuses on several key aspects such as the exploration of dyeing cotton fabrics from natural dyes such as Eclipta alba (leaves), Marking Nut (seed), and Tamarind Bark (Barks) in response to the growing demand for environmental friendly products and aiming to achieve durable and vibrant colours. The cotton fabric underwent a pretreatment process using mordants like Alum and pomegranate rind aimed at improving dye absorption and ensuring uniform colouring. The dyeing parameters such as the temperature of the dye bath, the concentration of dyestuff and time were optimized to achieve a good colour fastness. This chapter provides a thorough understanding of the performance of sustainable natural dyes in textile applications by integrating the analysis of various properties such as fastness and antimicrobial properties. The findings are intended to support sustainable practices in the textile sector by aiding in the development of environmental friendly dyeing methods.

The fourth chapter is to unlock the sustainable textile colour: Phycocyanin Pigments for Eco-Friendly Textile Dyeing. Textile dyeing industries are identified as one of the most polluting industries in the world. The impact of the effluent on the water pollution is significant. The use of complex dyes and carcinogenic chemicals in the dyeing process limits the waste's capacity to decompose naturally and exacerbates its harmful impact on the environment. Using natural dyes from various sources is one of the sustainable ways of dyeing textiles. The main drawbacks of naturally coloured fabric are its dull colours and poor wash fastness. On the contrary, algae were capable of producing bright colours naturally. Hence, attempts were made to evaluate the potential possibilities of using algae-based pigments in textile dyeing. In this research, Phycocyanin extracted from *Spirulina platensis* algae was utilized for dyeing cellulosic fabric. The use of different mordants like cream of tartar, alum, tannic acid, and ferrous sulphate was evaluated for the application. The results of this study showed better colour fixation on textile fabric. Objective evaluation of colour value of dyed fabrics showed a stronger colour with tannic acid and ferrous sulphate combination. The research results also evaluated the durability of the algae pigment-coloured textiles using washing fastness, fastness to rubbing, perspiration, and dry cleaning. The results were promising but still required improvement compared to a commercial dye. The pollution analysis performed on the final effluent obtained from the dyeing process showed that all pollution parameters (like BOD, COD, pH, Colour, TDS, and TSS) were under the standards. Phycocyanin pigments offer a sustainable colouration of textiles, whereas future studies must focus on the economical part of the dye extraction process.

The fifth chapter deals around the critical analysis on the various dyeing techniques used for the colouration of textiles using natural dyes. Natural dyes are equipped with medicinal rich qualities, sustainable, comfortable to human skin and biodegradable. In the ancient times, the clothes were dyed with natural colourants only. Later the emergence of synthetic dyes, their availability and their consistency in shade reproducibility shifted the dyers towards synthetic dyes. Even though various colours can be produced using synthetic dyes, their effect on human skin is not skin-friendly and many people report skin allergy/ailments. Now people are very much conscious and expect ecofriendliness in whatever they use in their day to day activities and are paying more attention towards natural dyed clothes. Even though the colour fastness and the shade reproducibility are the challenging task in natural dyeing, efforts have been taken by the natural dyers and researchers to resolve these problems. In this paper, an attempt is made to list the various natural resources and the possible colours produced from them. A critical analysis was carried out pertaining to the various techniques used for the natural colouration of textiles using different types of mordants and the dyeing techniques employed without the use of mordants. Detailed discussion regarding the sustainability aspect, and the future of this exciting field was also carried out. The other specialty of this paper involves in the exploration of the research works carried out with regard to use of remnants generated in natural/ayurvedic dyeing process as organic fertilizer.

The sixth chapter explains about the sustainable colouration of cotton fabric using Mexican mint leaves. For many years, natural dyes have been utilized to dye fabrics, offering rich colours and serving as an environmentally friendly substitute for synthetic dyes. Derived from plants, minerals, and sometimes insects, these dyes present a diverse palette of colours, each with distinct attributes and qualities. Natural dyes are well-known for their aesthetic appeal and ecofriendliness. However, ensuring long-lasting colour retention demands attention to detail. Aspects like choosing the right mordant, controlling dye concentration, managing pH levels, and applying post-dye treatments greatly influence dye durability. Additionally, maintaining natural dyed fabrics involves gentle washing and protecting them from extended sunlight exposure to prolong their lifespan. Mordanting plays a vital role in the natural dyeing process by fixing the colour to the fibres and enhancing colour fastness. Mordants form a chemical bond between dye molecules and fibres, ensuring long-lasting and vibrant colours. Various mordants, such as alum, iron, copper, and tannin-rich substances like oak galls and myrobalan, can be utilized in natural dyeing. Each mordant type imparts unique effects on the final colour, so it's crucial to select the appropriate mordant for the desired result. Before mordanting, it's essential to thoroughly scour the fabric to remove any impurities and prepare the fibres to accept the dye evenly. Mordanting not only helps to improve colour fastness but also allows for a greater range of colours to be achieved from natural dyes. By experimenting with different mordants and dyeing techniques, rich and unique visual effects can be created. Mexican mint leaves also known as Indian Borage are known for their medicinal properties. When consumed it protects from cold and cough and cleanses the body and promotes digestion. It is said to have wound healing properties also. The leaves have a strong odour. From these leaves, natural dye can be extracted and used for dyeing the fabrics. In this chapter, cotton fabric was mordanted using three different mordants and dyed using Mexican mint leaves extract. The mordants used were myrobalan, ferrous sulphate, and copper sulphate. Pale yellow colour was obtained when myrobalan and copper sulphate were used as the mordant. Light brown was obtained from the ferrous sulphate mordanted sample. The colour fastness properties were better for the dyed sample which was mordanted using copper sulphate. A handkerchief was developed with the dyed sample which was mordanted using copper sulphate.

The seventh chapter delves into the fusion of chemistry and technology to enhance colour fastness and achieve the desired hues, emphasizing the innovative use of foam finishing techniques. Textile production requires many steps from the fibre to final product. Imparting colour through dyeing and printing is critical in achieving both functional and aesthetic requirements. The colouration process, specifically dyeing, has been recognized as using great amounts of water, energy, and chemicals during the processing. Greenhouse gasses are emitted during the process contributing to climate change, excessive water use is problematic in water stressed regions of the world, and the dyeing effluent, containing chemicals used in achieving the desired colour, is a major contributor to water pollution. There is increased attention to the environmental impact created during the dyeing process and legislation around the world, specifically in the EU, is being implemented to address these issues. The industry is examining processing technologies that result in cleaner production systems including reducing energy, water and chemical use. There have been various processes developed in addressing these issues including foam dyeing. Foam dyeing was first introduced in the late 1970s and recent innovations have improved the process. Foam dyeing uses significantly less water when compared to traditional dyeing methods. As less water is used, less energy is used in heating the dye bath and drying the fabric. A review of foam dyeing, the innovative changes that have been made to improve its performance, and results of research looking at the impact of dye class and the use of pretreatments on colour uniformity, colourfastness and reduction in processing time (which results in lower energy used and lower GHG emissions) are included. The study compared benchtop dyeing to scaled up processing of the fabrics and results are reported. Information on the optimum processing conditions (percent pretreatment used, dye class, processing variables) to achieve the desired performance (colour, colourfastness) was determined. Various statistical methods were used in the analysis including ANOVA, T-Tests, and Tukey HSD, to identify significant groups. Through testing, it was found that the dye type and level of pretreatment has an impact on some of the colourfastness and washfastness scores. Results also showed there was no statistical significance between the L*A*B readings, colourfastness, and washfastness scores between benchtop dyeing, and foam dyeing.

The eighth chapter explains the significance of organic dyes. In recent years, the increased use of synthetic dyes in textile, paper, paint, cosmetic, food, and pharmaceutical industries has caused to increase interest in the field of dye wastewater treatment for the global society. The degradation of synthetic dyes is very difficult and these dyes are generally stable to light, oxidizing, and heat owing to their complex aromatic structure. Furthermore, another problem about synthetic dyes should be noted that these dyes have toxic, mutagenic, or carcinogenic compounds; they will have a huge impact on aquatic ecosystem and human health. Therefore, in the use of synthetic dyes, effective purification of wastewater becomes very essential. Nowadays, many techniques are used to purify the wastewater such as membrane separation, ion exchange, flocculation, chemical oxidation, electrolysis, filtration, microbiological degradation, and photo catalysis degradation. As far as the textile industry is concerned, dyeing and finishing process is an essential step so as to gain aesthetic properties to textile materials. The dyeing of textile materials is carried out in an aqueous solution at high temperature. For this reason, the dyeing process of textile material demand high consumption of energy, water, dyes, and textile auxiliaries. Recently, the dyeing industry has been seeking for alternative process in order to reduce toxic effluents and environmental pollution. Therefore, the interest in natural dyes in textile industry has risen in as natural dyes are biodegradable, non-toxic, and eco-friendly. The demand for eco-friendly materials and textiles has seen a significant increase in recent years, leading to a greater emphasis on sustainable clothing. It is a well-known fact that textile industries worldwide have been contributing to environmental pollution. Consequently, it has become imperative for textile industries for a transition to sustainable dyes and materials to minimize their environmental impact and promote ecofriendliness. Textile industries rank second globally in terms of polluting clean water sources. The process of textile dyeing alone releases around 72 toxic chemicals into water, some of which are difficult to remove. The various processes in the textile industry, from dyeing to rinsing, consume vast amounts of water. Dyeing factories often find it easier to dispose of effluent water rather than treating and reusing it. This practice results in the disposal of gallons of wastewater by mills each year, which contains harmful chemicals such as formaldehyde, chlorine, lead, and mercury. These chemicals pose risks to both living beings and the environment. Many chemicals are used throughout the cotton-to-cloth transformation process, including those for dyeing and fixing dyes. The responsibility for manufacturing sustainable textiles does not solely rest with textile mills. Every organization involved in the process must support this cause. Brands must collaborate closely with manufacturing factories to adopt clean technologies and processes to prevent further environmental damage. Achieving the goal of manufacturing eco-friendly clothing using sustainable dyes requires collective efforts from all stakeholders in the textile industry's supply chain. In this study, the use of natural dyes in textile dyeing process was investigated. Furthermore, chemical structure, origins, dyeing methods, advantages, disadvantages, toxicological, dermatologic, and antimicrobial effects and future of natural dyeing were reviewed.

The ninth chapter deals with the Natural Dyes for Sustainable Colouration of Textiles. Colourants are substances used to impart specific colour to textiles, paints, ceramics, plastics, and so on. Two types of colourants are used for textile applications: dyes and pigments. Both dyes and pigments could be derived from natural sources. The limitations of the natural colourants are the availability of resources and the poor colour fastness of the coloured textiles. Traditionally, natural colourants have been derived from specific plant, animal, and mineral sources. Recent research studies have identified dyes from various waste sources which is a circular economical approach. Additionally, the waste or biomass is huge in quantity that ensures resource adequacy. This chapter discusses natural dyes and pigments from various sources and their application on textiles.

The tenth chapter focusses on the Natural Dyeing of Cotton Fibres by Radiation Technologies. Cotton, a cellulose-based fibre, is the most commonly used natural fibre in the textile industry. Its well-known chemical structure with numerous hydroxyl groups enable cotton fibre to be highly absorbent and easily dyeable with several types of synthetic dyes, where large quantities of dyestuff and auxiliaries consumed with resulting toxic effluent. The environment and health issues, arising from synthetic dyestuff utilization during the colouration process of textiles, recently proximite to natural dyes. Even though natural dyes stand out in their environmentally friendly characteristics, they have significant challenges of poor adsorption and poor fastness properties in comparison with synthetic dyes on cellulose-based textiles. Radiation technologies, such as plasma treatment, microwave technology, ultraviolet irradiation, ultrasound technology, and infrared radiation, have been proven to possess not only the potential to enhance the colour and fastness properties of cellulose-based textiles, but also the tendency to reduce required energy, water, and chemical consumption during process. In this regard, combining these technologies with natural dyestuff use would be a promising alternative method to the conventional textile dyeing process, paving the way for the transition to sustainable textile dyeing processes. In the literature, it is revealed that irradiation technologies have been used not only on the fabric surface but also in the powder and extract of natural dyestuff in cotton dyeing process. Several types of natural dyes, such as curcumin, harmala seeds, cocklebur, acacia catechu, acacia cyanophylla, marigold flower, haar singhar flower, green tea, coffee extract, pomegranate rinds, and insect-based dye, cochineal have been utilized to dye cotton fabrics with irradiation technologies. On the other hand, various process-related parameters are of great importance in the efficiency of radiation technologies in the natural dyeing of cotton. This chapter delves into the principles of radiation technologies for textile surface treatment, with particular attention to cotton fabric applications. Additionally, a comparative study on the effects of the proses-related parameters on colour and fastness properties, as well as other attributes such as antibacterial activity and UV protection.

The eleventh chapter explains about the Ancient (Pre-historic and historic) Indian and European dye pigments with the help of a case study. Textile and Fashion industry is considered to be one of the most environmentally polluting industries. There are many processes involved in making a fashion garment, out of which colouration has adverse effects on environment and human health. Fashion industry is dynamic and hence colouration is inevitable. Most of the recent developments concentrate for sustainable process and techniques in the area of dyeing and printing. India being a land of cultural heritage is well known for its natural dyeing and techniques and hence sustainability is not a new phenomenon. Pre-historic period of Bimbedka caves in India and Altamira caves in Spain are considered to be one of the oldest cave in the world as per archaeological findings. It is evident that the pigments used in Cave painting are derived naturally without causing any harm to the environment and has been eco-friendly and sustainable. The commonly found colours in the cave painting are red, black, white, and brown. During historic period, paintings such as Kalighat are done on fabric also called 'patas'. One of the major concerns in modern day natural dyeing is the fixation, whereas the cave paintings withstand thousands of years beside all climatic changes. This chapter will provide a detailed analysis on the different colour pigments used on caves and fabric from different natural sources like roots, flowers, leaves, flowers, fruits, tree barks, animal—tusks, horns, insects, and minerals. It addresses the composition of dye pigments used in India and Europe during historic and prehistoric period to understand the differences, insights on the challenges and drawbacks of using it in the present day dyeing process. This chapter also emphasis on the evolution of ancient dye pigments used by our ancestors and to apply the knowledge on current sustainable colouration process in fashion industry.

The twelfth chapter deals with the Optimisation of Knit Dyehouse. It provides us with the Pathways towards 'Zero' Water Requirements. The textile and apparel industry has been playing an important role in the development of Bangladesh's economy. Textile dyeing is considered the backward linkage of the apparel industry and is considered the most significant in terms of water, energy, and chemical consumption, reprocessing, and productivity. Thus, the optimisation of any dyehouse for textile processing has been a holy grail for many dyehouse managers. Textile dyeing can be divided into many types, such as continuous, semi-continuous, or batch methods. According to the product, it can also be classified as knit dyeing, woven dyeing, yarn dyeing, denim dyeing, and garment dyeing/washing. Amongst these, woven and denim dyeing is more technology-driven as it needs a huge machine setup with major investments. On the other hand, knit, yarn, and garment washing are much more skill-oriented. In this chapter, based on the extensive experiences in the Bangladesh knit dyeing industry, various practical optimisation techniques for dyeing are discussed, a path that will lead to the least amount aka 'Zero' water requirements.

The thirteenth chapter covers the Sustainable Innovations in Textile Colouration to address the environmental challenges. Natural textile colourants have historically been sourced from insects, plants, and animals, with their use dating back to 3500 BC. Until W.H. Perkins discovered synthetic dyes in 1856, and these natural dyes were used to colour cellulosic and protein-based fibres. Today, the textile industry poses a significant environmental threat by producing vast amounts of fabric and consuming large quantities of water annually. During the dyeing process, approximately 10–15% of the unfixed dye is released into water bodies as wastewater, significantly contributing to pollution. Currently, 90% of our clothes are dyed using synthetic dyes. Addressing the environmental challenges related to textile dyeing and treatment and exploring new sustainable technologies are crucial. Textile dyeing depletes freshwater resources and continuously pollutes the surface water. For decades, researchers have developed waterless and pollution-free dyeing technologies. These efforts have led to the creation of waterless and air dyeing methods and the use of bacteria for dyeing, which begins by autoclaving textiles to prevent contamination. These methods offer numerous advantages over conventional techniques such as reduced waste and greenhouse gas emissions, shorter operating times, and lower energy consumption. Additionally, they exhibit good diffusivity and are readily available, reusable, non-flammable, and non-toxic, making them both economically viable and environmentally friendly. This study summarizes the status of sustainable research and development in textile material colouration with efforts to address these pressing issues.

I take this opportunity to thank all the contributors for their earnest efforts to bring out this book successfully. I am sure readers of this book will find it very useful.

With best wishes,

Dr. Subramanian Senthilkannan Muthu

Evaluation of Post-consumer Textile Waste Management Systems: The Case of Take-Back Systems in Mexico City

Daniela Ximena Rodríguez-Hernández and Alejandro Padilla-Rivera

Abstract Mexico City discards approximately 66,000 tons of post-consumer textiles annually, producing toxic leachates and greenhouse gases in landfills, which exacerbate climate change and its associated economic and social impacts. The textile waste problem in Mexico City is compounded by the lack of effective waste management systems, insufficient processing capacity, and inadequate regulations. These issues contribute to environmental degradation and pose significant challenges to sustainability efforts. Selective textile collection systems have emerged as a circular solution to address this issue. However, the efficiency and sustainability of these systems are not well understood. To shed light on this, we conducted a comparative Life Cycle Assessment (LCA) to evaluate the environmental impacts of various management alternatives for these wastes, including reuse, recycling, and landfill disposal. The study also explored the main social impacts of these systems using qualitative social interest indicators.

Here, we show that textile recycling and reuse can cause a rebound effect, increasing environmental impacts if the use of virgin fiber textiles is not entirely replaced. This increase is due to the higher use of materials, electricity, and fossil fuels, which elevate environmental impact in all categories except climate change. Notably, landfill disposal presents the highest environmental impact when recycled and reused products fail to replace new materials effectively. Our findings highlight several deficiencies within Mexico City's textile waste management framework. These include limited processing capacity, weak regulatory enforcement, and a lack of incentives to promote optimal waste management practices. Additionally, there is a critical need for increased public awareness about textile consumption and disposal. Socially, raising public awareness is crucial, as recycling and reuse options could paradoxically promote hyperconsumption, challenging sustainability.

These results emphasize the need for consolidating existing management systems to mitigate the environmental and social impacts of textile waste effectively.

D. X. Rodríguez-Hernández · A. Padilla-Rivera (✉)
Instituto de Ingeniería, Universidad Autónoma de México, México City, Mexico
e-mail: apadillar@iingen.unam.mx

S. S. Muthu (ed.), *Sustainable Coloration of Textiles*, Sustainable Textiles: Production, Processing, Manufacturing & Chemistry,
https://doi.org/10.1007/978-3-031-91217-7_2

By implementing these policies, Mexico City can significantly reduce the environmental footprint of its textile waste and promote a more sustainable future.

Keywords Post-consumer textiles · Circular economy · Waste · Take-back system · Life cycle assessment · Social analysis · Sustainability · City

1 Introduction

The textile and clothing industry is essential, currently providing people with items used in their daily lives, such as clothing. The textile industry is valued at 1 trillion dollars, represents approximately 7% of global exports, and employs 35 million people worldwide (Desore & Narula, 2018). However, as much as the industry plays a significant role in the global economy, it is also one of the major drivers of pollution, environmental degradation, and labor injustice worldwide. It is estimated that the fashion industry annually produces between four and five million tons equivalent of carbon dioxide (CO_2 eq), which accounts for approximately 10% of global CO_2 emissions (Yan, 2021). Additionally, it has a significant impact on water bodies due to the amount of water it uses, the more than 15,000 chemicals used in the production process, and pollution from microplastics to the oceans (Yan, 2021)—and this is only related to its production and manufacturing. Nevertheless, fast fashion has led to an exponential increase in clothing consumption in recent years. Thus, the more it is consumed, the more it is discarded.

Textile waste is classified according to the stage of its life cycle when it becomes waste: (i) production waste, referring to fibers, threads, leftover fabrics, and samples generated in the manufacturing process at factories, which due to their homogeneity, have a greater potential to be recycled; (ii) pre-consumer waste, referring to unsold and damaged products from the retail sector; approximately 35% of the total production becomes pre-consumer waste; and (iii) post-consumer waste, which are garments or other domestic textile items discarded by their owners because they no longer want them (due to wear or damage, or because they are no longer fashionable) (Yalcin-Enis et al., 2019). In proportion, more post-consumer textile waste (65%) is produced than production and pre-consumer waste (35%) (Yalcin-Enis et al., 2019).

The problems associated with the premature disposal of textiles are environmental, social, and economic. The decomposition of textile fibers in landfills produces greenhouse gases (e.g., CO_2, methane, and sulfides) and acid leachates from all the chemicals and dyes used in their manufacture. Additionally, in the case of synthetic fibers, there is the problem of long decomposition periods, which means they occupy permanent space in landfills or illegal dumps. This is also part of the social and economic problem that generates the opening of new landfills for the final disposal of waste from cities. The odors, the effects on the health of surrounding communities, and the loss of soil value are impacts that usually affect vulnerable populations, with environmental burdens and health risks transferred to minority and low-income communities.

The issue of textile waste is not new but has been exacerbated by factors derived from the low prices for acquiring new garments, poor fabric or seam quality, and fast fashion as a driving force for increasingly frequent clothing purchases. There is a global consensus that textile waste is a growing problem that demands a higher level of attention to be remedied. Although this issue has been more studied and addressed for textiles in the Global North, it is a worldwide trend.

In Mexico, the annual production of textile waste nationwide increased from 537.6 thousand tons in 2008 to 642.4 thousand tons in 2017 (Instituto Nacional de Estadistica y Geografia INEGI, 2023); that is, the annual production increased by 104.8 tons in less than a decade. Therefore, as in other Latin American countries, the incorporation of textile waste into the public agenda is still emerging. While there are some isolated actions to divert post-consumer textiles from landfills, the government has not developed initiatives to promote the separation and recovery of textiles as is done with other recyclable inorganics such as paper and cardboard, plastic, aluminum, and glass. A study by the Metropolitan Autonomous University indicates that between 3.26% and 4.27% of the total urban solid waste that arrives at transfer stations are textiles (Vásquez-Morillas et al., 2020) to end their life cycle in landfills.

In Mexico, although textiles are classified in the regulations as recoverable waste, various factors (cultural, technological, and fabric composition) prevent proper management, so 95% end up in landfills or open-air garbage dumps. However, some initiatives are beginning to aim to reverse this trend. On July 23, 2024, the Secretary of Environment of Mexico City (Sedema) announced the intention to propose a regulation for the comprehensive management of textile waste that seeks to gradually reduce the generation of textile waste by extending the life span of these products through the implementation of strategies based on circular fashion, eco-design, productive chaining, creation of return points, and the promotion of circular markets (Secretaria de Medio Ambiente, Sedema, 2024). The problem studied here must consider a wide network of actors and factors so that as a City, progress can be made toward a socially just circular economy. Therefore, it is important to provide an initial overview of the current landscape, evaluating the sustainability of emerging circular alternatives and the predominant waste management system.

In this sense, this chapter focuses on providing a general overview of the impacts of post-consumer textile management in Mexico City. It offers a general view of the management of textile waste within the framework of the circular economy, including some strategies currently adopted worldwide, success stories, and the environmental and social benefits of their implementation. Later, the particular case of post-consumer textile management is addressed, presenting the usual end-of-life scenario and alternative strategies that have emerged from the private sector and civil society. In Sect. 3, a case study is presented in which a comparative assessment with environmental and social criteria was carried out between the Urban Solid Waste Management System of Mexico City and a selective collection system of textile waste for future use (take-back system), considering the latter an alternative designed under the framework of the circular economy; the methodological tool of Life Cycle Assessment is used for environmental analysis. Finally, under the presented urban context, a proposal for a future system for the comprehensive management of post-consumer textiles in the City is made, addressing the environmental and social problems examined throughout the chapter.

2 Foundations of the Circular Economy Contextualized with Post-consumer Textiles

Clothing is an essential element for people. Besides serving as a tool for protecting the human body against various environmental conditions, in community life, clothing communicates identity, status, values, beliefs, and aspirations of its wearer (Newell, 2015). Thus, while the increase in clothing consumption in recent decades is inherent to the importance of fashion in society, the business model of fast fashion, characterized by mass production, variety, and affordability of garments (Bukhari et al., 2018), has led to a state of permanent hyperconsumption.

Fast fashion follows the linear economy production model, which naturally leads to the wastage of large quantities of non-renewable resources, the shortening of clothing lifecycles, and the production of trash. The study and dissemination of the environmental, social, and economic impacts generated by the overproduction and wastage of textiles have led governments, consumers, and companies to start considering circular alternatives.

The logic adopted from the circular economy to prevent textile waste can be approached from two fundamental principles. The first is from the design phase, where the choice of fibers, timelessness, quality of manufacture, and versatility of garments aim to extend the usage period of the garments or increase the feasibility that at the end of their life, they join the recycling or reuse chains (Koszewska, 2018). On the other hand, strategies to prevent waste seek to extend the useful life of products through the recovery, repair, and remanufacturing of resources that are already in circulation but have reached the end of their first useful life (Koszewska, 2018). Selective collection is key to making this last set of strategies possible.

2.1 Existing Strategies in the World

As the environmental and social problem of textile waste grows, so has consumer concern and applicable regulations. This, in turn, has led to an increase in alternatives for disposing of post-consumer textiles. Beyond disposal as mixed household waste, textiles (specifically clothing) can be donated, exchanged, sold, repaired, redesigned, or recycled (Chen et al., 2021).

For example, in Mexico, the second-hand market has experienced no growth. According to Statista Consumer Insights, 25% of respondents stated that they had purchased second-hand clothing in the last 12 months (Statista, 2023). It was also calculated that approximately 9% of respondents' closets are second hand, but according to trends, this could increase to 18% by 2030. This indicates that there is a market for the strategy of selling second-hand clothing. Examples of those who do this are seen from the formality (e.g., Go Trendier, Closter, Levi's, Megapaca, ReChido, and many more) and from informality (swap markets, online bazaars, street market stalls, bales, etc.).

On the other hand, textile recycling is less conventional. In Mexico, only 1% of textile waste is recycled, and it mostly belongs to the production waste classification (Secretaria de Medio Ambiente y Recursos Naturales, SEMARNAT, 2017). Recycling is defined as "the reprocessing of textile waste before or after consumption for use in new textile or non-textile products" (Sandin & Peters, 2018) and is classified according to the type of process, the quality of the final product, and the circuit it follows. Table 1 provides the classifications of recycling routes described

Table 1 Recycling routes

Classification	Route	Definition	Example
By process	Mechanical	Involves shredding textile waste into short fibers, reducing its quality, before manufacturing new materials	Non-woven materials for fillings or sound insulation, products are of lower value than the original product
	Chemical	Involves the depolymerization of polymers. The monomers are repolymerized to create new textile fibers or products	Usually mixed with virgin fibers to achieve quality and durability of the material
	Thermal	Fibers are spun by fusion using the same process as the original thermoplastic fibers. This technology improves the performance of recycled textiles	Used for thermoplastics, such as polyester
By quality	Downcycling	Refers to when the value and quality of the recycled product are lower than the original products	Manufacturing of non-woven products such as insulating materials (thermal or sound)
	Upcycling	Refers to when the value and quality of the recycled product are higher than the original product	Production of 100% recycled textile fibers
By circuit	Open	Processes in which the material from one product is recycled and used in another product	Textiles are not created from textiles
	Closed	When the material from one product is recycled and used in a product more or less identical	Textiles are created from textiles

Source: Own elaboration

by Sandin and Peters (2018) in their review about the environmental impact of textile recycling and reuse.

One of the main complications of textile recycling is the difficulty of recycling fabrics with a mix of synthetic and natural fibers, as they require separation (Stanescu, 2021), and the heterogeneity of the post-consumer textiles represents a high technical challenge (Niinimäki et al., 2020). However, the development of new technologies for the processes of classification and recycling is emerging (e.g., see https://www.youtube.com/watch?v=yMS1CRPlN1w).

Take-back programs are an emerging initiative that was born from companies to manage their waste in a way that advances the demands of their consumers to take responsibility for the environmental impacts generated by the brand and to create value through discarded items. The success of this strategy depends on the joint effort of companies and consumers. This work will evaluate this disposal system and its environmental and social impact in comparison with the solid waste management system of the municipalities of Mexico City.

Locally, actions are carried out to manage used clothing to prevent it from becoming waste. There are second-hand resale models, clothing repair workshops, upcycling enterprises, home collection systems for textile waste, and collection in containers located in clothing stores. The various options are managed by various actors from civil society, the private sector, and the public sector who support each other. For example, there are collaboration networks between the private sector (private-private), such as the observed collaboration between the collection and classification company re.colecto and brands like Robert's and C&A (re.colecto.mx, 2023). On the other hand, there are also collaborations between the public and private sectors, such as the collaboration of re.colecto with the Municipal Environmental Directorate of the government of Guadalajara, in which the collecting company was responsible for providing ten collection points for the reception and future classification of textiles, to ultimately allocate the collected items for donation to people in vulnerable situations and for recycling.

However, initiatives driven by civil society aimed at extending the useful life of discarded garments through resale, rental, repair, redesign, or donation are isolated actions from the formal or informal sector with a scope intended for a certain sector of the population. In this sense, actions categorized in this category are considered isolated, that is, they do not belong to an integrated system.

2.2 Impacts

Muthu (2014) mentions that, from the perspective of life cycle assessment, we must differentiate two concepts: impact and benefit or "credit." All end-of-life treatments have a negative impact, but they can also generate benefits, and these must be measured. In the case of textiles, and specifically clothing or household textiles, their benefit is calculated assuming that the manufacture of new products with virgin fibers is avoided, so low substitution rates can eliminate the benefits (Sandin & Peters, 2018).

Numerous documents recognize that prevention is the best option in waste management; however, this requires a paradigm shift and collective efforts: changes from the industry to build new business models, from society to change its consumption habits, and from decision-makers to modify the legislation and regulations applicable to waste generation (Niinimäki et al., 2020). This challenge is complex, meanwhile, alternatives that follow the waste hierarchy have been evaluated. In a review of 41 articles related to the environmental impact assessment of reuse and recycling, Sandin and Peters (2018) found that the related literature affirms that these two alternatives are preferable to incineration or final disposal in landfills.

It has been calculated that the environmental benefits of reuse are greater than those of recycling (Sandin & Peters, 2018). It is estimated that for every kilogram of fabric that is reused, the emission of 15 kg of CO_2 eq is avoided (Muthu, 2014). For its part, Farrant (2008) calculated that the activities of collection, treatment, and transport have an insignificant impact compared to that produced by the consumption of a new garment. On the reuse of 100 t-shirts (60% cotton and 35% polyester), he estimated a reduction of 14% in global warming potential, 15% in the use of natural gas and oil, and 30% in the reduction of waste generated, among others. However, Zamani et al. (2017) conducted a study of a clothing rental system which is considered a mode of reuse and concluded that the emissions generated by the logistics and transportation of people to the store can diminish the environmental benefits of reuse (Zamani et al., 2017) coinciding with the conclusions of the research conducted by Fortuna and Diyamandoglu (2017) on the reuse of clothing and transportation distances.

In the case of reverse logistics or take-back systems, benefits such as customer satisfaction for feeling that they contribute to environmental care, savings in the consumption of raw materials or new products, reduction of risks of handling hazardous materials, environmental footprint (due to the reduction of GHG emissions, energy use, and water), and the flow of waste to landfills are expected (Roy et al., 2023). For example, Roy et al. (2023) modeled a take-back system to compare the environmental impacts of producing 1 kg of cotton fabric with different percentages of recycled cotton against the production of 1 kg of virgin cotton fabric. As a result, they obtained that their alternative scenario could reduce GHG emissions by between 2.2% and 8.6%, the water footprint by between 0.6% and 24.5%, air pollution by between 1.4% and 11.6%, and land use by between 3.1% and 25.5%. The reverse logistics modality has established itself as a corporate sustainability strategy that gained strength in European countries, in China, the USA (Vilariño & Carenzo, 2023) and is beginning to expand to other cities such as Mexico City.

On the other hand, authors such as Östlund et al. (2015) show that recycling processes carried out with fossil energies can increase the impact of the process and the final product in the category of climate change potential if compared with a garment made of cotton fiber, which does not have a major impact in this category. On energy use, Allwood et al. (2006) discussed how, although recycling can be a way to avoid cotton cultivation in other regions, it increases the impact due to energy use in the country that manages the waste.

The social benefits of textile reuse and recycling through reverse logistics or take-back systems have been little studied. In general, the most cited social repercussion of circular economy activities refers to job creation. It is expected that a shift toward repair, reuse, and recycling activities will bring benefits in terms of resources and create new job perspectives, often at the point of consumption, in line with the trend to adopt more local approaches (Clube, 2022).

In the case of the reuse and recycling of post-consumer textiles, Repp et al. (2021) identified that an advantage of carrying out these strategies locally is that it increases employment in the sector in situ and requires less intensive labor. For example, Zamani (2016) identified that the reuse and recycling industry of textiles generates jobs along its value chain in transportation, distribution, cleaning, repair, and transformation services. There are also operational and technical jobs. In the case of the collection and classification systems of textiles developed by for-profit companies, Persson and Hinton (2023) question how these business models support the transition to a socially just circular economy. In their comparison with non-profit foundations that promote sustainability in its three main pillars, take-back systems driven by private companies have a capitalist character that drives consumerism and environmental degradation, in addition to being a threat to the livelihood of foundations since they document that these have suffered a decrease in the quantity and quality of the products they receive since it is more profitable for consumers to dispose of their garments where they receive an economic reward (Persson and Hinton, 2023).

3 Textile Waste in Mexico

In Mexico, textiles account for 1.4% of the total urban solid waste, which amounts to 642,000 tons of textile waste per year (SEMARNAT, 2019), with approximately 66,000 tons coming from Mexico City (CDMX). However, these thousands of tons become trash as they are incorporated as urban solid waste ending up in landfills. In CDMX, although NOM-024-AMBT-2013 classifies clothing and textiles in the "recyclable inorganic" waste fraction, the percentage of recycled textile waste is between 4% and 5%, much lower than other waste in the same fraction, such as paper and cardboard (39%), plastic (30.7%), and glass (16.6%) (SEMARNAT, 2017).

To calculate the post-consumer textiles flow for this work, only waste from household collection in the city halls was considered (assuming these are related to clothing/bedding), using an average of the percentages estimated in the most recent composition studies, which is 3.61%. Thus, it is considered that the amount of post-consumer textiles generated daily is 290.80 tons.

3.1 *Urban Solid Waste Management System in Mexico City: End-of-Life in Landfills*

In Mexico City, the comprehensive management of household waste is carried out by what in this work we will call the Urban Solid Waste Management System in Mexico City. It is a system articulated by various government institutions, the private sector, and the informal sector. Through different phases, the Urban Solid Waste Management System in Mexico City is responsible for collecting, transferring, and selecting the City's household waste to finally direct it to its end-of-life in beneficial processes (e.g., composting, recycling, co-processing) or in landfills.

The final destination of the City's household waste is mostly in landfills. Due to the weak culture of separation from the source, 86.1% of the waste generated in CDMX reaches final disposal (SOBSE, 2021). The final disposal in landfills represents the most common end-of-life scenario for post-consumer textiles in Mexico City and the country.

Despite initiatives to encourage circularity mainly in clothing, in Mexico, on average, only 5% of textile waste is destined for recycling, and an even smaller percentage is allocated to used clothing sales, barter markets, repair workshops, or rental stores (SEMARNAT, 2017). The rest of the garment waste is disposed of as mixed urban solid waste and is collected through the Urban Solid Waste Management System in Mexico City.

In Mexico City, the RSU collection process (Fig. 1) consists of four phases described below:

Curbside Solid Waste Collection

- It is the process of receiving household RSU door-to-door or on the sidewalks of avenues and streets; generators do not need to move to dispose of their waste.

The Regulation of the Solid Waste Law states that garbage must be separated into organic, recyclable inorganic, non-recyclable inorganic, and special management waste. However, the lack of adequate infrastructure and environmental education limits the success of this measure. Thus, only 40% of the citizenry adequately separates their waste (SEDEMA, 2021). As a result, post-consumer textiles discarded tend to join the mixed waste fraction from this phase of the system, which compromises their recovery capability.

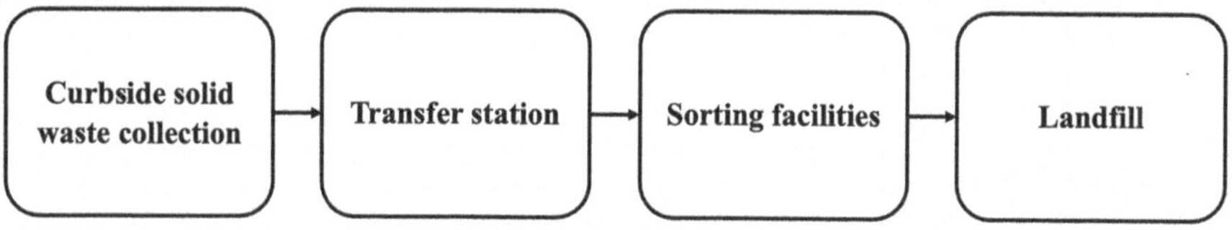

Fig. 1 Phases of the urban solid waste management system in Mexico City. (Source: Own elaboration)

Transfer Station

- The transfer station is a reception center for solid waste from curbside solid waste collectors to large-tonnage cargo vehicles (tractor-trailers) to send them to the next stage of their management: (i) organic waste: composting (16.1%); (ii) recyclable waste: sorting facilities (19%); (iii) limited use waste: compaction (7.6%); (iv) final disposal in landfills (57.3%). Textiles are usually directed to the sorting facilities or sent directly to final disposal.

Sorting Facilities

- If textiles are mixed within a load of waste with a large amount of mixed recyclables, they are directed to the sorting facilities. The main objective of the sorting facilities is to reduce the amount of waste that reaches final disposal in landfills and to save economic resources (Government of CDMX, 2022). At the sorting facilities, materials are sorted and separated based on their physical and chemical characteristics to be valued and reincorporated into production chains. Materials such as aluminum, PET containers, bags, paper, cardboard, multi-layer containers, and other metals are prioritized for recovery. Reports from the Ministry of the Environment and the Ministry of Works and Services do not report that textiles are recovered at the sorting facilities for recycling or resale, so this possibility is dismissed. The waste arrives at one of the four sorting facilities in the City in tractor-trailers and with auxiliary vehicles such as crawler excavators, forklifts, and front-end loaders on tires, and is deposited on conveyor belts for manual (one by one) or mechanical sorting of valuable waste. On average, 1.4% of the total materials that arrive at the facilities are recovered. Waste that is not recovered, termed "limited use," has the potential to be sent to compaction sites within the sorting facilities. Waste taken to this process is characterized by its high calorific power as it is compacted for later co-processing. Although textiles have these characteristics, it is not reported that they are used in this process.

Final Disposal

- This stage refers to the permanent deposit of waste in controlled sites to contain them in one place and thus prevent environmental and health impacts. Here, waste whose physical and chemical characteristics complicate their use arrives. Currently, there are 5 active landfills that receive waste from the City, none of which are within the City but are located in the State of Mexico (3), Morelos (1), and Puebla (1). That is, adverse effects on the environment and human health resulting from waste disposal, such as bad smells, aquifer contamination, explosions due to gas accumulation, land use, or disease transmission, are transferred outside the City, representing a significant case of socio-environmental injustice.

3.2 *The Valorization of Post-consumer Textiles: Take-Back Systems*

The increasing awareness and concern of consumers about the environmental crisis and the emergence of Extended Producer Responsibility (EPR) regulations (Hvass, 2014) have pressured brands to create concrete measures that demonstrate their commitment to the environmental issues associated with their industry (Choi et al., 2015). In this process, alternatives have been created from the design, selection of materials, use and choice of packaging, and management of post-consumer waste. Take-back programs emerged from this last item. In Mexico, as in other countries of the Global South, the government does not regulate the comprehensive management of post-consumer textiles (e.g., door-to-door collection, street collection containers, EPR, taxes, or incentives), so retail stores have leveraged their connection with consumers to promote the selective disposal of post-consumer textiles through take-back programs.

Take-back programs could be defined as a system of collection, classification, and redirection of clothing toward reuse or recycling alternatives (Corvellec & Stål, 2019); they may also include repair and sanitization. The success of this program largely depends on the relationship between the consumer and the brand (Hvass, 2014). In Mexico, take-back systems are an emerging innovation and involve various actors along the entire value chain; for their operation, they are associated with at least retailers, a professional collection and classification company, logistics services, recyclers, and second-hand sellers.

There are many examples of brands that have begun to implement take-back systems in their stores. For example, H&M started its used garment collection campaign in 2013, where consumers can bring garments (of any brand and in any condition) and in return, as a reward, receive a discount coupon on their purchases (Fig. 2). It was recorded that in its launch year, the program implemented in 54 countries collected 3047 tons of clothing (Shen, 2014). The items collected by a professional collection and classification company are redistributed or reprocessed to create commercial value (Shen, 2014). Some brands with a similar system in Mexico are C&A, Levi's, Robert's, and Pimperl.

The presence of companies specialized in the collection, classification, and redistribution of textiles is important as the recovery of the collected garments depends on their work and level of specialization (Nørup et al., 2019). In Mexico, there are two active companies and one that has permanently ceased activities (Table 2).

In contrast to the collecting companies in Mexico, there are non-profit foundations that align their values with environmental care and community support, such as Humana People to People in Spain. Humana collects, classifies, and recovers second-hand textiles and recirculates them through a 5-week system; its profits are allocated to social or environmental causes (Nørup et al., 2019).

In this sense, it has been questioned how business models for the recovery of post-consumer textiles support the transition to a socially just circular economy. For example, Persson and Hinton (2023) claim that take-back systems managed by private companies have a capitalist character that drives consumerism and environmental degradation by issuing coupons to buy new clothing. In addition, take-back

Fig. 2 Containers for the selective reception of textiles inside. (**a**) Container of startup re.colecto, located in the parking lot of a shopping mall, is the only collection company that has set up collection kiosks away from clothing stores in Mexico. On the walls of the container, they provide instructions on the textiles they receive and in what condition they should be delivered; (**b**) Container inside a retail clothing store, it is located in the checkout area and has a minimalist design with signage as shown in the image. (Source: Own elaboration)

systems of for-profit companies can be a threat to the survival of foundations since they document that these have suffered a decrease in the quantity and quality of the products they receive since it is more profitable for consumers to dispose of their garments where they receive an economic reward (Persson & Hinton, 2023).

Table 2 Examples of companies dedicated to the recovery and classification of textiles

Company	Collaboration	Description
Re.colecto	Collaborates with fast fashion sector companies and collects textiles gathered in their TB systems. Currently, it is the first company to enable a kiosk exclusively for textile collection in Mexico	A company that aims to close the lifecycle of garments and other textile products. It is based on strategies proposed from the circular economy. It has a system of collection, transportation, and classification. It associates with companies specialized in clothing recycling and second-hand stores. It is a for-profit company
RescatArte	Interacts directly with civil society. Its collection system is home-based (with an extra cost) or at its collection point	A project focused on promoting and disseminating sustainable themes and products in favor of the environment. It has a collection system and is responsible for selecting and classifying for resale or recycling through third parties. It is a for-profit company
Renovatex	Direct interaction with civil society through collection points	A non-profit company specialized in the collection of textile waste in the city of Cancún. Specialized in the collection of unwanted clothing, shoes, bedding, or industrial textiles. It sought the valorization of these articles for reuse and recycling. It had second-hand stores where people paid what they thought was fair for the clothing. It is permanently closed

Source: Own elaboration

4 Case Study

Objective To assess and compare the environmental and social impacts of two post-consumer textile waste management strategies in Mexico City, including reuse, recycling, and landfill disposal, using environmental life cycle assessment (LCA) and qualitative description of social aspects to inform future decisions for sustainable and circular management of these wastes in the city.

Methodology (LCA + Descriptive Social Analysis) A hybrid approach with a qualitative and quantitative method was used to assess the sustainability of textile waste management from an environmental and social perspective.

4.1 Comparative Life Cycle Assessment (LCA)

The methodology for assessing environmental impacts used in this study is the Life Cycle Assessment (LCA). LCA is a methodological tool that allows for the evaluation of potential impacts associated with a product or service through the collection of material and energy inputs and outputs throughout its lifecycle (Spathas, 2017). Its quantitative nature allows for the assessment, understanding, and monitoring of the negative impacts associated with the object of study.

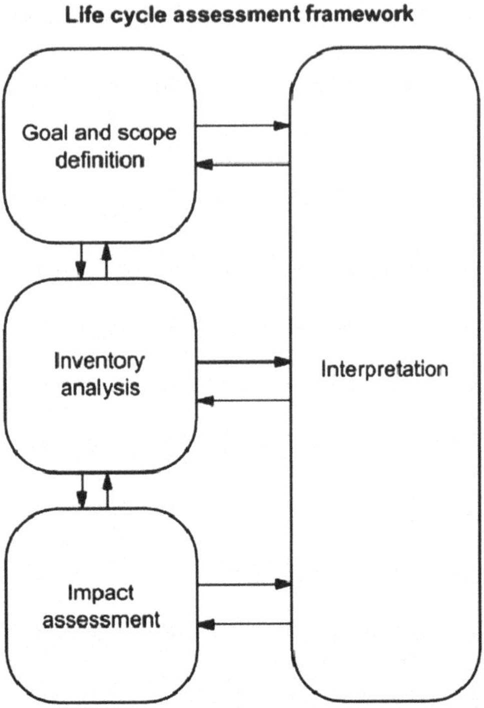

Fig. 3 Life cycle assessment stages according to ISO 14040:2006

LCA is an iterative process framed in four stages (Fig. 3), which are defined in ISO 14040:2006:

1. **Goal and Scope Definition:** Product or service to be evaluated is defined, and a functional unit for comparative assessment is established. System boundaries are defined at this stage to determine the processes included in the LCA.
2. **Life cycle Inventory Analysis:** Collection of input and output flows, referencing the functional unit.
3. **Life Cycle Impact Assessment:** Processing of the inventory in LCA software to assess potential environmental impacts. Impact categories and characterization models are selected at this phase, and the contribution of each life cycle process to the environmental impact of a product or service is identified.
4. **Life Cycle Interpretation:** Results of the assessment are evaluated and compared. General conclusions are formulated, recommendations proposed, and limitations described.

An LCA according to ISO 14044:2026 was conducted to evaluate the environmental impact of the post-consumer textile management systems. Two systems were analyzed: the Urban Solid Waste Management System in Mexico City and a take-back system in Mexico City. These scenarios were selected considering the most common final disposal method for post-consumer textiles in Mexico City, which is landfill, and compared with an emerging circular alternative employing

Table 3 Scenarios subject to evaluation

Scenario name	Processes	Avoided impacts
Functional unit Collection, transportation, classification, and end-of-life treatment of 1 ton of post-consumer textiles (100% cotton) produced in Mexico City		
Urban Solid Waste Management System of Mexico City	Household collection, transfer, selection, and final disposal in landfills	None
Take-Back System	Selective collection, classification, sanitization, repair, recycling, reuse, or landfill	Virgin raw material extraction. Premature final disposal. Garbage production

Source: Own elaboration

recycling and reuse methods, providing a representative image of post-consumer textiles management in Mexico City.

4.1.1 Goal and Scope

The study aimed to evaluate and compare the environmental impacts associated with the management of post-consumer textiles in Mexico City, using environmental life cycle assessment with an end-of-life approach. Two scenarios representing formally established and active post-consumer textiles management systems in the City were modeled and compared against each other (Table 3). The functional unit was defined through the service provided by a system in quantitative terms and allows for the comparison of different products with the same functionality (Zamani, 2014). Considering the system functions, the functional unit is the collection, transportation, classification, and end-of-life treatment of one ton of cotton fiber post-consumer textiles discarded in Mexico City.

The system boundaries assessed were the collection, transportation, distribution, classification, and final disposal are considered, including: (a) landfills; (b) recycling; and (c) reuse. The phases of production, distribution, sale, and use of the textiles collected are not included in this assessment (Figs. 4 and 5).

One of the limitations of this study is the lack of first-hand data specific to the evaluated systems, both in terms of energy and waste generation in the recycling process. As a result, secondary data were used, obtained through a bibliographic review of LCAs on textile end-of-life in other countries. This information was adapted to the specific context of Mexico, taking into account differences in energy sources and local recycling practices. Assumptions included the representativeness of data from other countries for the Mexican case and the homogenization of technical and operational parameters through the adaptation of available data. This approach may involve certain uncertainties, especially in the accuracy of estimates for the Mexican context.

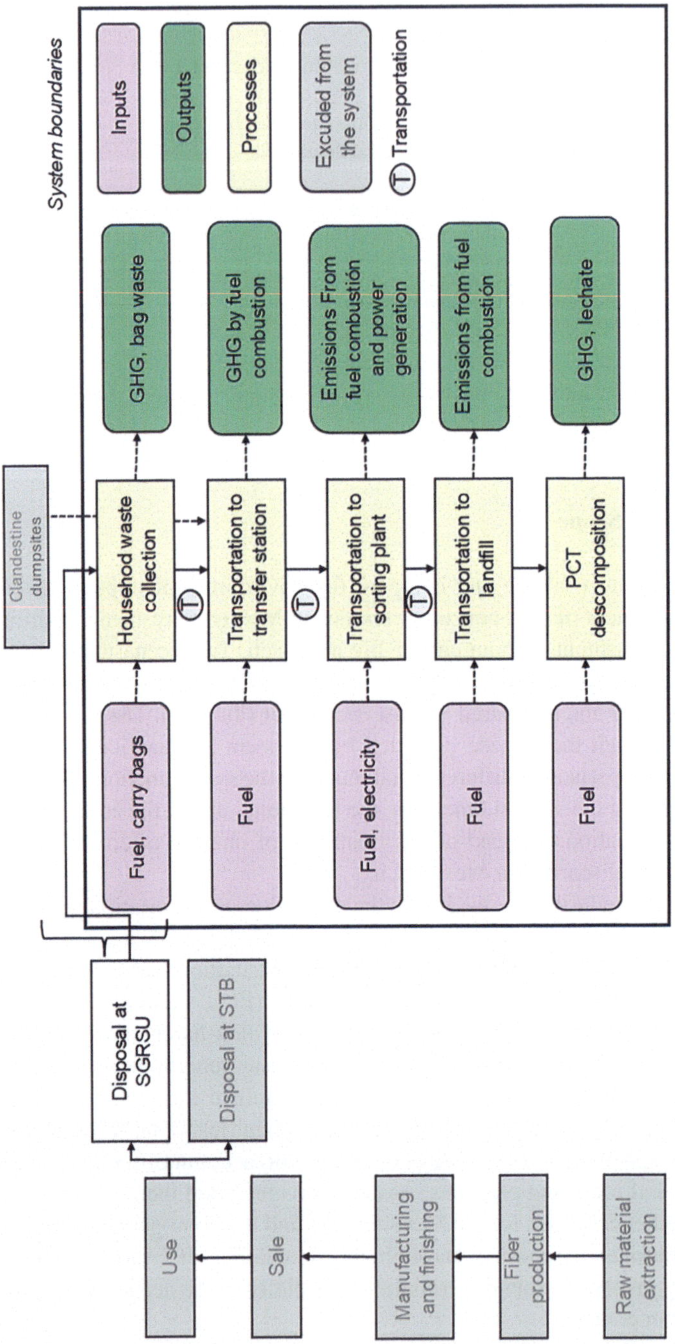

Fig. 4 BAU scenario, boundaries for the Urban Solid Waste Management System in Mexico City (SGRSU), indicating inputs and outputs. (Source: Own elaboration)

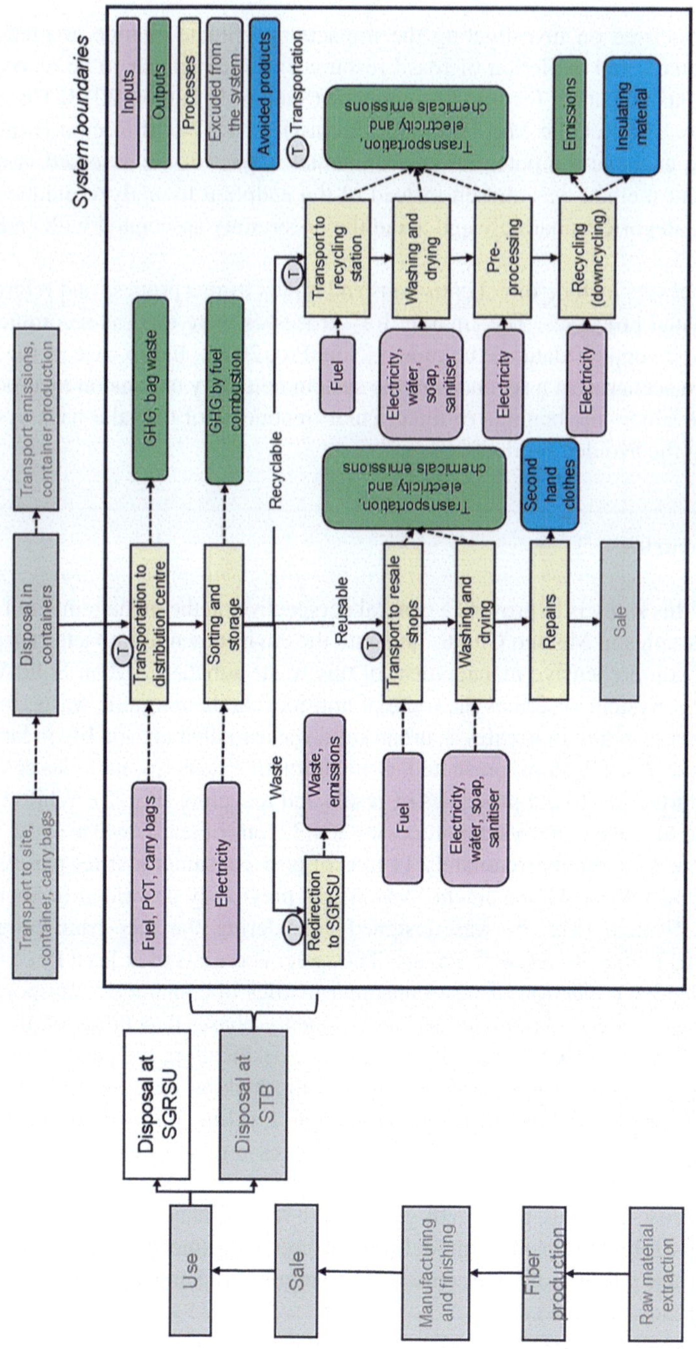

Fig. 5 Alternative scenario, boundaries for the take-back systems (STB), indicating inputs and outputs. (Source: Own elaboration)

4.1.2 Selection of Impact Categories and Assessment Method

The study focused on investigating the impacts on climate change, depletion of water resources, and depletion of fossil resources as they appear to be among the most relevant indicators for textile treatment (Sandin & Peters, 2018; Uncu Aki et al., 2020). The ReCiPe Mid Point (Hierarchist) V1.12/World Recipe H method was used to evaluate the potential environmental impact of the assessed systems. The midpoint method was chosen instead of the endpoint to analyze unique environmental categories separately and avoid the uncertainty associated with endpoint analysis.

Unit processes include direct emissions and inputs from a process and references to other related processes. The "market for" processes were used as recommended when specific supplier data are unknown (SimaPro, 2023). In the case of the take-back system scenario, in a second run, the system boundary expansion method was applied to evaluate the benefits of the circular treatments of the take-back systems considering the avoided products.

4.1.3 Scenarios

The aim of this study is to provide a general perspective on the management of post-consumer textiles in Mexico City to illustrate the environmental impacts caused by the lack of comprehensive management of this waste subclass. Urban Solid Waste Management System describes the route of non-recyclable inorganic waste, including from the collection of textiles as urban solid waste to their end-of-life in landfills (Fig. 4). Take-Back System, presented as an alternative scenario, includes selective collection of textiles to the processes of resale and recycling (Fig. 5). While Urban Solid Waste Management System processes 364.37 tons of textile waste daily, Take-Back System receives approximately 11 tons of post-consumer textiles per week.

Urban Solid Waste Management System was previously described in Sect. 3.1. Take-Back System (Fig. 5) was designed considering the take-back programs implemented in the city a few years ago. The system consists of at least three initial phases (selective collection of post-consumer textiles in containers; transportation to distribution centers; classification) and secondary phases depending on the classification of the item. In the case of reuse, after classification, the process involves transporting to resale stores, washing and drying, making necessary repairs, and selling. Conversely, when an item is classified for recycling, it involves transporting to recycling centers, washing and drying, preprocessing, and finally the recycling process.

For the reuse scenario, the sale of garments in second-hand stores is considered. These stores follow a business model that ensures consumers purchase the items and fulfill their purpose of being reused. The output of this management process is a garment ready for reuse.

A process of open-loop mechanical recycling is considered. The recycled fiber is used as input in other industries, such as the automotive industry. The products generated in this system are thermal and acoustic insulators for vehicles.

4.1.4 Inventory

Urban Solid Waste Management System

The environmental impacts of Urban Solid Waste Management System are associated with emissions generated by the transportation of waste and emissions during decomposition processes. The ton-kilometer (tkm) calculation was performed, and the vehicles using diesel and gasoline as fuel were identified. The average route distance (km) was calculated based on reports from the Secretary of the Environment of Mexico City in the Urban Solid Waste Inventory 2021 (Table 4).

During the transfer station phase, the contents of collector vehicles are emptied into larger capacity vehicles called tractor-trailers. This process involves the use of hoppers and gravity, hence no environmental loads are associated with it. However, emissions generated by the transfer of 250.14 t (85.8%) of post-consumer textiles passing through transfer stations to sorting facility and landfill sites must be considered. The tractor-trailers used are rigid trucks with a capacity between 26 and 31 tons (Gruasagudo, 2022), and diesel is the only fuel used (SEDEMA, 2022). The average distance traveled from transfer stations to sorting facility and landfill sites was calculated by creating a route in Google Maps from the city center to the destinations (Table 5).

During the sorting phase, the impact of transportation to the operational sorting facilities in the City and to the transfer station in the Azcapotzalco district, which has its own sorting center, was assessed. These facilities receive waste from transfer stations or directly from household collection (in the case of the Azcapotzalco facility). It has been estimated that post-consumer textiles arriving at the sorting facilities amount to a daily volume of 98 tons, of which 56.6 tons come from transfer stations, while the remainder arrives directly.

To manage the waste at the installations, auxiliary vehicles are used, which do not travel long distances, thus the calculation of the fuel used for their operation was made. According to McDougall et al. (2004), auxiliary loading vehicles use an average of 0.88 L/ton. The conveyor belts require 27 kWh/ton to operate (McDougall

Table 4 Collection stage inventory

Fuel type	Number of vehicles	(%)	Textile waste transported (t)	Average route distance (km)	(tkm)
Diesel	1458	87.2009569	0.872	30.49	26.59
Gasoline	214	12.79904	0.128	35	4.48

Source: Own elaboration based on data from *Inventario de Residuos Solidos de la Ciudad de México, SEDEMA, 2021*

Table 5 Transportation from transfer stations to landfills and sorting facilities

Landfills	Destination percent (%)	Textile waste transported (t)	Average km traveled per trip	(tkm)
Perseverancia (Cuautla)	4.034756917	0.03461821435	*114*	3.946476436
Naucalpan	7.180499612	0.06160868667	*37.3*	2.298004013
Bicentenario (Cuautitlan)	16.48095626	0.1414066047	*14.8*	2.09281775
Milagro (Ixtapaluca)	20.31055606	0.174264571	*51.1*	8.904919579
Cañada (Ixtapaluca)	20.37894174	0.1748513201	*49.4*	8.637655214
Sorting facility / CC	31.61428941	0.2712506031	*15.76*	4.274909505
Total				30.1547825

Source: Own elaboration based on data from, *SEDEMA*, 2022

et al., 2004). Finally, since textiles are not part of the recyclable waste recovered for valorization at this stage, it is assumed that 81% of the post-consumer textiles arriving at the sorting facilities are directed to their final destination in one of the five active landfills where the Urban Solid Waste (RSU) of Mexico City is deposited, the remaining 19% are remnants. The transport of 0.27 t of textiles to the various routes directing them to the RS involves 12.03 tkm transported in tractor-trailers.

Finally, in the decomposition process in landfills as final disposal sites, the calculation of biogenic biogas was performed considering methane and carbon dioxide emissions using the methodology proposed by the IPCC (2006). For this, activity data and specific model parameters were considered. The fraction of degradable organic carbon during the year of deposition is 0.40, considering that only natural fiber textiles contain organic carbon, for the case of cotton-polyester textile blends, it is 0.24 (Eggleston et al., 2006). The fraction of organic carbon that can decompose (DOCf) is 0.5 for natural fibers.

The methane generation index (k) was chosen considering the conditions of the three landfills with the greatest contribution, choosing a $k = 0.06$. Other parameters were chosen according to the model where OX = 0.10; MCF = 1, and $F = 0.5$ (Eggleston et al., 2006).

Cárdenas-Moreno (2014) calculated that for landfills, 23 L of leachate are produced per day. The leachate inventory was conducted according to the leachate composition presented in Table 6 (Güereca, 2006).

Take-Back System

Collection In this process, clothing is placed in selective collection containers for post-consumer textiles. This study assume that people typically travel about 1 km from their homes to the nearest shopping center to drop off clothes, this distance is considered negligible compared to other distances evaluated in this study.

Table 6 Leachate composition in landfill

Compound	(mg/L)
Aluminio	2.4
Amonio	210
Antimonio	0.066
Arsénico	0.014
Berilio	0.0048
Cadmio	0.014
Cloruro	590
Cromo	0.06
Cobre	0.054
Floruro	0.39
Fierro	95
Plomo	0.063
Mercurio	0.0006
Niquel	0.17
Zinc	0.68
AOX	2
DBO	1900
1,1,1-tricloroetano	0.086
1,2-dicloroetano	0.01
2,4-dicloroetano	0.13
Benzo (a) pireno	0.00025
Benceno	0.037
Clorobenceno	0.007
Cloroformo	0.029
Clorofenol	0.00051
Diclorometano	0.44
Endrin	0.00025
Etilbenceno	0.058
Hexaclorobenceno	0.0018
Isopropeno	0.076
PCB	0.00073
Pentaclorofenol	0.045
Fenol	0.38
Tetraclorometano	0.2
Tolueno	0.41
Toxaphene	0.001
Tricloroetano	0.043
Vinil cloruro	0.04
Dioxinas/furanos (ng/l)	0.32

Reprinted from "Desarrollo de una metodología para la valoración en Análisis de ciclo de vida aplicada ala Gestión integral de Residuos Municipales" by L. Güereca, 2006, *PhD Thesis*, Universidad Politécnica de Cataluña, p. 122. Retrieved from https://upcommons.upc.edu/bitstream/handle/2117/94157/01LPgh01de09.pdf?sequence=1&isAllowed=y

Consequently, the impacts associated with this process relate to the transportation of clothing from stores to the distribution centers and the use of carrying bags that contain the garments transported by users of the system.

An average distance of 26.39 km was calculated for the transportation of post-consumer textiles from flagship stores of the two major clothing brands using the take-back systems to the distribution centers.

The take-back systems require that garments must be packed in bags containing at least 3–5 items. However, the average weight contained in each bag is not specified. Moda.re, a Spanish company with a similar management system, estimated that one bag is used for every 4 kg of textiles. Considering that in Mexico, 10 out of every 14 carry-out bags provided at supermarkets are made of HDPE with an average weight of 7.5 g (Güereca et al., 2022), it is calculated that 1.875 kg of HDPE bags are used per ton of collected post-consumer textiles.

Transportation to Distribution Centers and Classification The impacts associated with this process relate to transportation. The average distance traveled from the destination distribution centers to the classification distribution centers is 276 km. The unit used to express the transportation of goods is tonne-kilometer (tkm), resulting in a calculated transport of 278.4 tkm. The dataset used for post-consumer textiles transportation was transport freight lorry 7.5–16 metric ton EUROS (GLO) running on diesel, with adjustments made for diesel use intensity (l/tkm) and engine emission contaminants.

From the classification distribution centers in Morelia, post-consumer textiles are transported according to their classification. This process utilized the transportation dataset, freight lorry, 3.5–7.5 metric ton EURO6, as the vehicles are recent models. The transportation distances are 111 km to the recycling facility and 0 km for reuse because the stores are located at the same site as the sorting center. The average distances were calculated using various possible routes.

Classification The classification process is manual, thus the electricity consumption of any automatic classification equipment, apart from the conveyor belts for garments, is not considered. According to Dahlbo et al. (2017), manual sorting with a conveyor belt advancing textiles consumes 1.27 kWh/t of textiles. From one ton of collected textiles, it is estimated that 42% is destined for resale in second-hand stores, 46% for mechanical recycling processes, and 2% for landfilling.

Preparation The inputs for washing and drying in the reuse and recycling processes include electricity, water, soap, and sanitizer for garments (Table 7). For preprocessing and repair, the electricity consumed by equipment used to remove buttons, zippers, and applications is considered, along with the amount of waste generated by this material filtering.

Table 7 Inputs in the reuse and recycling process for 1 ton of post-consumer textiles

Input	Quantity
Water (L/kg)	0.0196
Soap (kg/L)	0.0166
Washer electricity (MJ/kg)	1.26 × 1000
Dryer electricity (MJ/kg)	2.66
Sanitizer (L/kg)	0.006
Electricity (MJ)	1.77
Zippers, buttons, applications (waste) (kg)	0.366
Electricity recycling (MJ)	1.43
Textile waste (kg)	0.25

Adapted from *The Environmental Performance of High Value Recycling for the Fashion Industry*, by T. Spathas, 2017, *PhD Thesis*, Chalmers University of Technology, p. 24

Avoided Products

It is assumed that through reuse and recycling, an equivalent amount of products made from virgin materials is avoided. In the case of reuse, the avoided product is garments made from virgin fibers, and in the case of recycling (downcycling), it is high-density polyurethane used in automobiles and carpets.

The model for the production of textiles avoided due to reuse was calculated with 70% synthetic fiber (viscose) and 30% natural fiber (cotton), considering global fiber production statistics (Textile Exchange, 2021). The calculations used data from the production of these fibers available in the Ecoinvent database, which includes from raw material extraction to fabric weaving.

4.1.5 Life Cycle Impact Assessment

The SimaPro software version PhD 9.5.0.2 was used to model the management systems (https://www.simapro.mx/). The processes were modeled using specific data obtained from the Inventario de Residuos Solidos de la Ciudad de México (SEDEMA, 2022) as well as secondary data from bibliographic research and databases, primarily Ecoinvent 3.10.

4.1.6 Social Assessment

A mapping of the relevant social aspects for the sustainable operation of the studied systems was conducted, considering three groups of stakeholders. From this, a bibliographic review was carried out in scientific articles and gray literature to qualitatively describe 10 of the social impact categories relevant to the management

Table 8 Social indicators

Social impact category	Indicator	Definition	Associated SDG
Health and safety	Worker safety and protection	Workers equipped with personal safety gear at work	3 (good health and wellbeing)
Human rights	Child labor	Number of children working in the analyzed sector	8 (decent work and economic growth)
Employment quality (working conditions)	Fair wage	Worker's wage compared to minimum wage	
	Working/rest hours	Hours worked by employees per week	
	Employment benefits	Legal employment with benefits and job security	
Governance	Public commitment to sustainability issues	Actions taken with public funds related to waste management	11 (sustainable cities and communities)
Consumer/citizen satisfaction and participation	Consumer/citizen satisfaction	Satisfaction with the quality of service	
	Consumer/citizen participation	Citizens with access to the waste management system	
Value chain actor relationship	Transparency/corruption	Consumer awareness about the system	
		Presence of regular public reports by the organization	
Local development (socio-economic repercussions)	Development of environmental awareness and responsibility	% of users receiving environmental information from waste management	
	Local labor integration of formal workers from the informal sector	% of formal workers who were part of the informal sector	

Adapted from "Assessing the social performance of municipal solid waste management systems in developing countries: Proposal of indicators and a case study," by V. Ibáñez-Forés, M. D. Bovea, C. Coutinho-Nóbrega, & H. R. de Medeiros, 2019, *Ecological Indicators, 98*, p. 169 (https://doi.org/10.1016/j.ecolind.2018.10.031). Copyright 2019 by Elsevier

Table 9 Results by impact category of the evaluated treatments without considering the avoided products

Impact category	Unit	Landfill	Recycling	Reuse
Climate change	kg CO_2 eq	1231.704395	606.385942	303.368609
Depletion of water resources	m^3	0.357956939	11.1562399	10.5745667
Depletion of fossil fuels	kg oil eq	30.13073753	168.740603	97.480992

Source: Own elaboration

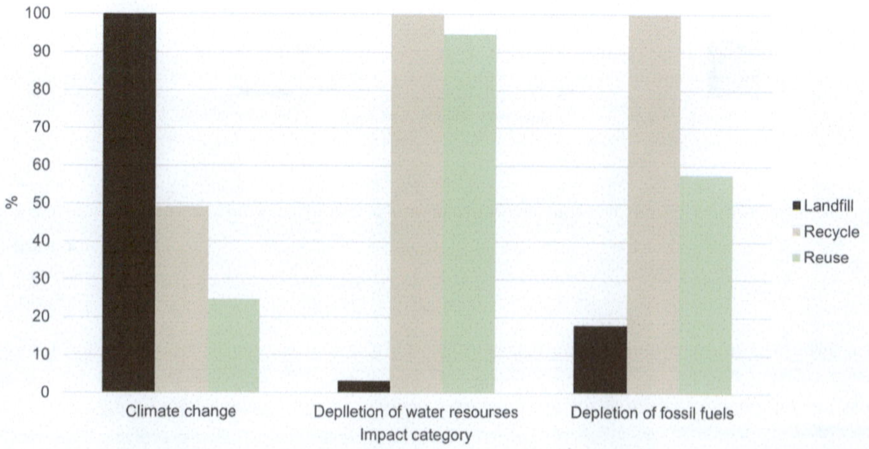

Fig. 6 Comparative graph of results for the three treatments without considering offsets for avoided products. (Source: Own elaboration)

Table 10 Impacts by category of the evaluated treatments considering avoided products

Impact category	Unit	Landfill	Recycling	Reuse
Climate change	kg CO_2 eq	1231.704395	−1860.8893	−8789.7653
Depletion of water resources	m^3	0.357956939	−48.173388	−619.20155
Depletion of fossil fuels	kg oil eq	30.13073753	−786.08946	−2032.4426

Source: Own elaboration

systems of solid urban waste in Latin America as proposed by Ibáñez-Forés et al. (2019) (Table 8).

5 Results

The results obtained for the disposal of 1 ton of post-consumer textiles without considering the avoided products are given in Table 9 (Fig. 6) and with avoided products given in Table 10 (Fig. 7).

In the analysis of emissions from climate change (Fig. 8), landfilling has the highest impact, generating 1231.7 kg CO_2 eq. The majority (98.36%) stems from the decomposition of textiles, while transportation and sorting operations contribute less than 2%. Methane emissions (45.8%) and biogenic methane (26.5%) are the main contributors, followed by carbon dioxide (27.6%).

For recycling, emissions amount to 606.38 kg CO_2 eq, where washing and drying are responsible for 63.47%, followed by cutting and recycling (23.3%). The electricity used in these processes is the major impact factor. Reuse generates 303.36 kg CO_2 eq, mainly attributable to washing and drying (85.8%).

Both treatments show significant impacts, yet the assessment does not include benefits derived from avoiding the production of new products. Compared to other

Fig. 7 Comparison of the environmental impacts generated by the treatment of 1 ton of post-consumer textiles with different methods considering offsets for avoided products. (Source: Own elaboration)

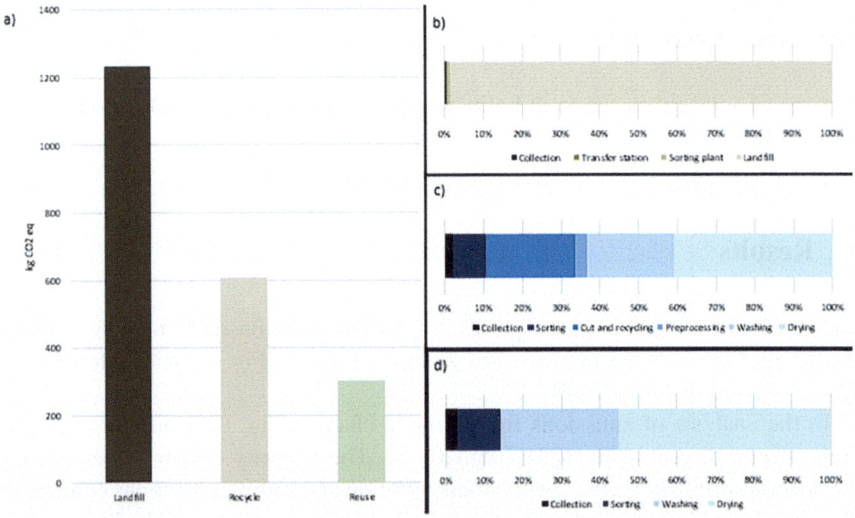

Fig. 8 Impacts and contribution by treatment on climate change. (**a**) Direct comparison between treatments; (**b**) impact contribution by landfill; (**c**) impact contribution by recycling; (**d**) impact contribution by reuse. (Source: Own elaboration)

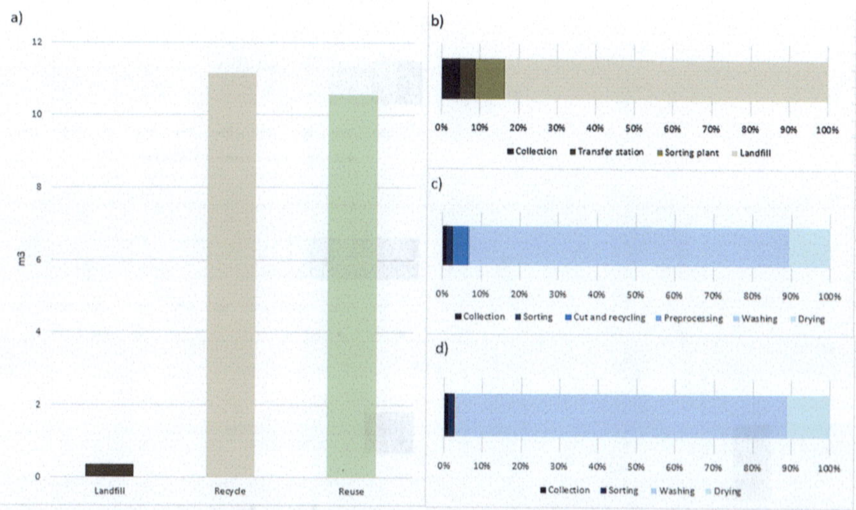

Fig. 9 Impacts and contribution by treatment on water depletion. (**a**) Direct comparison between treatments; (**b**) impact contribution by landfill; (**c**) impact contribution by recycling; (**d**) impact contribution by reuse. (Source: Own elaboration)

studies, emissions from these treatments are nearly ten times higher, highlighting the need for more sustainable strategies in post-consumer textile management.

In the case of water resource depletion (Fig. 9), recycling (11.15 m^3) and reuse (10.57 m^3) generate significantly higher impacts than landfilling (0.36 m^3), nearly 30 times higher. This is mainly due to the water used in washing garments, accounting for 82.2% and 86.1% of the impact in recycling and reuse, respectively. Drying contributes about 10%, with the rest of the processes adding only 1%. Approximately, 8.8 m^3 of water is used to wash a ton of garments. In landfilling, 83% of the impact originates from the final disposal, while activities like sorting, collection, and transfer contribute smaller percentages.

In the category of fossil fuel depletion (Fig. 10), recycling (168.74 kg crude oil eq) has an impact nearly ten times greater than landfilling (30.13 kg crude oil eq), while reuse reaches 97.48 kg crude oil eq. In recycling and reuse, washing and drying are the most energy-intensive processes, contributing 69.8% and 87.3% of the impact, respectively. Cutting and recycling contribute 17.2% in recycling, while sorting accounts for 9.1% in reuse.

The main factor for the impact of washing and drying is the energy derived from fossil fuels (95%), while detergents and disinfectants contribute 5%. In cutting and recycling, electricity accounts for 98.3% of the impact. In sorting and collection, diesel-powered transport is the major contributor. For landfilling, the decomposition of textiles generates 76.5% of the impact, followed by sorting (10.7%), collection

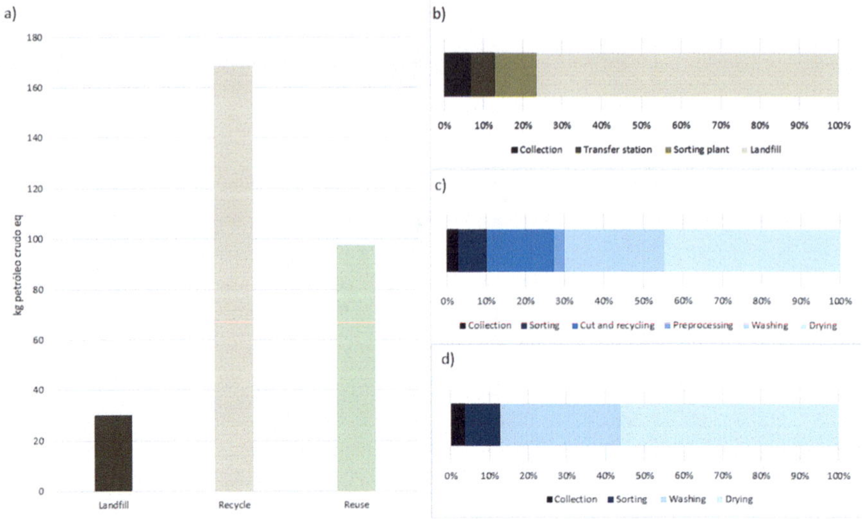

Fig. 10 Impacts and contribution by treatment on fossil fuel depletion. (**a**) Direct comparison between treatments; (**b**) impact contribution by landfill; (**c**) impact contribution by recycling; (**d**) impact contribution by reuse. (Source: Own elaboration)

(6.8%), and transfer (6%). This information highlights that intensive use of water and energy are key factors to address to reduce environmental impacts in textile waste management.

In the social aspect, important points stand out for each stakeholder group.

5.1 *Employees*

The social categories analyzed show key differences in the working conditions of employees in the urban solid waste management systems and take-back systems.

5.1.1 Child Labor

Although no employment of minors is reported in the formal urban solid waste management system (SEDEMA, 2022), records exist of child participation in informal activities related to waste separation and recycling at landfills (Azteca Noticias, 2021; Univisión Noticias, 2017), perpetuating significant risks to their health and well-being. In contrast, in the take-back systems, no minors were identified working directly, though their waste managed by municipal systems might indirectly involve this issue.

5.1.2 Working Conditions and Wages

In the urban solid waste management system, formal workers receive wages below the national average and the legal minimum wage, with notable gender inequalities (Secretaría de Economía, 2023). Volunteers involved in essential tasks lack regular income and benefits, accentuating their vulnerability (Guibrunet, 2019). Conversely, take-back systems employees have better earnings and access to basic benefits, although there is variability depending on the position and company (Indeed, 2023).

5.1.3 Safety and Labor Risks

Urban solid waste management system workers face multiple risks, such as exposure to hazardous materials and workplace accidents, with limited provision of personal protective equipment (PPE) Espinosa y Parra, 2020). Volunteers lack any protection, worsening their exposure (Guibrunet, 2019). In the take-back systems, although staff have PPE and better working conditions, risks related to handling poor-condition textiles include exposure to viruses, bacteria, fungi, or harmful fauna.

5.1.4 Labor Benefits

In the urban solid waste management system, formal workers have limited access to benefits like health care and vacations, while volunteers have no coverage at all (SEDEMA, 2022). In the take-back systems, benefits comply with Mexican labor legislation (Indeed, 2023), offering greater job security.

5.2 Consumers

5.2.1 Consumer/Citizen Satisfaction

Citizen participation and satisfaction with the textile waste collection systems are crucial. In Mexico City, although 75% of the citizens reported satisfaction with the general collection of solid waste (INEGI, 2023), the lack of textile separation limits their utilization. While municipalities do not manage textiles separately, the private sector leads this effort with collection programs (Wojnowska-Baryła et al., 2024) in stores like H&M and C&A, supplemented by tools like the Ecolana app, which helps locate collection points.

Although many citizens are willing to participate, barriers include the distance to collection points and lack of information. Deschamps et al. (2017) found in their exploratory study that in 2016, 78% of Mexicans felt there were not enough containers near their homes to properly dispose their post-consumer textiles. The same study showed that while 68% of the surveyed population mentioned using available recycling containers, more than half found it difficult or somewhat difficult to take their waste to the containers due to the distance and container conditions, affecting their participation in take-back programs. Incentives like discounts have increased textile turn-in, but interest seems more focused on the ease of disposing of clothing rather than personal benefits.

5.2.2 Participation and Awareness

The Mexican population frequently cleans out their closets, and the discarded clothing is generally donated to organizations, family, or friends. However, only half of the citizens express interest in having a place where their clothes can follow a closed circuit, and 59% would be willing to participate in collection points if it benefits the environment or vulnerable communities, while 39.8% would do so in exchange for economic incentives (Deschamps et al., 2017).

In Mexico City, there are no specific governmental programs for the collection of textiles. Although waste classification includes textiles, the majority of the population only separates into generic categories of organics and inorganics (AMAI, 2019). This results in about 4000 tons of textile waste ending up in landfills weekly, while take-back systems only process 11 tons in the same period. The main barriers to separation are lack of information, lack of infrastructure, and the time needed to transport (Deschamps et al.,2017).

Take-back systems, such as those implemented by C&A and H&M, have facilitated collection by being located in popular stores and offering incentives like coupons. For example, C&A collected 25.3 tons of clothing in 2023, showing a significant increase from the previous year (C&A, 2023). H&M reported a 40% growth in its global "Let's Close the Loop" program in 2019 (H&M, 2023). However, it has been observed that many consumers prioritize the convenience of disposing of their textiles over the economic benefits offered (Parcerisa, 2023).

Finally, although interest in acquiring second-hand clothing was limited in 2016 (Deschamps et al., 2017), perceptions have changed. Currently, second-hand clothing is considered a more sustainable, accessible, and unique option, which has expanded its market, especially among middle and upper classes (Arleo, 2023; Mondragon-Toledo & Morales-Tapia, 2022).

5.2.3 Development of Awareness and Environmental Responsibility

The Mexico City government has implemented programs that promote the culture of circularity, such as the Mercado de Trueque and Reciclatrón, focused on the separation and valorization of recyclable waste. However, textiles are not included in these efforts, despite their potential for utilization noted in regulations such as NADF-024-AMBT-2013 and PGIR. This contrasts with international initiatives, such as the Extended Producer Responsibility (REP) policy in France, which has successfully increased the collection of textiles through awareness campaigns and public funding (Bukhari et al., 2018).

In Mexico City, governmental campaigns focus on eliminating single-use plastics and promoting basic waste separation. Nevertheless, 65% of citizens are unaware of the final destination of their waste (AMAI, 2019), and there are no clear public options for the proper disposal of textiles. Authorities recommend individual actions, like upcycling or donations, but lack infrastructure for selective collection of textiles (Profeco, 2022).

On the other hand, retail companies have developed take-back programs that promote voluntary collection of used garments. Campaigns by brands like H&M, C&A, and Básicos de México highlight the benefits of textile recycling and offer incentives, such as discount coupons, although in some cases it is observed that consumers prioritize the convenience of disposing of their textiles over economic incentives (C&A, 2023; H&M, 2024). These initiatives aim to inform consumers about circular alternatives but often overlook the environmental impact of hyper-consumption and do not comprehensively address environmental awareness.

In conclusion, while the private sector has implemented strategies to encourage responsible disposal of textiles, the government has not developed specific actions for their management. This contrasts with international models, where more robust public policies have driven greater civic participation and valorization of textile waste.

5.2.4 Local Integration of Formal Employment from the Informal Sector

The transition to circular economy strategies, such as repair, reuse, and recycling, generates significant social benefits, particularly in the creation of local employment. These activities require less capital intensity and greater labor involvement, promoting job opportunities in sectors like transport, distribution, cleaning, repair, and transformation (Repp et al., 2021; Zamani, 2016). In the case of take-back systems, the subcontracting of specialized companies in collection, logistics, and recycling, such as Re.colecto, Marves, and Megapaca, fosters the development of small- and medium-sized enterprises, generating between 31 and 500 local jobs.

The success of these operations largely depends on the specific skills of the personnel, especially in manual textile sorting. Since composition labels are often lost or erased, proper identification of fibers and quality categories requires tactile experience and market knowledge, adjusting to fluctuations such as seasonality and fashion trends (Nørup et al., 2019; Botticello, 2012).

However, it has not been documented that take-back systems facilitate the formal integration of informal workers, as occurs in initiatives like PetStar, which trains PET collectors, improving their safety and income, though still without providing them full access to labor benefits. In Mexico, the PGIR 2015–2021 recognizes the role of informal recyclers and encourages their formalization through registrations and the creation of recycling cooperatives (SEDEMA, 2021). Thus, while reuse and recycling textile initiatives generate jobs and valorize local skills, the transition to formal labor inclusion of informal workers in these systems still presents challenges.

5.3 Government

5.3.1 Transparency

Transparency is measured by the availability and quality of information provided to consumers and other relevant actors. In the case of the urban solid waste management system, the Urban Solid Waste Inventory is published annually, accessible to the public, detailing its operation and waste flow. However, for textiles, specific data on generation, distribution, or final destination are not available, preventing a comprehensive analysis of their management. For instance, it is not known how much textile waste is generated per city hall or to which landfills it is directed.

Take-back systems include in sustainability reports data on the quantity of post-consumer textiles collected and general destinations (resale of second-hand clothing or recycling). However, they do not specify proportions, quantities of waste, or exact places of final disposition. A notable case was H&M in 2023, when an investigation by Aftonbladet revealed that clothing collected in Sweden was discarded in landfills in Nigeria, India, and Ghana, failing to fulfill commitments to extend the lifespan of garments (Lindberg, 2023). This lack of clarity can be interpreted as a deficiency in transparency and, potentially, as greenwashing, which undermines consumer trust.

5.3.2 Public Commitment to Sustainability Issues

Public commitment is reflected in the allocation of resources and specific actions to address the issue of textile waste. In Mexico City, no public funds have been allocated for the management of these wastes. However, in the state of Jalisco, pioneering initiatives have been implemented through the State Program for Integral Waste Management (SEMADET, 2023). These include permanent campaigns with fixed containers and collection centers, successfully collecting 6.8 tons of textiles in 2023 through collaboration between government, civil associations, and private initiative.

Companies operating take-back systems have not made public the figures regarding the economic resources allocated or the profits obtained from the management of textile waste. It should be noted that these initiatives are, to date, voluntary. While this may reinforce the positive perception of the brands, it is necessary to analyze whether these actions generate a positive environmental impact, as assumed.

6 Discussion

Textile waste management presents complex challenges from the perspectives of urban sustainability and the circular economy. On the one hand, the management of textile waste has critical implications for climate change, water management, and energy consumption. Landfills, while less demanding in water and energy, generate the highest greenhouse gas emissions due to the decomposition of textiles and methane release. On the other hand, alternatives such as recycling and reuse present significant environmental impacts, especially due to the intensive use of electricity and water in processes like washing and drying, which contribute to the depletion of water resources and fossil fuels. These figures considerably exceed those reported in other studies, indicating inefficiencies in the processes and a critical dependence on non-renewable energy sources.

From an urban sustainability perspective, it is fundamental to address these externalities through technological improvements such as optimizing washing and drying processes, integrating renewable energies, and water efficiency technologies; circular economy strategies that promote design for durability and recyclability, for example, according to Sandin and Peters (2018), the quality, choice of materials, and applications largely determine the viability of recycling. However, many brands do not align their design practices with recyclability requirements, limiting the potential impact of take-back programs by reducing the need for intensive washing and prolonging the product life cycle; and finally, the implementation of robust public policies that integrate textiles into municipal waste management programs, establishing infrastructure for their collection and valorization.

The social impacts of the management systems are unequal. In the Urban Solid Waste Management System, formal workers face low wages, gender inequality, and limited access to benefits, while volunteers work in vulnerable conditions, lacking any form of labor protection. In contrast, take-back systems offer better working conditions, though they do not comprehensively address the formalization of informal workers or the risks associated with handling poor-condition textiles.

The issue of environmental justice is also evident in the lack of transparency about the final destination of the collected garments varies considerably among brands, the programs that promise recycling or resale as a circular economy practice. In contexts where there are no strict regulations or clear standards for these systems, the post-consume textiles collected in take-back systems may often end up being exported to landfills in developing countries, exacerbating global inequalities and perpetuating pollution issues in vulnerable communities. That is, the problem is not solved but shifted.

In this sense, the circular economy, although powerful as a theoretical framework to operationalize sustainability, faces criticisms regarding its lack of concern for social justice and systemic transformation. Corvellec et al. (2022) criticize the empty appropriation of this concept, pointing out its frequent use in ecological modernization narratives based on capitalist growth. Similarly, Hobson and Lynch (2016) and Jaeger-Erben et al. (2021) highlight that many initiatives labeled as "circular" focus only on environmental aspects, neglecting essential social dimensions.

Take-back systems are not immune to these criticisms. While they aim to extend the value of products at the end of their life cycle, their implementation often prioritizes incentives that encourage consumption, such as discount coupons. Hvass (2014) notes that up to 87% of these coupons are redeemed in the European Union, reinforcing the mindset of discarding and consuming more, rather than promoting a cultural shift toward waste reduction.

It is considered that the lack of awareness and specific programs limits citizen participation in the proper management of textiles. While private companies lead initiatives such as voluntary collection programs, they are not aligned with broad public strategies, creating disconnection between individual actions and systemic goals. Effective governance should consider environmental education campaigns that help raise awareness among the population about the impact of textiles and the importance of reuse and recycling as strategies to extend the life cycle of garments, in addition to reducing consumption; accessible infrastructure where the expansion of collection points and access to responsible post-consumer textiles management programs are sought; design of policies that promote transparency and business commitment, avoiding greenwashing.

6.1 Proposal for Future Systems for the Proper Management of Post-consumer Textiles

The approach to developing an adequate post-consumer textile management system must be comprehensive, encompassing both formal management and informal efforts. The current post-consumer textiles collection and recycling systems, along with informal practices such as clothing bazaars, swap markets, buy-and-sell of used clothes in street markets, and businesses of repair and upcycling, are complementary but are currently not effectively articulated. This lack of integration prevents these efforts from adequately countering the negative effects of fast fashion and hyperconsumption, issues that perpetuate the large-scale generation of textile waste.

In a future post-consumer textile management system, separate collection of textiles is essential to prevent them from mixing with urban solid waste and thus preserve their quality. If proper separation is not carried out, textiles can be damaged by leachates or other liquids from the waste. Selective collection can be carried out through various strategies, such as the installation of containers in public thoroughfares, as seen in several countries in the global north, or through take-back systems, swaps, donations, repair workshops, and even governmental collection initiatives.

Once collected, textiles must be classified according to their quality and utility. The management of this classification may depend on the entity in charge of the collection. It is crucial to follow the waste hierarchy when allocating them, prioritizing practices like reuse, upcycling (creative recycling), renting, swapping, or resale, before recycling and, ultimately, disposal in landfills. This approach not only maximizes the value of textiles but also contributes to a more efficient and sustainable circular economy model. For this scheme to work, collaboration from governmental institutions, private enterprise, NGOs, and society will be required.

As highlighted earlier, environmental education is crucial for the success of any comprehensive recycling system. Currently, many people do not separate their waste properly due to a lack of knowledge, the perception that there are not enough incentives, the absence of ingrained habits, or even the belief that their efforts will have no impact since the waste will be mixed in collection processes. This lack of awareness and commitment to recycling represents a significant challenge to increasing recycling rates in the country.

To move toward an effective post-consumer textiles management system, it is essential to change attitudes and behaviors regarding waste separation from the home. This can be achieved by implementing strategies that have been successful in the past in promoting the recycling of other valuable materials, such as glass, paper, plastic, or aluminum. Examples of these strategies include governmental programs like the "Mercado de Trueque" or "Reciclatrón," which have encouraged the separation of recyclable waste in different cities in the country.

In addition to educational campaigns, partnerships with the private sector can play a key role in strengthening reward programs. An example of this is the initiative of the civil association Ecolana, which has promoted reward programs for users who participate in recycling activities, incentivizing the collection of recyclable waste, including textiles. Through collaboration between government and businesses, more robust incentive systems can be created that encourage active citizen participation in post-consumer textiles collection, in addition to ensuring transparency in the processes.

It is important to recognize that volunteers play an essential role in the amount of waste that is recovered for reuse and recycling. In many cases, informal collectors are responsible for a significant part of the collection and sorting of recyclable materials (Guibrunet, 2019). For this reason, it is essential to promote the formalization of this sector through the creation of financial incentives that can finance technology innovation projects in recycling or the installation of specialized recycling facilities for textiles. This would not only allow textiles to be considered a valuable waste but also create new job opportunities for informal collectors.

The implementation of a comprehensive system for post-consumer textiles management must address not only technical and logistical issues but also social, educational, and cultural aspects. It is imperative that all involved actors—governments, businesses, civil organizations, and the general public—work in a coordinated manner to ensure that efforts made are not isolated but are effectively complemented. Only through a collaborative approach and a transformation in consumption and recycling habits will it be possible to mitigate the impact of fast fashion and promote a circular economy that values post-consumer textiles as recoverable resources.

7 Conclusions

The environmental and social assessment of the management of post-consumer textiles post-consumer textiles shows that the environmental impact of landfills is greater compared to circular treatments, only if the latter effectively replace the use of products made with virgin raw materials. However, if the substitution is not realized, the Circular Treatment System (CTS) may present greater environmental impacts due to the high consumption of materials, electricity, and fossil fuels.

From a social perspective, both systems show strengths and weaknesses but lack transformative elements that drive a just transition toward a circular economy. Neither questions the current production and consumption model, reinforcing the status quo. This approach limits the commitment to environmental and social sustainability, prioritizing economic growth as a solution with a weak focus on equity and justice.

The active participation of users is crucial for the environmental benefits of recycling and reuse to materialize. Additionally, volunteer workers play a key role in the recovery and valorization of textiles, underscoring the need to dignify their work and facilitate their transition to formal employment.

To foster a more inclusive and effective system, public investment and the implementation of Extended Producer Responsibility schemes are required, increasing the diversion of textiles from landfills to sustainable alternatives. These schemes must integrate the participation of users, informal recyclers, and the private sector, promoting a model that prioritizes both environmental sustainability and social justice.

Acknowledgement This chapter is based on the master's thesis of Daniela Ximena Rodríguez Hernández, who completed the postgraduate program in Sustainability Sciences at the Universidad Nacional Autónoma de México. Acknowledgments are extended to the Life Cycle, Climate Change, and Sustainability Research Group (CIVICCS) and the Instituto de Ingeniería, UNAM, for providing the necessary software tools and databases to carry out this research. Likewise, we thank the Consejo Nacional de Humanidades, Ciencias y Tecnologías (CONAHCyT) for the scholarship granted to carry out this research (CVU: 1146384).

References

Allwood, J., Malvido de Rodríguez, C., Ellebæk Laursen, S., & Bocken, N. (2006). *Well dressed? The present and future sustainability of clothing and textiles in the United Kingdom.*

Asociación Mexicana de Agencias de Investigación de Mercado (AMAI). (2019). *Reciclaje en la CDMX*. 30–35.

Arleo, S. (2023). *Thanks, it's second-hand: Shifting values of second-hand thanks, it's second-hand: Shifting values of second-hand clothing in Mexico City* [Colby College]. https://digitalcommons.colby.edu/honorstheses/1419

Botticello, J. (2012). Between classification, objectification, and perception: Processing secondhand clothing for recycling and reuse. *TEXTILE, 10*(2), 164–183. https://doi.org/10.2752/175183512x13315695424356

Bukhari, M. A., Carrasco-Gallego, R., & Ponce-Cueto, E. (2018). Developing a national programme for textiles and clothing recovery. *Waste Management & Research: The Journal of the International Solid Wastes and Public Cleansing Association, ISWA, 36*(4), 321–331. https://doi.org/10.1177/0734242x18759190

C&A. (2023). Reporte de sustentabilidad 2023. http://cyafototeca.com/REPORTE_SUSTENTABILIDAD_2024_FINAL.pdf

Cárdenas-Moreno, P. R. (2014). Estimación de biogás y lixiviados generados en sitios de disposición final de residuos sólidos urbanos de Estado de México. Instituto Politécnico Nacional.

Chávez, D. (2023, julio). [Entrevistado por D. X. Rodríguez].

Chen, X., Memon, H. A., Wang, Y., Marriam, I., & Tebyetekerwa, M. (2021). Circular economy and sustainability of the clothing and textile industry. *Materials Circular Economy, 3*(1). https://doi.org/10.1007/s42824-021-00026-2

Choi, T.-M., Guo, S., Ho, S. S.-Y., & Li, W.-Y. (2015). Effects of used garment collection programs in fast-fashion brands. In *Sustainable fashion supply chain management* (pp. 183–197). Springer.

Clube, R. (2022). Is job creation a legitimate social benefit of the circular economy? *Resources, Conservation, and Recycling, 181*(106220), 106220. https://doi.org/10.1016/j.resconrec.2022.106220

Corvellec, H., & Stål, H. I. (2019). Qualification as corporate activism: How Swedish apparel retailers attach circular fashion qualities to take-back systems. *Scandinavian Journal of Management, 35*(3), 101046. https://doi.org/10.1016/j.scaman.2019.03.002

Corvellec, H., Stowell, A. F., & Johansson, N. (2022). Critiques of the circular economy. Journal of Industrial Ecology, 26(2), 421–432. https://doi.org/10.1111/jiec.13187

Dahlbo, H., Aalto, K., Eskelinen, H., & Salmenperä, H. (2017). Increasing textile circulation—Consequences and requirements. *Sustainable Production and Consumption, 9*, 44–57. https://doi.org/10.1016/j.spc.2016.06.005

Deschamps, T. C., Carnie, B., & Mao, N. (2017). Public consciousness and willingness to embrace ethical consumption of textile products in Mexico. *Textiles and Clothing Sustainability, 2*(1). https://doi.org/10.1186/s40689-016-0017-2

Desore, A., & Narula, S. A. (2018). An overview of corporate response towards sustainability issues in the textile industry. *Environment, Development & Sustainability., 20*(4), 1439–1459.

Dos Santos, P. S., & Campos, L. M. S. (2021). Practices for garment industry's post-consumer textile waste management in the circular economy context: An analysis on literature. *Brazilian Journal of Operations & Production Management, 18*(1), 1–17. https://doi.org/10.14488/bjopm.2021.004

Eggleston, S., Buendia, L., Miwa, K., Ngara, T., & Tanabe, K. (2006). *Directrices del IPCC de 2006 para los inventarios nacionales de gases de efecto invernadero.* https://www.ipcc-nggip.iges.or.jp/public/2006gl/spanish/pdf/0_Overview/V0_0_Cover.pdf

Farrant, L. (2008). *Environmental benefits from reusing clothes*. Technical University of Denmark.

Fortuna, L. M., & Diyamandoglu, V. (2017). Optimization of greenhouse gas emissions in second-hand consumer product recovery through reuse platforms. *Waste Management (New York, NY), 66*, 178–189. https://doi.org/10.1016/j.wasman.2017.04.032

Gruasagudo, P. (2022, mayo 5). *Cuántas toneladas carga un camión: Peso camión según modelo*. Agúdo. https://www.gruasagudo.es/cuantas-toneladas-carga-un-camion/

Güereca, P. (2006). *Desarrollo de una metodología para la valoración en Análisis de ciclo de vida aplicada ala Gestión integral de Residuos Municipales*. Universitat Politécnica de Catalunya.

Güereca, P., Padilla, A., García, X., Zamorano, S., & Rolon, I. (2022). *Análisis de los impactos ambientales de las bolsas de plástico de un solo uso en la Ciudad de México*. Instituto de Ingeniería, UNAM. https://www.youtube.com/watch?v=LGTDrkDnZ1Q

Guibrunet, L. (2019). What is "informal" in informal waste management? Insights from the case of waste collection in the Tepito neighbourhood, Mexico City. *Waste Management (New York, NY), 86*, 13–22. https://doi.org/10.1016/j.wasman.2019.01.021

Let's close the loop. (2024). H&M. https://www2.hm.com/es_mx/sustentabilidad-en-hm/our-work/close-the-loop.html

Hobson, K., & Lynch, N. (2016). Diversifying and de-growing the circular economy: Radical social transformation in a resource-scarce world. *Futures, 82*, 15–25. https://doi.org/10.1016/j.futures.2016.05.012

Hvass, K. (2014). Post-retail responsibility of garments–a fashion industry perspective. *Journal of Fashion Marketing and Management, 18*(4), 413–430. https://www.emerald.com/insight/content/doi/10.1108/jfmm-01-2013-0005/full/html

Ibáñez-Forés, V., Bovea, M. D., Coutinho-Nóbrega, C., & de Medeiros, H. R. (2019). Assessing the social performance of municipal solid waste management systems in developing countries: Proposal of indicators and a case study. *Ecological Indicators, 98*, 164–178. https://doi.org/10.1016/j.ecolind.2018.10.031

Indeed. (2023). *Sueldos en Industrias Marves en México*. Indeed. Recuperado el 15 de junio de 2024, de https://mx.indeed.com/cmp/Industrias-Marves/salaries

Instituto Nacional de Estadistica y Geografía. (2023). Sistema de Información Nacional para Gestión Integra de los Residuos. https://www.inegi.org.mx/temas/residuos/

Jaeger-Erben, M., Jensen, C., Hofmann, F., & Zwiers, J. (2021). There is no sustainable circular economy without a circular society. *Resources, Conservation, and Recycling, 168*(105476), 105476. https://doi.org/10.1016/j.resconrec.2021.105476

Juárez-Juárez, C. R. (2008). *Análisis de Ciclo de Vida del Sistema de Gestión de Residuos Sólidos de la Ciudad de México*. Instituto Tecnológico de Estudios Superiores de Monterrey.

Koszewska, M. (2018). Circular economy — Challenges for the textile and clothing industry. *AUTEX Research Journal, 18*(4), 337–347. https://doi.org/10.1515/aut-2018-0023

Lindberg, S. (2023, junio 13). Aftonbladet's investigation into H&M's recycling in 9 points. Aftonbladet. https://www.aftonbladet.se/nyheter/a/jlME1e/aftonbladet-investigation-into-h-m-srecycling-airtags-in-items

McDougall, F., White, P., Franke, M., & Hindle, P. (2004). Gestión Integral de Residuos Sólidos, inventario de ciclo de vida. *Procter & Gamble, 1*(1), 624.

Mena, M. (2023). *La ropa, el producto más popular en las compras de segunda mano en México*. Statista. https://es.statista.com/grafico/27425/articulos-usados-mas-comprados-por-los-mexicanos/

Moazzem, S., Wang, L., Daver, F., & Crossin, E. (2021). Environmental impact of discarded apparel landfilling and recycling. *Resources, Conservation, and Recycling, 166*(105338), 105338. https://doi.org/10.1016/j.resconrec.2020.105338

Mondragon-Toledo, B., & Morales-Tapia, D. (2022). From the Mexican "tianguis" to Instagram: Second-hand markets in Puebla-Tlaxcala. *JOMEC Journal*, (17), 100. https://doi.org/10.18573/jomec.226

Muthu, S. S. (2014). End-of-life management of textile products. In *Assessing the environmental impact of textiles and the clothing supply chain* (pp. 144–162). Elsevier, Amsterdam.

Newell, A. S. (2015). *Textile waste resource recovery: A case of study of New York State's textile recycling system*. Cornell University.

Niinimäki, K., Peters, G., Dahlbo, H., Perry, P., Rissanen, T., & Gwilt, A. (2020). The environmental price of fast fashion. *Nature Reviews. Earth & Environment, 1*(4), 189–200. https://doi.org/10.1038/s43017-020-0039-9

Nørup, N., Pihl, K., Damgaard, A., & Scheutz, C. (2019). Evaluation of a European textile sorting centre: Material flow analysis and life cycle inventory. Resources, Conservation, and Recycling, 143, 310–319. https://doi.org/10.1016/j.resconrec.2019.01.010

Azteca Noticias [@aztecanoticias]. (2021). #LosNiñosDeLaCanastal ¿Trabajo infantil o explotación? Youtube. Recuperado el 27 de noviembre de 2024, de https://www.youtube.com/watch?v=U_qZN3m_vYE

Univisión Noticias [@univisionnoticias]. (2017). Trabajo infantil y pagos de miseria en reciclaje para la gran industria refresquera. Youtube. Recuperado el 27 de noviembre de 2024, de https://www.youtube.com/watch?v=OE8R-nJcxVg

Östlund, Å., Wedin, H., Bolin, L., Berlin, J., Jönsson, C., Posner, S., Smuk, L., Eriksson, M., & Sandin, G. (2015). Textilåtervinning. *Tekniska möjligheter och utmaningar.*

Procuraduria Federal del Consumidor (PROFECO). (2022). Adiós a tu ropa, pero con responsabilidad. Procuraduria Federal del Consumidor (PROFECO). https://www.gob.mx/profeco/documentos/adios-a-tu-ropa-pero-con-responsabilidad?state=published

Persson, O., & Hinton, J. B. (2023). Second-hand clothing markets and a just circular economy? Exploring the role of business forms and profit. Journal of Cleaner Production, 390(136139), 136139. https://doi.org/10.1016/j.jclepro.2023.136139

Dale la vuelta. (2023). Re.colecto. https://recolecto.mx

Repp, L., Hekkert, M., & Kirchherr, J. (2021). Circular economy-induced global employment shifts in apparel value chains: Job reduction in apparel production activities, job growth in reuse and recycling activities. *Resources, Conservation, and Recycling, 171*(105621), 105621. https://doi.org/10.1016/j.resconrec.2021.105621

Roy, S., Chu, Y. Y. J., & Chopra, S. S. (2023). Life cycle environmental impact assessment of cotton recycling and the benefits of a take-Back system. *Resources, Conservation & Recycling Advances, 19*(200177), 200177. https://doi.org/10.1016/j.rcradv.2023.200177

Sandin, G., & Peters, G. M. (2018). Environmental impact of textile reuse and recycling—A review. *Journal of Cleaner Production, 184*, 353–365. https://doi.org/10.1016/j.jclepro.2018.02.266

Secretaría de Medio Ambiente. (2022). Plan de Gestión Integral de Residuos para la Ciudad de México 2021-2025. SEDEMA. https://pgir.cdmx.gob.mx/#/

Secretaría de Medio Ambiente. (2022). Inventario de Residuos Solidos de la Ciudad de México, 2021. SEDEMA. https://www.sedema.cdmx.gob.mx/storage/app/media/DGCPCA/residuos/InventariodeResiduosSolidos2021.pdf

Secretaría de Medio Ambiente y Desarrollo Territorial. (2023). *Realiza Guadalajara campaña de acopio de textiles.* https://guadalajara.gob.mx/gdlWeb/#/detalle/2019/Realiza-Guadalajara-campaa-de-acopio-de-textiles-

Secretaría de Medio Ambiente y Recursos Naturales. (2017). Informe de la situación del medio ambiente en México compendio de estadísticas ambientales. Indicadores clave de desempeño ambiental y de crecimiento verde. Edición 2017. SEMARNAT. https://apps1.semarnat.gob.mx:8443/dgeia/informe15/tema/pdf/Informe15_completo.pdf

Secretaría de Medio Ambiente y Recursos Naturales. (2019). Informe de la Situación del Medio Ambiente en México. Capítulo 7: Residuos Sólidos Urbanos. SEMARNAT. https://apps1.semarnat.gob.mx:8443/dgeia/informe15/tema/pdf/Cap7_Residuos.pdf

Secretaría del Medio Ambiente. (2024). Alista Sedema proyecto de Norma Ambiental para el Manejo Integral de los Residuos Textiles. Sedema. https://sedema.cdmx.gob.mx/comunicacion/nota/alista-sedema-proyecto-de-norma-ambiental-para-el-manejo-integral-de-los-residuos-textiles

Secretaría de Economía (2023). Recolectores de Desechos, Material Reciclable y otros Materiales. Data México. https://www.economia.gob.mx/datamexico/es/profile/occupation/recolectores-de-desechos-material-reciclable-y-otros-materiales

Shen, B. (2014). Sustainable fashion supply chain: Lessons from H&M. Sustainability, 6(9), 6236–6249. https://doi.org/10.3390/su6096236

Spathas, T. (2017). *The environmental performance of high value recycling for the fashion industry.* Chalmers University of Technology.

Stanescu, M. D. (2021). State of the art of post-consumer textile waste upcycling to reach the zero waste milestone. *Environmental Science and Pollution Research International, 28*(12), 14253–14270. https://doi.org/10.1007/s11356-021-12416-9

Textile Exchange (2021). Preferred Fiber & Materials, Market Report 2021. https://textileexchange.org/app/uploads/2021/08/Textile-Exchange_Preferred-Fiber-and-Materials-Market-Report_2021.pdf

Uncu Aki, S., Candan, C., Nergis, B., & Sebla Önder, N. (2020). Understanding denim recycling: A quantitative study with lifecycle assessment methodology. In *Waste in textile and leather sectors*. IntechOpen.

Vásquez-Morillas, A., Álvarez-Zeferino, J. C., Espinosa-Valdemar, R. M., Velasco-Pérez, M., Cruz-Salas, A. A., & Tecorralco-Bobadilla, A. L. (2020). Composición de los residuos de la estación de transferencia Azcapotzlco. *Tecnologías Sustentables.*

Vilariño, A., & Carenzo, S. (2023). Oportunidades y desafíos para la recuperación y reciclaje del descarte textil. In M. Saidón & S. Sorroche (Eds.), *¿Qué hacer con los residuos?*

(pp. 341–374). Teseo. https://www.teseopress.com/quehacerconlosresiduos/chapter/capitulo-10-textiles-oportunidades-y-desafios-para-la/

Wojnowska-Baryła, I., Bernat, K., Zaborowska, M., & Kulikowska, D. (2024). The growing problem of textile waste generation—The current state of textile waste management. *Energies, 17*(7), 1528. https://doi.org/10.3390/en17071528

Yalcin-Enis, I., Kucukali-Ozturk, M., & Sezgin, H. (2019). Risks and management of textile waste. In *Environmental chemistry for a sustainable world* (pp. 29–53). Springer.

Yan, M. (2021). *A life cycle assessment to quantify environmental impacts of a Spanish clothing reuse and recycling project, Moda re*. Imperial College London.

Zamani, B. (2014). Towards Understanding Sustainable Textile Waste Management: Environmental impacts and social indicators [Chalmers University of Technology]. https://publications.lib.chalmers.se/records/fulltext/204502/204502.pdf

Zamani, B. (2016). The challenges of Fas fashion. In *Environmental and social LCA of Swedish Clothing Consumption*. Chalmers University of Technology.

Zamani, B., Sandin, G., & Peters, G. M. (2017). Life cycle assessment of clothing libraries: Can collaborative consumption reduce the environmental impact of fast fashion? *Journal of Cleaner Production, 162*, 1368–1375. https://doi.org/10.1016/j.jclepro.2017.06.128

Eco-Friendly Dyeing of Cotton Fabrics with Natural Pigments

M. Bhuvaneshwari and M. Maanvizhi

Abstract The rich color palette is achieved from the natural resources which may be used extensively for coloring textile materials through printing, dyeing, and various other techniques. In general, human emotions and qualities are connected to colors that are also obtained from natural resources like plants, minerals, and animals. Various studies are done by researchers to investigate the new coloring resources that enhance the properties of fabrics rather than coloration. This study investigates the key aspects such as the dyeing of cotton fabrics from natural resources such as Eclipta (leaves), marking nut (Seed), and Tamarind bark to achieve durable and vibrant colors aiming for environmentally friendly products. The cotton fabrics were pre-treated to remove the dirt and to enhance the uniform dye uptake. Mordants like alum and pomegranate rind were used in this study to improve the color fastness properties and dyeability. The parameters such as the type of dye, the concentration of dyes (% Wt.), and the type of mordants were optimized in the dye bath at a constant temperature and time to analyze the good color fastness. This research provides a holistic understanding of the performance of sustainable natural dyes in textile applications with specific properties to color fastness. The findings are intended to support sustainable practices in the textile sector by aiding the development of environmentally friendly dyeing methods.

Keywords Natural dye · Mordant · Color fastness · Sustainability · Eco-friendly

M. Bhuvaneshwari · M. Maanvizhi (✉)
VIT Fashion Institute of Technology (VFIT), Vellore Institute of Technology (VIT) - Chennai Campus, Chennai, India
e-mail: maanvizhi.m@vit.ac.in

S. S. Muthu (ed.), *Sustainable Coloration of Textiles*, Sustainable Textiles: Production, Processing, Manufacturing & Chemistry,
https://doi.org/10.1007/978-3-031-91217-7_3

1 Introduction

Nature provides an eco-friendly way of coloring by utilizing natural resources which is a sustainable approach to overcoming harm caused by the application of synthetic dyes. Synthetic dyes are the leading dyeing industry owing to various colors availability the cost efficiency (Atav et al., 2020). However, the wider usage of synthetic dyes has played a crucial role in environmental degradation. The textile industry is a significant contributor to the global water pollution. Approximately, 20% of the global water and soil pollution arises from the textile dyeing process (Tiwari, 2022). Dyes from natural sources such as plants and animals are considered as sustainable dyes to replace these harmful synthetic dyes (Tiwari, 2022). The most available predominant natural dye colors are yellow, orange, red, blue, green, brown, and black (Samanta & Konar, 2011).

Natural dyes are grouped into three main categories: Vat dyes and mordant dyes, which need auxiliary compounds to create affinity with the fibers. The next category is substantive dyes which require no mordant. A few rare techniques include rust dyeing, mud dyeing, and painting. Substantive dyes are applied by taking double the quantity of the weight of the material and processing it below the boiling temperature for an extended period. Turmeric dyes, onion skins, walnut husks, and tea are a few examples of substantive dyes. It is a familiar fact that vat dyes are applied on the fiber surface in soluble form and then converted into an insoluble form of dyes. The most common natural vat dye is an indigo dye which requires a vatting process to form a bond between the fiber and the dye color.

In the current world's focus on sustainable goals, natural colorants have gained significant attention. The benefits of the natural dyes are shown in Fig. 1.

The research studies show that the quantity of anthocyanins in the petals of fresh Delonix regia red flowers is 8.3 g per 100 g of dry flowers (Samanta & Konar, 2011). This organic method of extracting the required compounds from natural sources has given a way to protect and promote the ecosystem and reduce water and soil pollution. The natural dyeing process includes a wide range of works right from research on natural dye sources, the extraction processes, dyeing techniques, research on mordents, color depth analysis and their sustainability (Haji et al., 2024). Moreover, recent studies have also shown the advancements in the extraction of dyes from natural resources, widening the color ranges and dyeing application procedures and replacing the metal mordent with bio-mordents to make the process more sustainable.

The progress of extraction techniques has been transformed by the incorporation of modern technologies such as ultrasonic assistance, supercritical extraction, and microwave-assisted extraction. It was found that the usage of microwave radiations was improving the process of extraction of natural dyes from plants such as madder roots and oak and observed the impact of the same on the enhancement of light and wash fastness for wool and silk fabrics (Raza et al., 2024). It is noticed that the chemicals used in the dye's extraction process create minimum impact on the environment and the selection of the extraction methods is decided by many factors such

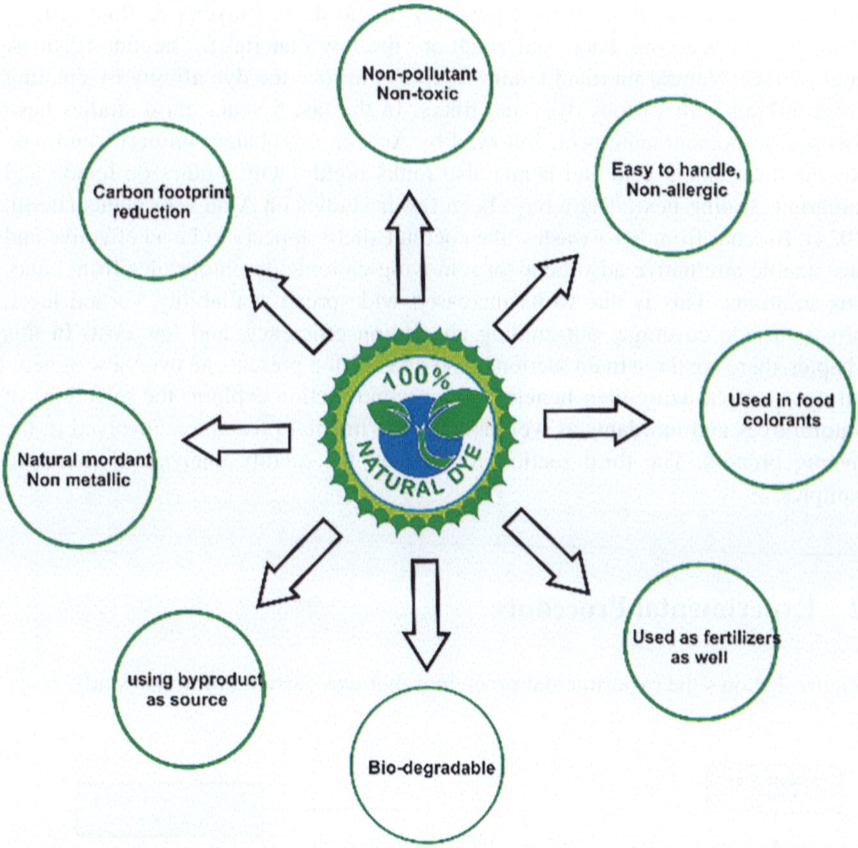

Fig. 1 Benefits of natural dye

as availability, economy, and efficiency (Teja et al., 2023). It is proved that the extraction methods of the dyes from natural resources influence the absorption wavelength of the dyes (Aslan, 2024). The impact of time, mordent, temperature, and pH are the crucial factors deciding the effect and the color quality of the dyed fabrics. It was found that the alkali treatment that is given for the cotton fabrics made changes in the conjugate systems and ionization capacity and caused yellow dye when dyed with gardenia on cationic cotton fabric (Wang et al., 2024). On the other hand, it was found that the wool fabric required acidic conditions for gaining good results in UV protection of the material with or without mordents (Dulo et al., 2022). Dye affinity and color fastness of natural dyes are enhanced when the fabric is treated with metal mordents due to which the effluents are discharged with metal compounds that are harmful to the environment. Requirements of bio-mordants could be a sustainable substitute to conventional metal mordents (Rahman et al., 2023). Thus, bio-mordants are crucial to enhance the color fastness of the biodegradable dyes (Reazuddin Repon et al., 2024). Metallic mordants, oil mordants, and tannins are the three main categories of mordents (Saxena & Raja, 2014). Pomegranate peel, oak bark, and cutch are the raw material for tannins (Pisitsak et al., 2018). Natural mordant tannins help to improve the dye affinity by creating cross-linking with various dyes and fibers. In the last 5 years, most studies have focused on pomegranate peel, followed by Acacia, myrobalan, turmeric, and oak. Research on tannic acid and henna also ranks highly, with studies on lemon and tamarind coming next. There have been fewer studies on Aloe vera plants (Benli, 2024). Biochar from agro-wastes like coconut shells appears to be an effective and sustainable alternative adsorbent for removing cationic dye molecules from aqueous solutions. This is due to its increased widespread availability, carbon level, broad surface coverage, outstanding absorption efficiency, and low cost. In this chapter, there are three main sections. The first section presents an overview of natural dyes, emphasizing their benefits. The second section explains the selections of natural dyes and mordants as well as the experimental procedures involved in the dyeing process. The third section focuses on the quality analysis of the dyed samples.

2 Experimental Procedure

Figure 2 shows the experimental procedure that was carried out in this study.

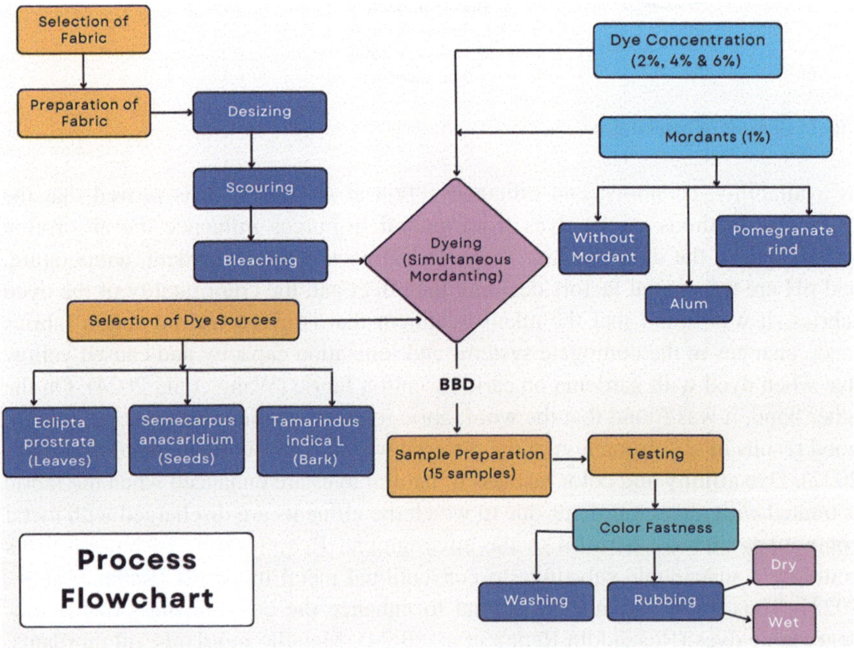

Fig. 2 Experimental procedure

2.1 Selection of Fabric

Cotton is the most common plant fiber used in the textile industry. It is well known for its sustainability. This cellulosic fiber is preferred for its compatibility with natural dyes and eco-friendly properties. Staple cotton fibers have been used in textile manufacturing for centuries due to their advantages such as strength, durability, and versatility. They can be spun or woven into all sorts of construction, ranging from light to heavy durable fabrics. Cotton can absorb moisture rapidly and dry fast, and also cotton fabrics have good resistance to alkalis and high temperatures during ironing which makes the fabric suitable for many applications. Considering these factors the cotton fabric is selected for the study that will be accepted by consumers worldwide.

2.2 Preparation of Fabric

The cotton fabric needs a series of preparatory processes such as desizing, scouring, and bleaching to optimally prepare the fibers for dye uptake. Desizing is the first process that involves the removal of sizing agents that might affect the penetration of dyestuffs. It is carried out in the desizing bath having a wetting agent at a temperature of 60 °C to 90 °C for 1.5 h and with pH maintained between 3 and 4. The unbleached cotton fabric is thoroughly washed after the desizing process to remove all the residual agents and dried and it is carried to ensure uniform dyeing.

The desized fabric is then scoured to remove waxes, natural oils, and other impurities that may affect dye absorption. The composition of the scouring bath includes caustic soda (3%), soda ash (2%), and wetting agent (1%) at a material-to-liquor ratio of 1:20 maintained at boiling temperature for 1–1.5 h. The scouring process ensures the cotton fabric to be dyed is free from impurities and makes the fabric ready for the subsequent process by improving the hydrophilicity of the fabric and results in better uptake and even application of the dye.

The final preparation stage is bleaching. Bleaching is necessary if a light color is desired after dyeing. This process involves treating the fabric in a bath containing hydrogen peroxide (4%), soda ash (2%), and caustic soda (1%) at a material-to-liquor ratio of 1:20. To ensure even penetration of the bleaching agent, the process is carried out at a temperature of 90–100 °C for 1–1.5 h. The bleached fabric is then washed thoroughly and dried in shade to achieve whiteness. All these preparation steps are necessary to ensure that the cotton fabric is ready for dyeing, thus ensuring a high-quality product that will have durable color fastness properties.

2.3 Selection of Dye Sources

The selection of natural dye sources is motivated by considerations that are ecological and involve nontoxic chemicals for application in cotton fabric. Recent studies emphasize the importance of sustainable and safe dyeing processes (Pizzicato et al., 2023). Highlight advancements in natural dyes, focusing on sustainable extraction techniques and the preparation of substrates. Research stated a chemical-free dyeing approach using functional natural dyes that minimize environmental risks (Rehman et al., 2022). According to the researchers, selected dyes should cause no danger in handling and are environmentally friendly. Hence this study investigates the natural dyes acquired from different parts of plants such as leaves, seeds, and bark based on their availability and their potential for dyeing. Eclipta (*Eclipta prostrata*)—leaves, Marking Nut (*Semecarpus Anacardium*)—seeds, and Tamarind Bark (*Tamarindus Indica* L) were selected for the study. These selected plants also offer antimicrobial properties due to their natural medicinal qualities in addition to the dyestuffs.

The leaves of *Eclipta prostrata* (Fig. 3), often called Eclipta, a flowering plant belonging to the daisy family—Asteraceae are used for dyeing. This traditional plant is valued for medicinal purposes and is native to the tropical regions of India, Korea, China, and Thailand. It grows well in a moist environment and warm conditions. The plant contains bioactive compounds such as steroidal alkaloids and has properties such as antibacterial, anti-hemorrhagic, antiviral, and antifungal. With these properties, Eclipta is an excellent and eco-friendly source for dyeing cotton fabrics.

Marking Nut (Fig. 4), botanically called *Semecarpus antheridium*, a dye source used in this study is derived from its seeds. This deciduous tree grows in the outer Himalayas and has historical importance in Indian medicine. The seeds of the Marking Nut produce a very strong dye that can give a rich dark brown color to cotton fabrics. It is also known for its medicinal uses, including anti-inflammatory and antibacterial properties.

Tamarind Bark (Fig. 5), derived from the plant *Tamarindus indica* L., is the other source of dye used in this study. The origin of Tamarind Bark is Asia and tropical

Fig. 3 *Eclipta prostrata*. (Adapted from Widipedia, 2025)

Fig. 4 *Semecarpus anacarldium*. (Adapted from Indiabiodiversity, 2025)

Fig. 5 *Tamarindus indica* L. (Adapted from flikr, 2025)

Africa, and it has numerous beneficial compounds like citric and tartaric acids thus, it is considered a natural dye. The medicinal properties of tamarind bark add extra ecological value that it can provide for dyeing. The use of these natural dye sources aligns with contemporary demands for sustainable and safe textile practices.

2.3.1 Extraction of Dyes from Selected Natural Resources

The preparation of the natural dyes from *Eclipta prostrata*, *Semecarpus anacarldium* and *Tamarindus indica* L requires a careful procedure to ensure the quality and efficacy of the dyes.

1. ***Eclipta prostrata*:** Mature leaves of the Eclipta plant are collected and dried in the shade for 2 days. After drying, the leaves are powdered finely using a mortar and kept in a cool dry place for half an hour and then stored in a container for use as a dye. This method helps retain the bioactive compounds of the leaves, which makes the dye effective and eco-friendly.
2. ***Semecarpus anacarldium*:** The harvested marking nuts are dried in the shade, and then powdered finely. The powder is kept in a glass bottle for later use as a dye. The color produced by the Marking Nut powder is deep and very durable and can be used for textile purposes.

3. ***Tamarindus indica* L:** The bark of tamarind trees that are more than 40 years old are selected to produce high-quality dyes. The collected bark is dried in the shade and powdered finely by a mixer. The powder then gets swelled in water for a day and is filtered and boiled in water at 90 °C for 10 min. The final powder obtained from the mentioned procedure is considered a natural dye of superior quality and used for dyeing textiles.

The extraction process of natural dyes from *Eclipta prostrata*, *Semecarpus anacardium*, and *Tamarindus indica* L. yields distinct colors with Eclipta producing green, Marking Nut resulting in navy blue, and Tamarind Bark generating black. These potential natural dyes demonstrate the highlight of eco-friendly and sustainable textile dyeing and emphasize their role in adopting sustainable practices and ensuring the safety and ecological compatibility of the dyeing process.

2.4 Sample Selection Using Box-Behnken Design (BBD)

Box-Behnken Design (BBD), the statistical and mathematical method is applied to find the best experimental setup with the least number of possible experiments. It explores the relationship between independent input variables and measurable responses and effectively maximizes the results (Hassan et al., 2020). BBD is used for regression modelling to prevent extreme condition trials and unnecessary outcomes. Thus, this study employed BBD for performing regression correlation and optimization. Minitab (version 16) is used for developing a three-level factorial design. BBD interprets the data, analyzes and generates the results. The parameters for BBD are assigned in three levels with minimum (−1), mid-value (0), and maximum value (1). Factors A, B, and C represent the dye source, dye concentration (weight in %), and type of Mordant, respectively (Table 1). Fifteen sample combinations (Table 2) based on BBD were prepared, and their results were analyzed.

2.5 Dyeing Process

A dyeing process with simultaneous mordanting was explored in this study using various dyestuffs and mordants. Dyestuffs include Eclipta, marking nut, and tamarind bark, with dye concentrations of 2%, 4%, and 6%. The mordants used were

Table 1 Test factors and levels

Factor	Dye source	Dye concentration	Mordant (1%)
Code	A	B	C
−1	1—*Eclipta prostrata* (EP)	2%	1—No mordant (NO)
0	2—*Semecarpus anacardium* (SA)	4%	2—Pomegranate rind (PR)
1	3—*Tamarindus indica* L (TI)	6%	3—Alum (AL)

Table 2 Coded values and actual values of BBD

Sample	Coded values			Actual values		
	A	B	C	A	B	C
1	−1	−1	0	EP (1)	2	PR (2)
2	1	−1	0	*TI* (3)	2	PR (2)
3	−1	1	0	EP (1)	6	PR (2)
4	1	1	0	*TI* (3)	6	PR (2)
5	−1	0	−1	EP (1)	4	NO (1)
6	−1	0	1	*TI* (3)	4	NO (1)
7	1	0	−1	EP (1)	4	AL (3)
8	1	0	1	*TI* (3)	4	AL (3)
9	0	−1	−1	*SA* (2)	2	NO (1)
10	0	−1	1	*SA* (2)	6	NO (1)
11	0	1	−1	*SA* (2)	2	AL (3)
12	0	1	1	*SA* (2)	6	AL (3)
13	0	0	0	*SA* (2)	4	PR (2)
14	0	0	0	*SA* (2)	4	PR (2)
15	0	0	0	*SA* (2)	4	PR (2)

alum and pomegranate rind (1%). The dyeing is carried out in the dye bath at 60 °C for 1 h with a material-to-liquor ratio of 1:20. Fifteen samples were prepared based on the Box-Behnken Design (BBD) results, covering different combinations of dyestuffs and mordants. The coded values (−1, 0, 1) for variables A (dyestuff), B (dye concentration), and C (mordant) corresponded to specific actual values (EP for Eclipta, TI for tamarind, SA for marking nut; concentrations of 2%, 4%, 6%; and NO for no mordant, PR for pomegranate rind, AL for alum, respectively). Samples 13, 14, and 15 represent the replicas as per BBD to avoid errors. After dyeing, samples were rinsed thoroughly in cold water and air-dried at room temperature. This methodology aimed to investigate the effectiveness of simultaneous mordanting in the dyeing process regarding evaluating the fastness properties and color yield.

2.6 Color Fastness Test

Color fastness of a dyed/printed fabric refers to its resistance to color changes under specific conditions. Consequently, color fastness must be quantified in terms of these changes and expressed as a magnitude as specified by Prabhavathi et al. (2014). A fabric that maintains its color during care and use is considered color-fast. Standardized tests, including washing, dry rub, and wet rub, were conducted to determine the color fastness of the natural dyed samples. The specific tests included as mentioned by Jabar et al. (2020) and Soundri (2024).

- Shade change after washing: Test method AS 2001.5.4.
- Color fastness test for rubbing: AATCC Test Method 8—Dry and Wet Crocking.

Color Fastness to Washing The laundromat was used to assess the color fastness of washing specimens of sizes 10 cm × 5 cm are cut from the sample and sandwiched between a pair of desized and bleached samples in similar dimensions. A soap solution weighing about 2 g/L was prepared and the test sample was immersed in that solution for 30 min at a temperature of 60 °C (Zhao et al., 2022). Then the sample was removed and rinsed in cold water, squeezed and dried. The color change and staining of the samples were graded on a gray scale and the values were noted.

Color Fastness to Wet Rub Test samples measuring 25 cm × 10 cm were cut from the dyed sample and mounted on the base of the crockmeter. A piece of damp white fabric was placed under the scrubber, and the sample was rubbed 11 times. The white fabric was then evaluated using a gray scale to determine the fastness (Yaman et al., n.d.).

Color Fastness to Dry Rub Test samples measuring 25 cm × 10 cm were cut from the material and mounted on the base of the crockmeter. A piece of white fabric was placed under the scrubber, and the sample was rubbed 11 times. The white fabric was then evaluated using a gray scale to determine the fastness (Yaman et al., n.d.).

3 Results and Discussion

3.1 Color Fastness Test

The color fastness to washing and rubbing on both dry and wet conditions on different samples is given in Table 3.

The EP extract obtained as powder was green in color. The literature suggests that it contains alkaloids, tannins, and phenolic compounds, terpenoids, flavonoids, saponins, and anthraquinone glycosides (Titus et al., 2024). The extraction methods used in this work favor all the mentioned compounds. Figure 6a, b—evaluation of the EP dyed fabrics—shows that the PR mordanted fabric had a lighter shade than the AL mordanted fabric. The same trend has been observed for SA and TI and mentioned in Fig. 6c, d and Fig. 6e, f, respectively. Most of the literature suggests that extraction of dyes using acidic solvents such as methanol and ethanol is not sustainable.

3.2 Color Fastness to Washing

The study shows that the impact of mordanting and dyeing temperature on the wash fastness exposes that the absence of mordent results in a low-grade scale rating of 3, irrespective of the temperature, as shown in Fig. 5. The same scenario was absorbed

Table 3 Color fastness test report

Sample	Coded value			Actual values			Color fastness to washing		Color fastness to rubbing	
	A	B	C	A	B	C	Color change	Staining on cotton	Dry rub	Wet rub
1	−1	−1	0	EP (1)	2	PR (2)	3	3	4	4
2	1	−1	0	*TI* (3)	2	PR (2)	3	3	4	3
3	−1	1	0	EP (1)	6	PR (2)	5	3	5	4
4	1	1	0	*TI* (3)	6	PR (2)	4	5	5	4
5	−1	0	−1	EP (1)	4	NO (1)	3	4	4	4
6	−1	0	1	*TI* (3)	4	NO (1)	4	3	4	4
7	1	0	−1	EP (1)	4	AL (3)	4	5	5	5
8	1	0	1	*TI* (3)	4	AL (3)	3	4	4	4
9	0	−1	−1	*SA* (2)	2	NO (1)	3	4	3	3
10	0	−1	1	*SA* (2)	6	NO (1)	4	4	4	3
11	0	1	−1	*SA* (2)	2	AL (3)	3	3	3	3
12	0	1	1	*SA* (2)	6	AL (3)	5	5	4	4
13	0	0	0	*SA* (2)	4	PR (2)	4	4	4	4
14	0	0	0	*SA* (2)	4	PR (2)	3	3	3	3
15	0	0	0	*SA* (2)	4	PR (2)	4	4	4	3

for all three herbs EP, SA, and TI. Table 3 reveals that the cotton fabric dyed with 6% EP (1) with mordant PR shows good color fastness and poor staining. This highlights the significance of bio-mordants that are PR in enhancing dye fixation and improving the durability of the color on the fabric, as the mordants also aid the formation of stable complexes with the dye molecules and fibers, preventing color loss during washing.

The dyed fabric gray scale rating with the mordents was 3 and 4 for the rest of the alum and PR combinations. The fabric dyed gray scale rating with TI neither AL nor PR showed a rating 5, this shows the color fixation of TI has scope for further studies for the improvement with some bio-mordent or metal mordent. The gray scale rating for the fabric dyed with SA and AL shows 5, which indicates that the mordant plays a crucial role in deciding the color fastness and color bleeding for the dye SA.

3.3 *Color Fastness to Dry Rubbing*

Results on the impact of mordant on color fastness to rubbing of bio-mordant both dry and wet rubbing are mentioned in Table 3. They showed that the EP with bio-mordent PR, and TI with mordant PR reveal good dry rubbing results, however, the wet rubbing is poor. Furthermore, EP with Al mordent shows good at both dry rubbing and wet rubbing of the dyed fabric. The same trend was produced by the researcher (Nambela & Haule, 2024).

a. EP treated with AL mordent

b. EP treated with PR mordent

c. SA treated with AL mordent

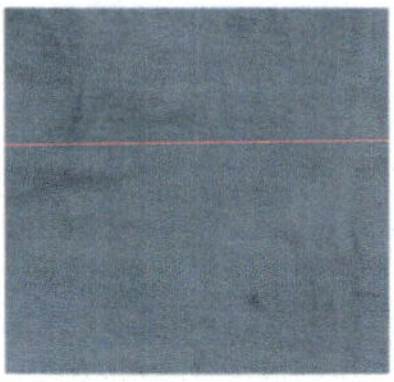

d. SA treated with PR mordent

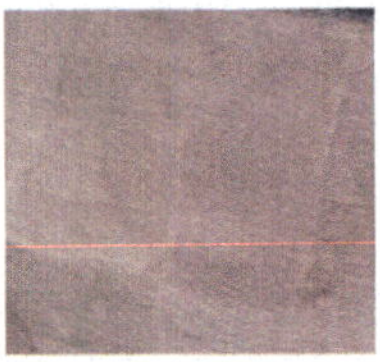

e. TI treated with AL mordent

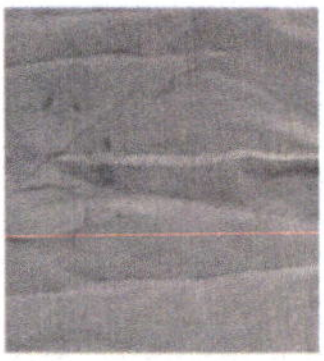

f. *TI* treated with PR mordent

Fig. 6 Fabric samples with AL and PR mordants. (**a**) *EP* treated with AL mordent. (**b**) *EP* treated with PR mordent. (**c**) *SA* treated with AL mordent. (**d**) *SA* treated with PR mordent. (**e**) *TI* treated with AL mordent. (**f**) *TI* treated with PR mordent

4 Conclusion

The dyes extracted from natural resources like Eclipta (leaves), marking nut (seeds), and tamarind bark have shown vibrant colors with good color fastness. This sustainable approach with natural resources offers a rich color palette to the textile industry and aligns with environment-friendly products by providing a better alternative to synthetic dyes which may be harmful to the environment and creates some health impacts. The study includes the pre-treatment of cotton fabrics to ensure uniform dye uptake by making the fabric ready to absorb the natural dyes effectively obviously resulting in good colors intensity or fastness. Also, the study demonstrates the use of alum and pomegranate rinds as the mordants for the improvements in the color fastness properties and help to fix the dyes onto the fabric and prevent color fading over time. The finding shows the comparisons between alum and pomegranate rind as mordants where the colors achieved are similar in both the mordants stating that pomegranate rind (bio-mordant) is a sustainable alternative.

The research highlights the optimization of mordants and dyes extracted from natural resources and their quantity of use in cotton fabrics. Even though the current results are encouraging, still there is much more to explore in terms of identifying different fabrics and their interactions with the identified natural dyes. In the future, studies may focus on optimizing the types of mordants used, the concentration of the dyes, and dyeing conditions, and their possible outcomes were analyzed.

This research focuses on the potential of natural dyes obtained from natural resources to transform the textile industry with eco-friendly dyeing methods, reducing environmental footprints and promoting consumer-friendly products. As the global market is moving toward sustainable practices, the adoption of natural dyes and bio-mordants as per the study may play a crucial role in shaping the dyeing process ensuring high-quality products with vibrant colors. Overall, this research underscores the responsibility of the textile industry about the importance of sustainable practices followed without compromising the quality.

References

Aslan, F. (2024, February). New natural dyes extracted by ultrasonic and soxhlet method: Effect on dye-sensitized solar cell photovoltaic performance. *Optical and Quantum Electronics, 56*(4). https://doi.org/10.1007/s11082-024-06294-x

Atav, R., Güneş, E., Çifçi, D. İ., & Güneş, Y. (2020, November). Comparison of wool fabric dyeing with natural and synthetic dyes in view of ecology and treatability. *AATCC Journal of Research, 7*(6), 15–22. https://doi.org/10.14504/ajr.7.6.3

Benli, H. (2024, February). Bio-mordants: A review. *Environmental Science and Pollution Research International, 31*(14), 20714–20771. https://doi.org/10.1007/s11356-024-32174-8

Dulo, B., De Somer, T., Phan, K., Roosen, M., Githaiga, J., Raes, K., et al. (2022, December). Evaluating the potential of natural dyes from nutshell wastes: Sustainable colouration and functional finishing of wool fabric. *Sustainable Materials and Technologies, 34*, e00518. https://doi.org/10.1016/j.susmat.2022.e00518

flikr. [Photo]. Retrieved January 9, 2025, from https://www.flickr.com/photos/35334437@N05/9307612433/

Haji, A., Vadood, M., Öztürk, M., Yigit, İ., Eren, S., & Eren, H. A. (2024, March). Prediction of colour strength in environmentally-friendly dyeing of polyester fabric with madder using supercritical carbon dioxide. *Coloration Technology, 141*, 44. https://doi.org/10.1111/cote.12757

Hassan, M. Z., Roslan, S. A., Sapuan, S. M., Rasid, Z. A., Mohd Nor, A. F., Md Daud, M. Y., et al. (2020, June). Mercerization optimization of Bamboo (Bambusa vulgaris) fiber-reinforced epoxy composite structures using a Box–Behnken design. *Polymers, 12*(6), 1367. https://doi.org/10.3390/polym12061367

Indiabiodiversity. [Photo]. Retrieved January 9, 2025, from https://indiabiodiversity.org/species/show/231099

Jabar, J. M., Ogunmokun, A. I., & Taleat, T. A. A. (2020, December). Color and fastness properties of mordanted Bridelia ferruginea B dyed cellulosic fabric. *Fashion and Textiles, 7*(1). https://doi.org/10.1186/s40691-019-0195-z

Nambela, L., & Haule, L. V. (2024, April). Extraction of natural dyes from Delonix regia flowers for cotton fibres. *Tanzania Journal of Science, 50*(1), 126–133. https://doi.org/10.4314/tjs.v50i1.10

Pisitsak, P., Tungsombatvisit, N., & Singhanu, K. (2018, February). Utilization of waste protein from Antarctic krill oil production and natural dye to impart durable UV-properties to cotton textiles. *Journal of Cleaner Production, 174*, 1215–1223. https://doi.org/10.1016/j.jclepro.2017.11.010

Pizzicato, B., Pacifico, S., Cayuela, D., Mijas, G., & Riba-Moliner, M. (2023, August 1). Advancements in sustainable natural dyes for textile applications: A review. *Multidisciplinary Digital Publishing Institute (MDPI), 28*. https://doi.org/10.3390/molecules28165954

Prabhavathi, R., Devi, A. S., & Anitha, D. (2014). Improving the colour fastness of the selected natural dyes on cotton (Improving the sunlight fastness and wash fastness of the eucalyptus bark dye on cotton), *1*(4), 27–30. Retrieved December 1, 2024, from https://www.iosrjournals.org/iosr-jpte/papers/Vol1-issue4/F0142730.pdf

Rahman, M. M., Kim, M., Youm, K., Kumar, S., Koh, J., & Hong, K. H. (2023, January). Sustainable one-bath natural dyeing of cotton fabric using turmeric root extract and chitosan biomordant. *Journal of Cleaner Production, 382*, 135303. https://doi.org/10.1016/j.jclepro.2022.135303

Raza, A., ul-Wara, K., Adeel, S., Mia, R., Hosseinnezhad, M., Rather, L. J., et al. (2024, March). Modern ecofriendly approach for extraction of luteolin natural dye from weld for silk fabric and wool yarn dyeing. *Sustainable Chemistry and Pharmacy, 39*, 101554–101554. https://doi.org/10.1016/j.scp.2024.101554

Reazuddin Repon, M., Dev, B., Rahman, M. A., Jurkoniene, S., Haji, A., Alim, M. A., et al. (2024, March). Textile dyeing using natural mordants and dyes: A review. *Environmental Chemistry Letters, 22*(3), 1473–1520. https://doi.org/10.1007/s10311-024-01716-4

Rehman, A., Irfan, M., Hameed, A., Saif, M. J., Qayyum, M. A., & Farooq, T. (2022, March). Chemical-free dyeing of cotton with functional natural dye: A pollution-free and cleaner production approach. *Frontiers in Environmental Science, 10*. https://doi.org/10.3389/fenvs.2022.848245

Samanta, A. K., & Konar, A. (2011). Dyeing of textiles with natural dyes. *Natural Dyes, 3*(30–56), 212–222.

Saxena, S., & Raja, A. S. M. (2014). Natural dyes: Sources, chemistry, application and sustainability issues. *Textile Science and Clothing Technology*, 37–80. https://doi.org/10.1007/978-981-287-065-0_2

Soundri, S. G. M. (2024). Dyeing of cotton with natural dye extract from coconut husk and purple cabbage. *JETIR*. www.jetir.org

Teja, A. S., Srivastava, A., Satrughna, J. A. K., Tiwari, M. K., Kanwade, A., Yadav, S. C., et al. (2023, February). Optimal processing methodology for futuristic natural dye-sensitized solar cells and novel applications. *Dyes and Pigments, 210*, 110997–110997. https://doi.org/10.1016/j.dyepig.2022.110997

Titus, A., Sangeetha, K. S., Anitha, P., Sindhu, P. V., & Pradheep, K. (2024). Phytochemical screening of solvent extracts from Eclipta prostrata L.: A preliminary investigation. *Research Journal of Agricultural Science, 15*(6), 1261–1265.

Tiwari, K. (2022, September). A potential application of microbial pigment: An alternative to synthetic dye & colourants. *International Journal of Current Science Research and Review, 05*(09). https://doi.org/10.47191/ijcsrr/v5-i9-07

Wang, L., Hu, H., Du, Y., Mi, X., Zhu, Q., Chen, Q., et al. (2024, January). Eco-friendly and sustainable application of gardenia yellow extraction as natural dye source for dyeing and bio-functional finishing of cotton fabric. *Cellulose, 31*(4), 2583–2601. https://doi.org/10.1007/s10570-023-05724-5

Widipedia. [Photo]. Retrieved January 10, 2025., from https://en.wikipedia.org/wiki/Eclipta_prostrata

Yaman, N., Ozdogan, E., & Seventekin, N. (n.d.) *Improvement fastnesses and color strength of pigment printed textile fabric*. http://www.jeffjournal.org

Zhao, X., Xiong, M., Jiang, L., Yang, Q., Zhou, C., & Liu, J. (2022, December). The effect of washing parameters on the quantity of dye discharge from clothes. *Fashion and Textiles, 9*(1). https://doi.org/10.1186/s40691-022-00289-6

Unlocking Sustainable Colour: Phycocyanin Pigments for Eco-Friendly Textile Dyeing

R. Rathinamoorthy and K. Rajesh kumar

Abstract The textile dyeing industry is recognized as one of the most polluting industries globally, with significant contributions to water pollution. The use of complex dyes and carcinogenic chemicals in dyeing processes hinders the natural decomposition of waste and amplifies its harmful environmental impact. Employing natural dyes from various sources presents more sustainable alternative for textile dyeing. However, the main drawbacks of natural dyes are their dull colours and poor wash fastness. In contrast, algae are capable of naturally producing vibrant colours, prompting efforts to explore algae-based pigments for textile dyeing.

In this research, phycocyanin extracted from *Spirulina platensis* algae was used to dye cellulosic fabrics. Various mordants, including cream of tartar, alum, tannic acid, and ferrous sulphate, were evaluated for their effectiveness in improving dye fixation. The results indicated that tannic acid combined with ferrous sulphate provided superior colour strength on textile fabrics. The objective colour evaluation showed stronger colour intensity with this combination. The study also assessed the durability of algae-pigment-dyed textiles through tests for washing fastness, rubbing fastness, perspiration resistance, and dry cleaning. While the results were promising, improvements are still needed to match the performance of commercial dyes. Pollution analysis of the final effluent from the dyeing process revealed that all key parameters (BOD, COD, pH, colour, TDS, and TSS) were within acceptable standards. Phycocyanin pigments offer a sustainable alternative for textile colouration, though future research should focus on optimizing the economic feasibility of the dye extraction process.

Keywords Algae pigment · Phycocyanin · Sustainability · Cotton fabric · Colour fastness to washing · Environmental friendly

---

R. Rathinamoorthy (✉) · K. Rajesh kumar
VIT-Fashion Institute of Technology, Vellore Institute of Technology, Chennai Campus, Chennai, India

S. S. Muthu (ed.), *Sustainable Coloration of Textiles*, Sustainable Textiles: Production, Processing, Manufacturing & Chemistry,
https://doi.org/10.1007/978-3-031-91217-7_4

1 Introduction

Although often overlooked, the majority of garments we wear daily have undergone various treatments, such as dyeing and finishing, during the fibre, yarn, fabric, or garment phases. The harmful effects of these treatments on humans, animals, and the environment are frequently disregarded. A significant industrial revolution in textile production occurred during the 1800s, particularly with the development of natural dyes from sources like flowers, leaves, vegetables, and insects. While natural dyeing is a time-consuming and complex process, especially given the scale of modern textile production, it was initially favoured. However, natural dyes could not consistently produce the desired colours, creating a gap that synthetic dyes successfully filled.

Synthetic dyes have since experienced remarkable growth due to their wide colour range, large-scale production, efficiency, and quick application. These advantages have made synthetic dyes the dominant choice, despite their eco-unfriendly nature and potential health risks, such as allergies and carcinogenic effects (Zerin et al., 2020). Today, 90% of our clothing is dyed using synthetic dyes (Willis, 2023). The textile industry is the largest consumer of synthetic dyes, accounting for two-thirds of the world's dye (Doble & Kumar, 2005). The use of synthetic dyes in textiles are associated with numerous environmental and health concerns. The production process, which involves pre-washing, dyeing, and post-processing, consumes vast amounts of water. This results in the creation of large volumes of contaminated aqueous waste, posing a serious threat to biodiversity and planetary resilience (Bisht et al., 2020). According to reports, the textile dyeing sector is the second-largest global contributor to water pollution (ILO, 2021).

Dyeing textiles involves the use of significant quantities of potentially hazardous chemicals, but awareness and transparency regarding these substances are often lacking. The complexity of supply chains and the segmented nature of distribution channels contribute to a lack of information about the specific chemicals used in textile production. Nearly 30% of the chemical substances and compounds employed in textile manufacturing and wet processing remain undisclosed, allowing suppliers to use potentially harmful substances that pollute the environment and pose risks to consumers (Swedish Chemicals Agency, 2016). As Wicker (2023) notes, neither professionals nor consumers are typically aware of the full list or quantities of chemicals used in the production of the garments they wear, from fabric manufacturing to final finishing (Wicker, 2023).

Synthetic dyes are not only used in the textile industry but also in sectors such as food, leather, pharmaceuticals, and cosmetics. Globally, the demand for dyestuffs reaches nine million tonnes, with azo dyes accounting for 70% of the total. These dyes are derivatives of aromatic hydrocarbons such as benzene, phenol, naphthalene, and aniline. Notably, azo dyes are often linked to various cancers and allergies (ILO, 2021), including the carcinogenic effects of aromatic amines (Licina et al.,

2019). Benzidine, a compound associated with azo dyes, has been specifically linked to bladder cancer (Lyon, 2010). The European Union has already banned many of these azo dyes, which are still widely used in fast fashion apparels. Human skin absorbs azo dyes easily, partly due to their water solubility, which can lead to skin and eye irritation.

Given these concerns, the textile processing industry is increasingly encouraged to shift towards natural dyes as an alternative to synthetic dyes. However, this shift brings its own set of challenges. Questions arise about whether natural dyes can truly serve as a breakthrough solution. Issues such as achieving vivid colours, a wide range of shades, and minimizing colour fading are problematic. Additionally, natural dyes bond poorly with regenerated and synthetic fibres, and their production requires significant resources, from harvesting to processing. After-treatment of naturally dyed fabrics also presents difficulties (Ranson, n.d.). While natural dyes are free from toxic chemicals, they are not without implications. The water usage in natural dye processing is nearly the same as that in synthetic dye processes. Though natural dye substances themselves do not pollute the environment, the fixatives and additives used to enhance dye adherence can. Mordants, which often contain metallic compounds, are particularly problematic and can have similar environmental impacts to synthetic dyes. Moreover, there is limited evidence regarding the dye-yielding potential of many natural colourant species, with only a small fraction of these species explored for their efficacy in dye extraction from natural resources (Ajmal et al., 2014; Khan et al., 2014; Rahman Bhuiyan et al., 2018). Natural colouring substances offer significant advantages: they are rapidly biodegradable, non-carcinogenic, and non-toxic to both the environment and humans (Shahid et al., 2013). Additionally, when combined with various mordants, polygenetic natural dyes can produce a wide range of colours—a notable limitation of synthetic dyes.

Natural dyes are also renewable, with some species growing on barren land, making them cost-effective. Certain natural dyes are anti-allergenic, making them suitable for skin-contact garments, while others are moth-resistant, ideal for children's apparel (Samanta & Konar, 2011). Nature provides abundant resources for various industries, particularly the dye and pigment sectors. Dyes can be extracted from diverse sources such as plants (leaves, stems, bark, roots, fruit pulp, and skins), animals and biological origins (insects, snails), and minerals (ochre, clay, red soil) (Jihad, 2014). Some natural dyes require the addition of a mordant to enhance colour intensity, while others, rich in tannins, do not need mordants and are known as substantive dyes. Examples of substantive dyes include annatto seeds, turmeric, sappanwood (Cæsalpinia), and safflower (Saxena & Raja, 2014). This chapter is divided into three sections. Firstly, it provides an overview of existing literature on natural colourants, their limitations, and the untapped potential of seaweed-based textile dyes. The second section reviews the literature on the utilization of various seaweeds and algae-derived pigments. Finally, the third section presents an experimental evaluation of Phycocyanin Pigments in textile dyeing and colouration, followed by a quality analysis of the dyed textiles.

2 Natural Colourants and Their Issues

Bechtold et al., investigated the differences in dye substances from Canadian goldenrod plants sourced from various locations and collected over different periods from 1999 to 2003 (Bechtold et al., 2007). Plant components such as buds, stems, and leaves were collected, air-dried at ambient temperature, and stored in a dark space. The dye solution was extracted using the boiling water method, and used to dye the wool yarn in hank form. The photometric results and colour strength of the dyeing process did not meet the expected standards. It was also observed that samples sourced from different regions showed minor variations in colour strength and intensity, though the procedure used was deemed appropriate. Other researchers effectively utilized agro-industrial residues and by-products as sources of natural dyes, with the primary benefit being that agro-waste does not occupy land directly involved in crop production, avoiding competition with other uses. The supercritical fluid extraction method, which retrieves dye substances without water, remains underutilized in natural dye extraction (Pizzicato et al., 2023).

In another study, researchers extracted yellow dye from the petals of the Cassia alata flower using a conventional boiling technique. Cotton, tussar silk, and wet-blue goat leather (chrome tanned) were dyed at varying temperatures, with a standard dyeing time of 10 min for all samples. Among all methods tested, ultrasonic extraction and ultrasonic dye bath techniques were found to be superior. The increase in the rate of dye exhaustion was directly proportional to the dye concentration in all samples. Cotton fabric exhibited good colour fastness to rubbing and washing, while its fastness to perspiration was relatively good. However, after perspiration testing, all samples faded to a lighter shade due to the higher pH. Light fastness was poor to moderate across all samples (Muruganandham et al., 2023). Msaddek et al. isolated a dye substance called indicaxanthin from *Opuntia ficus-indica* fruit and applied it to modified acrylic fabrics. To compare dye uptake efficiency, both conventional and ultrasonic methods were used. The ultrasonic method showed around 49.62% higher dye uptake compared to the conventional technique. The colour fastness to washing, rubbing, and light was rated fair to good (Guesmi et al., 2013). The importance of mordanting methods was assessed and found to significantly improve the colour fastness of dyes extracted from *Clitoria ternatea*, *Tagetes erecta Linn* flowers, and the peel of *Punica granatum* using the aqueous extraction method (Kanchana et al., 2013).

Use of eucalyptus leaf to dye wool fabric yielded yellow to brown shades as a results based on the concentration of the extract. Addition of ferrous sulphate as mordant had better impact and developed a darker shade of brown on wool fabric. Usage of copper sulphate, stannous chloride, and alum as mordants have shown better dye results, however post-mordanting with alum have evidenced lower colour strength (Mongkholrattanasit et al., 2011). In a study, extracted tannins from tamarind seeds to use as a mordant for dyeing cotton, wool, and silk fabrics with turmeric and pomegranate rind extracts. The results showed an increase in colour value for pre-mordanted samples compared to other samples. Good colour fastness to washing was reported for all three fabrics (Prabhu & Teli, 2014). The use of date leaves (*Phoenix dactylifera Linn*) to dye cotton and silk fabrics was analysed using pre, meta-, and post-mordanting methods. The results showed acceptable colour intensity and fastness characteristics without the need for additional mordants or tannins. The dye extract utilized the inherent polyphenolic compounds in the date leaves as a self-mordant. The addition of mordants led to variations in shades, and increased colour strength was observed. Metallic salt mordants improved the colour fastness to washing, particularly on silk fabrics (Hossain et al., 2022).

Dye extraction from the tender stems and leaves of *Sesbania aculeata* was carried out using the aqueous extraction method. Three mordanting methods and five mordants were tested on Seri silk and a polyester-cotton blend (67–33%). Ferrous sulphate mordant produced the darkest shades, while alum mordanted samples resulted in the lightest shades. Bright colours were achieved on silk fabrics with metallic mordants, compared to polycotton fabrics. Pre-mordanting on both selected fabrics resulted in a colour fastness rating of 4–5, while other mordanting methods produced average to satisfactory results. Light fastness was also found to be satisfactory across all samples (Swami et al., 2012).

In summary, various studies have explored natural dye extraction and mordanting methods to improve colour fastness. Based on geographical location and cultural practices, several plant species have been treated for their colour-yielding properties. While the literature discussed here is not exhaustive, it highlights some rare studies that utilized unique plant sources for dye extraction. Researchers have also used agro-industrial by-products for natural dye extraction, emphasizing the untapped potential of supercritical fluid extraction (Pizzicato et al., 2023). Studies using *Cassia alata* flower petals found ultrasonic extraction to be superior in achieving higher dye exhaustion, with good colour fastness to rubbing and washing but poor light fastness (Muruganandham et al., 2023). Guesmi et al., demonstrated that ultrasonic dyeing improved dye uptake in modified acrylic fabrics compared to conventional techniques (Guesmi et al., 2013). Other studies underscored the importance of mordanting, with tamarind seed tannins and metallic salt mordants enhancing dyeing results in cotton, wool, silk, and other fabrics (Hossain et al., 2022; Prabhu & Teli, 2014; Swami et al., 2012). But the overall results also evidenced that the suitability of the mordant is again subjective to the specific dye and no standard mordant or mordanting process was reported.

3 Potential of Seaweed and Algae as a Dye

Seaweed and algae, abundant marine resources, have gained attention for their potential as sustainable and eco-friendly dye-yielding materials for textile applications. These marine organisms contain a diverse range of bioactive compounds, including pigments, polysaccharides, and proteins, which can be harnessed to create natural dyes. Due to the rich pigment content (e.g., chlorophyll, carotenoids, and phycobiliproteins), abundance, sustainable source of production, and lower environmental impact, seaweed and algae were focused more in this area. Additionally, the development of unique colours is one of the main reasons why researchers are always eager about seaweeds and algae. For instance, brown algae are rich in fucoxanthin and tannins, providing earthy tones, while red algae contain phycoerythrin, yielding red or pink shades. Green algae are rich in chlorophyll, giving green tones. The variety of pigments allows for a diverse palette of natural colours. In addition to being eco-friendly, seaweed and algae dyes have been shown to possess other advantageous properties, such as antimicrobial and UV-protection qualities. These added functionalities make them appealing for use in high-performance textiles, particularly in sportswear, medical textiles, and fashion. Both of these aquatic organisms, often found in marine and freshwater ecosystems, have been historically overlooked for dye production in comparison to plants, but their unique chemical compositions make them valuable for modern textile applications.

El-Khatib et al., extracted green pigments from the algae family *Chlorella Spirogyra* to dye wool fabrics (El-Khatib et al., 2016). Chitosan and tannic acid were used as auxiliaries to enhance the dye affinity in wool. Wool fabrics pretreated with this eco-friendly method exhibited higher colour intensity compared to non-pretreated fabrics. Additionally, the pretreated fabrics demonstrated lower fungi and bacteria growth due to the antibacterial properties of the dye pigment. Seaweed (*Sargassum* sp.), in combination with concentrated sodium alginate flour containing 5% moisture, was used to dye Bima woven fabric. It was found that the concentration of sodium alginate was directly proportional to the intensity of the colour obtained (Wiraningtyan et al., 2020). A study on dye substances extracted from green algae (*Cladophora*) revealed another promising possibility for natural colourants. Cotton fabric dyed with *Cladophora* exhibited stronger colour values. The findings also showed that the 80% methanol extract of the colourant achieved higher colour intensity at a lower cost (Ali, 2007).

The non-toxic cyanobacterial pigment phycocyanin, derived from *Arthrospira (Spirulina) platensis*, has been used as a hair dye. The dye's affinity for hair remained stable even after five washes with shampoo (Kraseasintra et al., 2022). A study explored the use of *Spirulina* extract for dyeing fibres and bioplastics using a liquid dye setup. Based on the concentration of filtered *Spirulina* powder and the precipitation method employed, a vivid blue-green hue was achieved. An equal proportion of *Spirulina* powder and alum mordant influenced colour absorption and stability. The dye was also applied to alginate bioplastics, bio yarns, and bio films, resulting in a range of green to blue shades. However, during the evaluation of long-term colour fastness, the dye was found to be affected by oxidation and humid conditions in bio films. Additionally, the robustness of the dyed material faced challenges, particularly when exposed to light (Delgado et al., 2024).

Caulerpa lentillifera, a green seaweed, was analysed for its potential as a dye for colouring textiles. The extracted dye was applied to silk fabric using various mordants to fix the colouring agent. Among the mordants used, iron produced darker shades of brown in both extraction methods, while alum resulted in lighter shades of yellow. In terms of colour strength, the extraction with boiling water and iron mordant yielded darker colours, whereas the ammonium hydroxide technique produced lighter colour strength values. The fabric demonstrated good colour fastness to washing and perspiration, with ratings of 4–4.5. However, colour fastness to rubbing and crocking resulted in a rating of 3, while colour fastness to light was poor (Kadir et al., 2023). Similarly, dyeing cotton and wool fabrics using different iron mordants significantly increased the *K*/*S* value and colour fastness, employing various natural colourant powders derived from plants and insects (Zarkogianni et al., 2011). Rani (2020) investigated the use of a marine brown seaweed, *Cystoseira indica*, for dyeing 100% cotton single jersey knitted fabrics (Rani, 2020). The extracted dyes were applied using tannic acid as a bio-mordant and sodium chloride, sodium carbonate, and formic acid as exhausting agents. The fabric was exposed to varying percentages of fixing agents and different dyeing durations. The lowest percentage of formic acid showed optimal dye absorption with a maximum dyeing duration of 120 min. The colour fastness to light was evaluated as moderate to good. However, the presence of chromophores in the natural dye led to photochemical degradation, resulting in poor to moderate colour fastness to light, which aligned with the findings of Zarkogianni et al. (2011).

Pigments from *Chryseobacterium* sp., *Serratia marcescens*, and *Chromobacterium violaceum*, such as yellow-orange, red, and purple, were used to dye a variety of natural and synthetic fabrics, including acrylic fibre, silk, polyester, cotton, polyester microfiber, and viscose. The results of the study showed that acrylic, silk, polyester microfiber, and polyester exhibited higher colour fastness, while cotton and polyester had average colour fastness to washing, light, and perspiration (Chander, 2022). Kadir et al. dyed cotton and silk fabrics with a pigment extracted from *Sargassum* sp. seaweed, using a surface modifier of PAMAM dendrimer to analyse dyeability (Kadir et al., 2023). These surface modifiers are composed of subdivided monomers containing various primary amino groups, which enhances the dyeability of fabrics, particularly with natural dyes. Methanol-extracted *Sargassum* sp. dry powder was applied to PAMAM dendrimer-pretreated fabrics. Alum, iron, and vinegar were used as mordants in an exhaustive dyeing method, and it was found that the *K*/*S* value of pretreated fabrics increased, and the colour fastness was rated as good to excellent. However, the light fastness rating was poor, with a rating of 3–5.

Red, green, and brown strains from *Kappaphycus alvarezii* (red seaweeds—Rhodophyta) were obtained using two extraction methods: the boiling water method and the microwave method. The extracted substances were evaluated with a UV–Vis spectrophotometer to determine optimal absorbance capacity and identify factors contributing to the shades attained. Bamboo and silk fabrics were dyed using the exhaustive dyeing method, with ferrous sulphate and potassium aluminium sulphate as mordants. The results showed that the microwave extraction method produced darker stains on the samples compared to the boiling method. Colour fastness to washing was rated good to excellent, while colour fastness to rubbing (crocking) was acceptable, but light fastness was found to be poor (Ab Kadir et al., 2014a). Four marine algae (*Iyengaria stellata*, *Sargassum muticum*, *Colpomenia sinuosa*, and *Laurencia obtusa*) were studied to assess their dyeing efficacy on cotton fabrics. Dye compounds were extracted using various mediums, including organic, alkaline, acidic, and aqueous solutions. A range of parameters, such as concentrated exhausting agents, pH levels, temperatures, pre- and post-mordanting processes, and dyeing durations, were applied. The dyed samples were then evaluated for colour strength, and it was found that the use of tannic acid and iron as mordants, in both pre- and post-mordanting processes, significantly enhanced the colour strength. The role of tannic acid during mordanting was particularly strong, as it formed an insoluble structure on the cotton surface, providing stability (Azeem et al., 2019). Table 1 consolidates the important research works that used seaweed and algae-based pigments to dye the textiles as reported by the literature (Mutaf-Kılıc et al., 2023).

Seaweed and algae have garnered attention as sustainable and eco-friendly dye sources for textile applications. A review of the literature reveals that brown algae (fucoxanthin and tannins) have been explored for earthy tones, red algae (phycoerythrin) for red or pink shades, and green algae (chlorophyll) for green hues. The research summary also highlights that seaweed and algae-based dyes have been used on cotton, wool, silk, and synthetic fibres. In addition to providing colour, these dyes offer unique functionalities, such as antimicrobial and UV-protection properties. The literature further notes the use of *Chlorella spirogyra* for dyeing wool fabrics, *Sargassum* sp. for Bima woven fabric, *Cladophora* for cotton fabric, and *Spirulina* extract for fibres and bioplastics. Mordants and auxiliaries such as chitosan, tannic acid, and alum have been shown to enhance dye affinity and colour fastness. However, challenges like poor light and abrasion fastness have been reported across studies. Additionally, no significant attempts have been made to standardize or regulate the dyeing process for scaling up to mass production. The cost-effectiveness of such extraction and application methods is also still under research to develop a viable model for bulk production.

Table 1 Summary of literature that used algae and seaweed pigments to dye the natural fiber-based textiles (Reprinted with permission from Mutaf-Kılıc et al., 2023)

Algae species	Pigment	Pigment extraction process	Fabric dyed	Mordant	Colour of dyed material	Reference
Caulerpa lentillifera	Chlorophyll	Boiling water extraction method, ammonia fermentation method	Silk	Ferrous sulphate, potassium aluminium sulphate, acetic acid	Yellowish green, brown	Ab Kadir et al. (2014b)
Caulerpa lentillifera, *Sargassum* sp.	Chlorophyll, fucoxanthin	Boiling water extraction method	Silk	Ferrous sulphate, potassium aluminium sulphate, vinegar	–	Ab Kadir et al. (2016)
Sypirogyra sp.	Chlorophyll	–	Wool fiber	Chitosan, tannic acid	–	El-Khatib et al. (2016)
Arthrospira platensis, *Gracilaria vermiculophylla*	Phycocyanin, phycoerythrin	Thawing and sonication	Cotton, wool	Cream of tartar, ammonium sulphate	–	Ferrandiz et al. (2016)
Sargassum sp.	Fucoxanthin	Maceration extraction with methanol	Cotton, silk	Ferrous sulphate, potassium aluminium sulphate, vinegar	Yellowish, brownish	Ismail et al. (2016)
Sargassum sp.	Fucoxanthin	Methanol extraction in water bath at 60 °C	Polyester	Ferrous sulphate, potassium aluminium sulphate, vinegar	Reddish yellow, greenish yellow	Ab Kadir et al. (2017)
Gracilaria sp.	Phycoerythrin	Thawing and magnetic stirring	Cotton	Cream of tartar	Light red	Moldovan et al. (2017)
Cladophora glomerata	Chlorophyll	Carotenoid aqueous, alkaline, organic, ethanolic/KOH, methanolic/ KOH, acidic extraction	Cotton	Iron sulphate, copper sulphate, aluminium sulphate, tannic acid	Creamy, yellow, brown	Mir et al. (2019)

(continued)

Table 1 (continued)

Algae species	Pigment	Pigment extraction process	Fabric dyed	Mordant	Colour of dyed material	Reference
Spirulina platensis, *Turbinaria conoides*, *Halymenia dilatata*	Phycocyanin, fucoxanthin, phycoerythrin	Freeze thawing, mechanical shaking, Soxhlet extraction	–	Tannic acid, potassium aluminium sulphate (alum), mango kernel residue, fresh lemon juice, *aloe vera* gel	Creamy, yellow, light blue, pink	Mona and Sekarenthiran (2019)
Lyengaria stellata, *Sargassum muticum*, *Colpomenia sinuosa*, *Laurencia obtusa*	Fucoxanthin, phycoerythrin	Aqueous, alkaline, acidic, and organic extraction	Cotton	Copper sulphate, iron sulphate, aluminium sulphate, tannic acid	Light-dark brown, creamy white, greyish white	Azeem et al. (2019)
Cystoseira indica	Fucoxanthin	NaOH extraction	Cotton	Tannic acid, sodium chloride, sodium carbonate	Reddish yellow	Rani (2020)
Stoechospermum marginatum	Fucoxanthin	Aqueous, acidic (acetic acid), alkaline (sodium hydroxide)	Silk/Jersey, cotton	Ferrous sulphate, aluminium sulphate	Reddish yellow	Rani et al. (2020)

4 Phycocyanin Pigment from *Spirulina platensis*

Phycocyanin is one of the main protein components of *Spirulina platensis*, known for its antioxidant and radical-scavenging properties (Hoseini et al., 2013). Its medicinal properties have been shown to effectively protect the liver and kidneys during detoxification and help activate the immune system. The immune-boosting abilities of phycocyanin demonstrate potential in suppressing cancer cells and inhibiting the development of viral infections (Moorhead & Capelli, 1993). Additionally, *Spirulina platensis* is renowned for its vibrant blue-green pigment, primarily derived from phycocyanin. This pigment, along with other bioactive compounds, provides unique structural and functional properties, making *Spirulina* an attractive natural alternative to synthetic dyes. The basic chemical structure of phycocyanin is presented in Fig. 1.

Attempts were made to develop a print paste using phycocyanin pigment, along with alginate and glycerine. The study aimed to extract the dye from dried algae by grinding it into a powder, followed by its extraction and application with varying concentrations of alginate and glycerine thickeners. The pigment was applied to cotton fabric via screen printing. The results showed poor adhesion of the colour to the fabric, although the parameters for print paste consistency were optimized. While the study demonstrated the potential of developing a print paste with phycocyanin pigment, the fastness properties showed poor resistance to washing (Ciptandi et al., 2021). Another study explored the use of *Spirulina* as a dyeing agent for natural fibres and alginate-based bioplastics, aiming to optimize a zero-waste dyeing process. The extracted dye was applied to natural fibres and bioplastics, producing hues ranging from green to blue. The research indicated that phycocyanin can be used to dye fibres, yarns, fabrics, and bioplastics for packaging. However, the dyed materials exhibited poor fastness to light, indicating instability in colour fixation (Delgado et al., 2024). In a separate study, phycocyanin pigment was used to dye silk fabric with the bio-crosslinker genipin. Results showed that the use of a crosslinking agent increased colour depth, durability, and functionalities like UV-protection and antioxidant properties. The colour strength (*K/S*) of PC/GNP-treated silk improved at pH levels between 3.6 and 7 but decreased at higher pH levels. Compared to non-crosslinked samples, crosslinked samples exhibited better

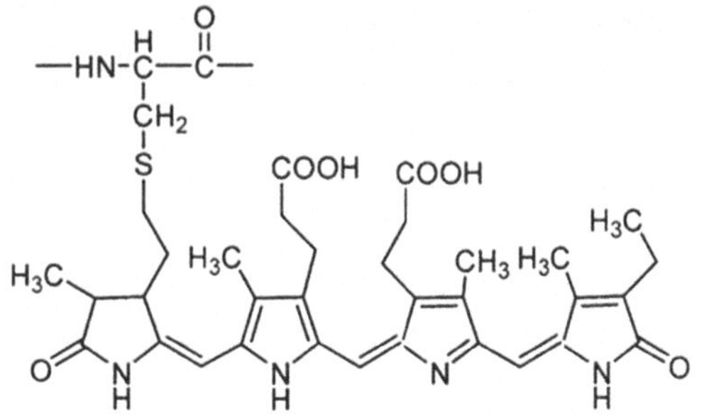

Fig. 1 Chemical structure of phycocyanin

colour fastness to washing and rubbing (>4) (Wu et al., 2023). An attempt was made to use phycocyanin as a dye source by incorporating it into polyvinyl alcohol (PVA) nanofibres. The resulting PVA fibres exhibited excellent colour strength of 0.2 with a light blue hue. The fibres demonstrated improved wash fastness, with values ranging from 3 to 4, and staining resistance scoring between 4 and 5 on a 5-point scale. While previous studies reported poor lightfastness for this dye, the current research indicated a good lightfastness value of 7 on a 9-point scale. Additionally, the dope-dyed nanofibres showed a threefold increase in strength as a result of the process (Ghaffar et al., 2023).

Beak et al. evaluated the dyeability of phycocyanin dye on cotton and silk fabrics. Different mordants, including copper sulphate ($CuSO_4$), calcium acetate ($Ca(CH_3COO)_2$), aluminium sulphate ($Al_2(SO_4)_3$), aluminium potassium sulphate ($KAl(SO_4)_2$), iron sulphate ($FeSO_4$), and tin chloride ($SnCl_2$), were tested for their performance with phycocyanin dye. The results showed that silk fabric exhibited significantly lower colour strength compared to cotton fabric. Moreover, the effect of mordants was more pronounced on cotton than on silk. Regarding fastness properties, the phycocyanin-dyed samples displayed poor light and wash fastness but relatively good fastness to rubbing. The poor dyeability of cotton fabric was attributed to the protein structure of the phycocyanin dye (Baek et al., 2022). Several approaches were already made to include phycocyanin pigments in the food as a food colourant (Imchen & Singh, 2023; Mao et al., 2024; Roda-Serrat et al., 2018).

Although some attempts have been made to use phycocyanin pigments for dyeing textiles, the literature indicates that these pigments exhibit poor adhesion and fastness properties on cotton fabric. Whereas, the performance was noted good with silk fabric. The use of mordants and crosslinking agents has been shown to improve colour depth and fastness. The primary challenges reported in the existing literature are the pigments' poor fastness, particularly to light exposure. Additionally, limited scalability and high sensitivity to process parameters are other significant issues highlighted in previous studies. Therefore, this research focuses on using phycocyanin pigments to dye cotton, linen, and silk fabrics through various mordanting methods (natural and metallic), including pre-mordanting, meta-mordanting, and post-mordanting techniques. The study also emphasizes the fastness properties of phycocyanin dyes against washing, perspiration, rubbing, and dry cleaning. The main objective is to enhance colour stability and improve fastness properties.

5 Materials and Methods

5.1 Materials

Phycocyanin dye power, mordants like alum, cream of tartar, tannic acid, and ferrous sulphate were sourced from Nice Chemicals India Pvt Ltd, Chennai. The sourced chemicals and dyes are of analytical grade and used as such without further modification.

Cotton, Linen and Silk fabrics were sourced from the local retail outlets in Chennai, India. Cotton fabric with, EPI—110, PPI—130, and GSM of 140; Linen fabric with EPI—90, PPI—110, and GSM of 160; and silk fabric with EPI—90, PPI—110, and GSM of 120 was used in this study. The sourced fabrics were pre-bleached and have undergone no pre-treatments.

5.2 Mordanting and Dyeing

The fabrics were treated with selected mordants using both pre-mordanting and post-mordanting methods. A 5% concentration of the mordants was used, based on the weight of the material or fabric involved in the dyeing process. The material-to-liquor ratio was maintained at 50:1. For both pre- and post-mordanting, the fabric was immersed in the mordant solution for 30 min at 60 °C, before or after the dyeing process.

For dyeing the mordanted fabrics, 5 g of phycocyanin powder was dissolved in 100 mL of water to prepare a dye bath. White or bleached fabrics were immersed in the dye bath for 40–60 min at 30–40 °C. After the dyeing process, the fabrics were either rinsed (pre-mordanted) and dried (meta-mordanted) or treated (post-mordanted) and then dried. The detailed combinations of treatments used in this research are provided in Table 2.

Table 2 Details of the experimental methods attempted to dye phycocyanin dyes on textiles

S. no	Sample type	Method used	Mordants used	pH	Treatment time (min)	Treatment Temperature (°C)
1	Cotton	Pre-mordanting	Alum	4	40	30
2		Post-Mordanting	Cream of tartar	5	50	35
3		Combination	Tannic acid Ferrous sulphate	6	60	40
4	Linen	Pre-mordanting	Alum	4	40	30
5		Post-Mordanting	Cream of tartar	5	50	35
6		Combination	Tannic acid Ferrous sulphate	6	60	40
7	Silk	Pre-mordanting	Alum	4	40	30
8		Post-Mordanting	Cream of tartar	5	50	35
9		Combination	Tannic acid Ferrous sulphate	6	60	40

5.3 *Characterization of Dyed Fabric*

5.3.1 Colour Strength Determination

The colour strength is to be determined by calculating the *K*/*S* value of the dyed fabric. This value is directly proportional to the amount of dye present in the material. *K*/*S* value can be found using UV spectrophotometer by using the following formula developed by Kubelka and Munk:

$$K/S = \frac{(1-R)^2}{2R}$$

where *K*—coefficient of adsorption, *S*—coefficient of Scattering, *R*—reflectance value of fabric at $\lambda_{\max}$.

5.3.2 Evaluation of Fastness Properties

The dyed fabric samples were evaluated for their colour fastness properties as per the standards mentioned below:

- Colourfastness to domestic and commercial laundering as per Standard UNE-EN ISO 105C06:2010
- Colourfastness to rubbing as per Standard UNE-EN ISO 105-X12:2003
- Colourfastness to perspiration as per Standard ISO 105 E04
- Colourfastness to dry cleaning as per BS EN ISO 105-D01

6 Results and Discussion

In the preliminary experiments, attempts were made to analyse the effects of dyeing temperature, treatment time, and different mordants. Based on the practical results, a treatment temperature of 40 °C with (and a treatment duration of) 60 min were selected for various mordant combinations. Consequently, all selected fabric types were subsequently dyed at 40 °C for 60 min. The dyeing process was conducted using different mordants on cotton, linen, and silk fabrics. The results indicated that cotton and linen fabrics exhibited poor colour adhesion of dyes, even after the application of various mordanting agents. The methods of application namely pre-mordanting, post-mordanting, and a combination of both did not yield appreciable results. However, the application of dye on silk fabric was found to be efficient, and the results are discussed in this section.

6.1 Colour Strength of Dyed Samples

The dyed fabrics were analysed for their colour strength values using a spectrometer, and the results were provided in Table 3. The colour absorption strength was reported as the *K*/*S* value, with the corresponding wavelength marked as λ_{max}. The colour coordinates are defined as follows: L^* represents lightness (positive side) and darkness (negative side); a^* denotes redness (positive) and greenness (negative); b^* reports yellowness (positive) and blueness (negative) values of the dyed silk fabric, respectively. Table 3 provides the colour absorption (*K*/*S*) values of cotton and linen fabrics. The results demonstrated that dye absorption on these fabrics was minimal, even after the mordanting process. Both cotton and linen exhibited very poor colour strength, resulting in low *K*/*S* values. These findings align with the previous results discussed in the literature (Baek et al., 2022). Based on these results, the samples were not further processed for evaluation.

With respect to the silk fabric, a better fixation was noted with a few mordants; where, tannic acid and its combinations resulted in very good colour adhesion. This was due to the more effective binding of the dye substance to silk fabric compared to cotton and linen. Similar results have been reported in the literature with phycocyanin pigment on cellulosic and silk fibres (Baek et al., 2022). The results are presented in Table 4, where it can be noted that most samples showed a negative b^* value, indicating the blue colour of the dye on the sample. Additionally, the a^* value of the silk fabrics was also noted as negative, indicating a green shade. Phycocyanin pigment is derived from blue-green algae, which tends to yield a bluish-green colour. Among the samples tested, sample 4, treated with tannic acid and ferrous sulphate, showed the highest colour strength with a dark brown shade. Compared to all other mordants, tannic acid produced better colour strength with the phycocyanin dye powder.

Table 3 Colour strength (*K*/*S*) values of cotton and linen fabric

Method used	Mordants used	Cotton *K*/*S* at 400 nm	Linen *K*/*S* at 400 nm
Pre-mordanting	Alum	0.1	0.1
	Cream of tartar	0.078	0.1
	Ferrous sulphate	0.11	0.00
	Tannic acid	0.03	0.01
Post-Mordanting	Alum	0.00	0.2
	Cream of tartar	0.2	0.14
	Ferrous sulphate	0.00	0.3
	Tannic acid	0.5	0.04
Combination	Alum, cream of tartar	0.00	0.00
	Cream of tartar, tannic acid	0.11	0.05
	Tannic acid, alum	0.1	0.14
	Tannic acid, ferrous sulphate	0.002	0.1
	Alum, alum	0.00	0.10

Table 4 Maximum colour strength values of the samples

Samples (Pre and post-treatment of mordant)	λ_{max}	*K/S* value	*L**	*a**	*b**	Actual colour
1 (Alum, cream of tartar)	610	0.15	85.58	−5.87	−4.47	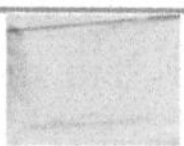
2 (Alum, alum)	620	0.7	84.75	−5.45	−3.42	
3 (Tannic acid)	400	0.31	40.4	1.64	−11.32	
4 (Tannic acid, ferrous sulphate)	600	3.9	−75.68	−13.35	0.89	
5 (Alum, tannic acid)	400	0.44	82.33	−5.81	1.63	
6 (Tannic acid, cream of tartar)	620	0.38	78.42	−12.06	−7.21	
7 (Tannic acid, alum)	620	0.15	77.57	−9.99	−5.15	

6.2 Colour Fastness to Washing

The silk fabric dyed with phycocyanin dye using various mordant combinations was evaluated for wash fastness with multifiber fabrics, and the results are provided in Fig. 2. From the results, it can be noted that the phycocyanin dyed silk fabric exhibited very poor colour fastness to washing. Even with the use of different mordants, the fastness rating ranged between 1 and 2. However, in the case of tannic acid, a slight improvement in fastness was observed. A higher fastness rating of 3–4 was assigned to the silk sample dyed with a combination of ferrous sulphate and tannic acid mordants. Additionally, in terms of chemical interaction, phycocyanin dye has a limited number of reactive groups and interacts with silk fabric primarily through hydrogen bonding and Van der Waals forces. However, the use of tannic acid significantly increased the colour depth due to its excellent crosslinking ability (Kimura, 1993). Previous studies that tried with four different brown and red algae to dye the cotton fabric also showed a similar results of tannic acid in dye fixation (Azeem et al., 2019).

The weak affinity of the phycocyanin dye was the primary reason the silk fabric could not achieve a deep colour shade. The poor interaction between the dye molecules and the fibre structure also resulted in lower wash durability of the colour (Wu et al., 2023). The *K/S* value of the samples ranged from 0.15 to 3.9. The proposed

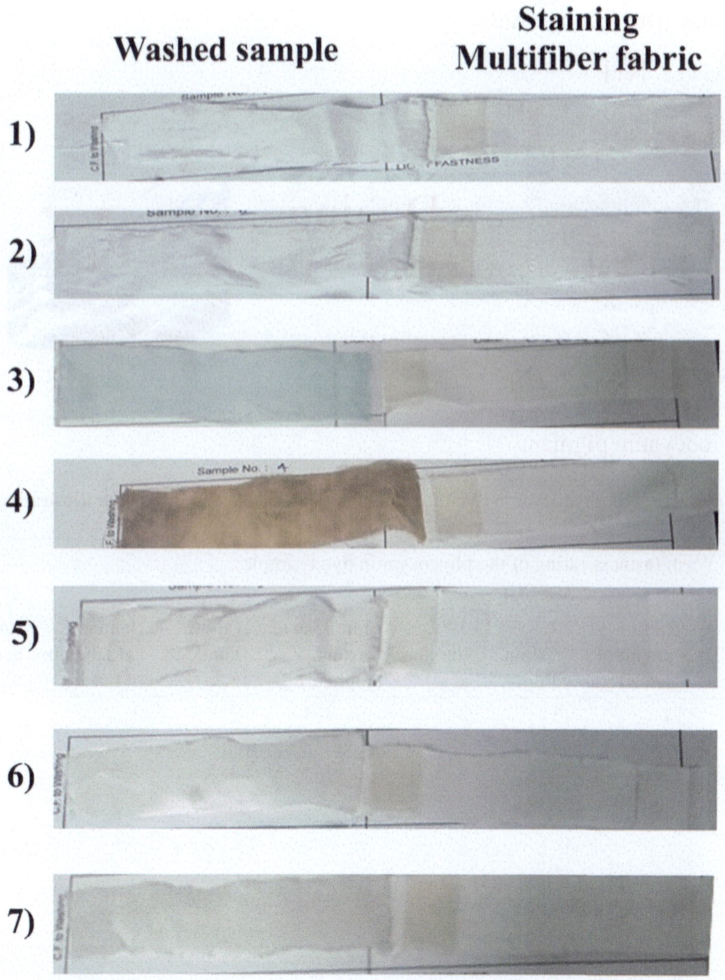

Fig. 2 Samples analysed for colourfastness to washing and their staining fabric

interaction mechanism between the phycocyanin dye and the silk fabric is illustrated in Fig. 3. However, the staining results were appreciable, with the staining level of the fabric marked in the range of 4–5 (Table 5), indicating minimal to no staining. This aligns with previous research (Ciptandi et al., 2021), which evaluated the potential application of phycocyanin paste as a printing medium and analysed its performance. The findings were similar to the current study as well showing good staining but poor fastness, with a rating of 1. This is also evident in current research, as shown in Fig. 2, which represents the multifibre fabric used for wash fastness testing.

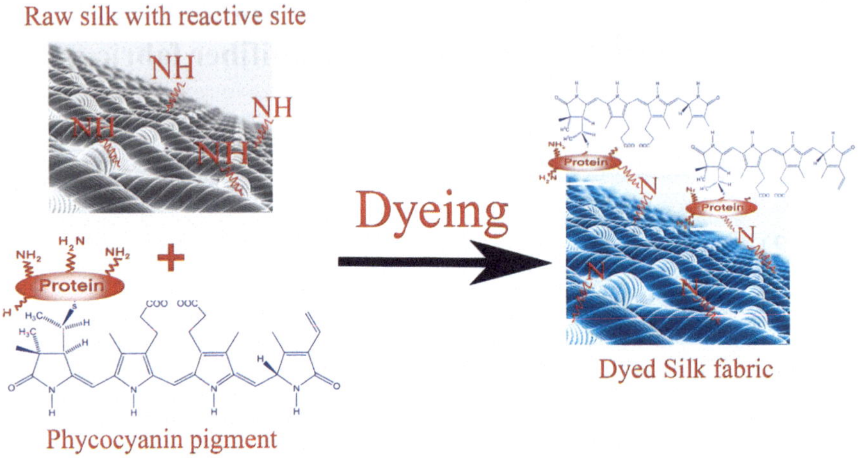

Fig. 3 Interaction of phycocyanin dye molecule with silk fabric. (Authors own illustration)

Table 5 Wash fastness rating of the phycocyanin dyed samples

Samples	1 (Alum, cream of tartar)	2 (Alum, alum)	3 (tannic acid)	4 (Tannic acid, ferrous sulphate)	5 (alum, tannic acid)	6 (tannic acid, cream of tartar)	7 (tannic acid, alum)
Colour change	1	1	2	3–4	2	2	1
Staining on acetate Cotton Nylon Polyester Acrylic Wool	4–5	4–5	4–5	4–5	4–5	4–5	4–5

6.3 *Fastness to Perspiration*

Silk fabrics are known for their poor perspiration fastness and often lose strength when exposed to sweat. Therefore, attempts were made to evaluate the durability of the phycocyanin-dyed silk fabric against perspiration. The detailed results for acid and alkaline media are provided in Table 6, and the corresponding staining ratings are shown in Fig. 4. The results indicate that the use of tannic acid significantly improved the performance of the dyed fabric. Samples 1 and 2, treated with alum and cream of tartar, exhibited very low durability compared to other samples

Table 6 Colour fastness rating of phycocyanin dyed silk fabric against perspiration

Sample	1 (Alum, cream of tartar)	2 (Alum, alum)	3 (Tannic acid)	4 (Tannic acid, ferrous sulphate)	5 (Alum, tannic acid)	6 (Tannic acid, cream of tartar)	7 (Tannic acid, alum)
Perspiration fastness against the acid medium							
Colour change	1	1	2–3	4	2	3–4	3
Staining on acetate Cotton Nylon Polyester Acrylic Wool	4–5	4–5	4–5	4–5	4–5	4–5	4–5
Perspiration fastness against alkaline medium							
Colour change	1	1–2	3	3–4	2	4	4
Staining on acetate Cotton Nylon Polyester Acrylic Wool	4–5	4–5	4–5	4–5	4–5	4–5	4–5

containing tannic acid. Specifically, samples 4 and 6 demonstrated excellent resistance to perspiration fastness, with ratings of 4 and 3–4, respectively. A combination of tannic acid with ferrous sulphate and cream of tartar resulted in a strong affinity between the dye and the silk fabrics. However, in the case of tannic acid combined with alum, and alum combined with tannic acid as pre- and post-mordants, the fastness ratings were 3 and 2, respectively.

For perspiration fastness in an alkaline medium, a higher rating of 4 was reported for tannic acid with cream of tartar as well as for the combination of tannic acid with alum. The second-highest rating, 3–4 for colour change, was observed for the tannic acid and ferrous sulphate combination, similar to the fastness results in the acidic medium. The findings were consistent with those for perspiration fastness in the acidic medium, where the use of tannic acid led to better adhesion and resistance to perspiration. In contrast, the combination of alum and cream of tartar exhibited poor fastness in the acidic medium, with a rating of 1.

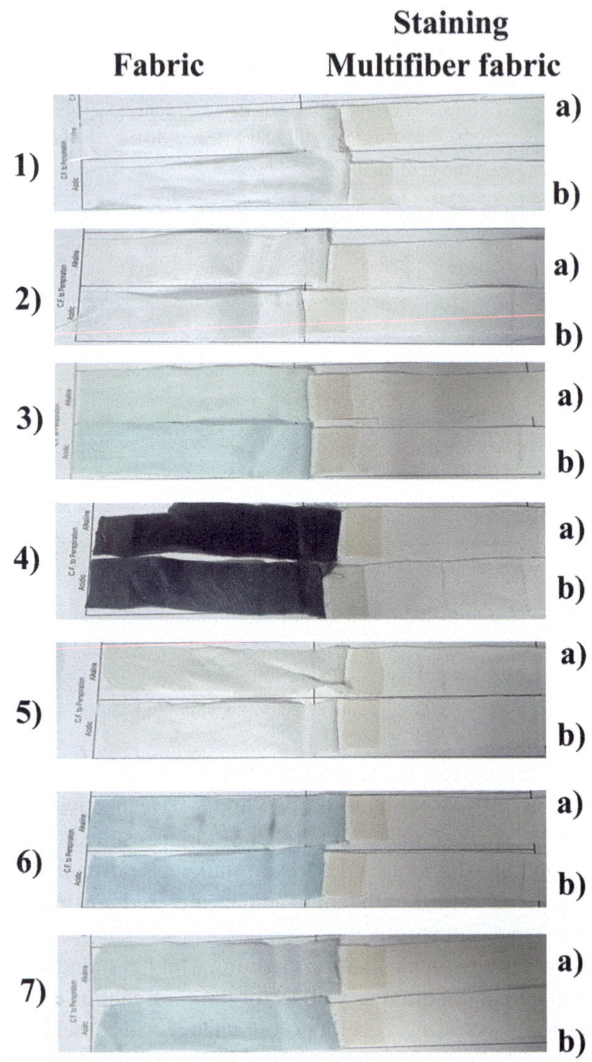

Fig. 4 Samples analysed for colourfastness to perspiration and their staining fabric. Fastness to (**a**) alkaline and (**b**) acidic side

6.4 Colour Fastness to Rubbing

Rubbing fastness simulates real-life conditions where the dyed fabric is subjected to abrasion. The experiment was conducted in both dry and wet states, with the results provided in Fig. 5 and the corresponding rating given in Table 7. The results showed that dry rubbing did not affect colour fastness, as all samples dyed with different mordants exhibited a high fastness rating of 5. However, in the case of wet rubbing, a slight difference was observed, with ratings reported as 4–5. The reduction in the

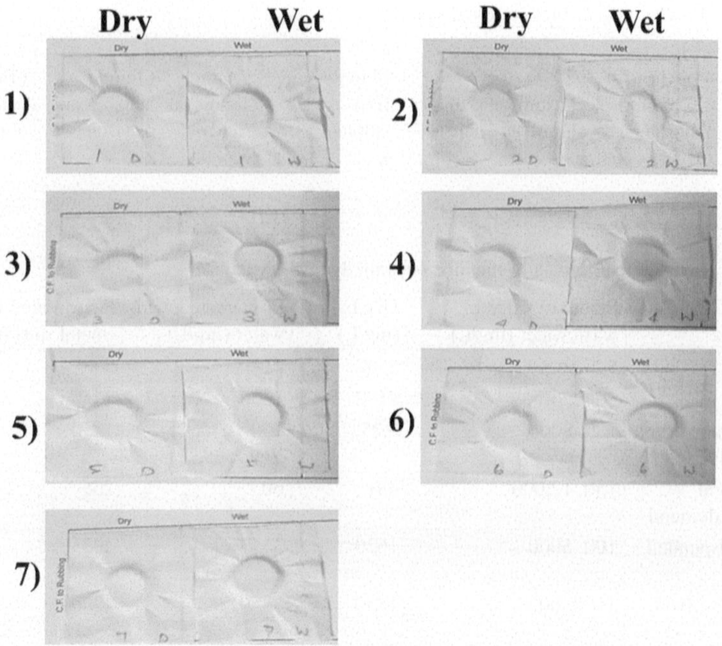

Fig. 5 Samples analysed for colour fastness to washing and their staining wet and dry rubbing

Table 7 Rubbing fastness of phycocyanin dyed samples

Sample		1 (Alum, cream of tartar)	2 (Alum, alum)	3 (Tannic acid)	4 (Tannic acid, ferrous sulphate)	5 (Alum, tannic acid)	6 (Tannic acid, cream of tartar)	7 (Tannic acid, alum)
Dry rub	Grade	5	5	5	5	5	5	5
Wet rub	Grade	4–5	4–5	4–5	4	4–5	4–5	4–5

rating in the wet state was primarily due to the poor adherence of the dye molecules to the silk fabric.

6.5 *Colour Fastness to Dry Cleaning*

Silk fabrics are often dry cleaned due to their higher cost and to prevent damage that can occur during regular washing. In this research, phycocyanin-dyed silk fabrics were evaluated for their fastness to dry cleaning, with the results presented in Table 8. Compared to regular laundering, the dry cleaning process had little impact on colour fastness. All the samples showed minimal colour changes. As previously

Table 8 Dry cleaning fastness of dyed samples

Sample	1 (Alum, cream of tartar)	2 (Alum, alum)	3 (Tannic acid)	4 (Tannic acid, ferrous sulphate)	5 (Alum, tannic acid)	6 (Tannic acid, cream of tartar)	7 (Tannic acid, alum)
Change in shade	4	4	5	5	5	5	4

Table 9 Environment impact of the phycocyanin dyeing wastewater

Effluent parameter	Standard value of wastewater (mg/L)	Dye bath (mg/L)	Post-treated acid water (mg/L)	Post-treated water (metal salt) (mg/L)
pH	5.5–10.5	5.16	4.16	5.89
Colour	50–2500	2100	1000	2350
Chemical oxygen demand	150–8000	2.25	700	1250
Biological oxygen demand	150–12,000	481	60	208
Total suspended solids	100–5000	1620	3545	3975
Total dissolved solids	1500–6000	1850	3850	4200
Alkalinity	500–800	–	459	567

mentioned, the tannic acid-mordanted samples exhibited higher colour fixation with a rating of 5, compared to other mordants, which had a rating of 4, as given in Table 8.

6.6 *Wastewater Analysis from the Dyeing Process*

Apart from the colour evaluation, the study also analysed the environmental impact of the effluent that was released from the dyeing process. In this aspect, the effluent from various stages of the dyeing process namely, from dye bath, after acid treatment for neutralization and effluent after post-mordant treatment was collected and checked for values of pH, alkalinity, chemical oxygen demand, biological oxygen demand, colour, total dissolved solids, and total suspended solids. From the results reported in Table 9, all the selected parameters of the effluent were under the limits of standards prescribed by the Central Pollution Control Board, India. As per the analysis, a higher concentration of biological oxygen demand was reported with the dye bath effluent in the range of 500 mg/L due to the use of dye in the process. Whereas the other parameters are concerned namely chemical oxygen demand and total dissolved solids, a higher concentration was noted in the post-treated effluent obtained from the final process. However, the values were under the limit. The results showcased that, the use of metal mordant is one of the major causes of higher

COD and TDS in the post-treated effluent. Despite the adherence to the norms, future studies should make efforts to reduce the value further, and hence the effluent will be free from all pollutants and the process will become eco-friendly.

6.7 Discussion on the Results

Despite the brilliant blue colour of the phycocyanin pigment, a major issue identified in this research is its poor affinity for textile fibres. Preliminary analysis revealed weak adherence to cotton and linen fabrics. Among all the tested fabrics, silk performed better. However, in the case of cellulosic fibres, the interaction with the phycocyanin molecule was not permanent, and the dye instantly lost its adhesion when exposed to water. Previous studies have attributed the poor fastness properties of the dye on cellulosic fabrics to its protein components (Baek et al., 2022). In this research, efforts were made to dye silk fabric, and the fastness results were moderate. Although both silk and the dye contain protein components, their interaction was weak due to low chemical affinity. The dyed silk fabric exhibited good fastness when treated with a combination of tannic acid and ferrous sulphate for pre and post-mordanting, respectively. Despite good staining scale ratings, the fastness was improved in only a few combinations with tannic acid. The primary reason for the improved fastness in these cases is the effective chemical interaction between the dye and the textile material (Burkinshaw & Kumar, 2009). Regarding perspiration fastness, the dyed silk fabric performed well, especially in combinations of tannic acid with ferrous sulphate, cream of tartar, and alum, achieving a rating of 4 and above. The staining ratings were also satisfactory. Similarly, the dye showed effective performance in both wet and dry rubbing, as well as dry cleaning tests, with excellent fastness results, particularly in dry cleaning and rubbing evaluations.

The main limitation of this study is its exclusive focus on colour extraction from phycocyanin powder using methanol, without evaluating other solvents. Methanol was selected based on its effectiveness across various plant species, but future research could explore different extraction mediums, which may yield better dyeability and more durable adhesion to fabrics. Additionally, no dye-fixing agents or crosslinking agents were used in this study to enhance the dye's fixation on textiles. Based on the performance of tannic acid in this analysis, it is plausible that a crosslinker could significantly improve the dye's affinity for silk fabric. This approach could also be extended to cellulosic textiles. Neither this research nor previous studies (Baek et al., 2022) have attempted crosslinking on either cellulosic or silk fabrics. However, a recent study demonstrated that the use of the bio-crosslinker genipin improved both dye affinity and wash durability (Wu et al., 2023). Future efforts should focus on testing various crosslinking agents with both cellulosic and protein fibres to better understand the effective fixation of phycocyanin dyes. The effluent analysis results indicated that the dyeing method used was effective in reducing pollution. However, a higher level of COD and TDS was observed (though still within acceptable limits), likely due to the use of metal mordants. While these mordants were effective in this study, from an environmental perspective, alternative mordants should be identified for phycocyanin dyeing across various textile materials.

7 Conclusions

The use of natural dyes is a sustainable approach from an environmental perspective. In this project, an extract of *Spirulina platensis* was utilized as a sustainable dye for textile materials in an economically feasible manner. The dyeing process was carried out using several mordants, with tannic acid showing better performance as both a pre- and post-mordant when combined with ferrous sulphate. While the dye produced a light colour on silk fabric, its fixation on cotton and linen fabrics was less effective. The tannin-mordanted process demonstrated higher colour strength and better performance against washing. However, other mordants resulted in very poor fastness ratings, despite their good staining properties. Phycocyanin-dyed silk fabric exhibited better fastness against perspiration, wet and dry rubbing, and dry cleaning. Although the study demonstrated the potential of phycocyanin dyes in textile applications, the overall performance of the dye was suboptimal and did not meet commercial standards. Future studies should focus on improving dye fixation, particularly on cellulosic textiles. While several mordants were evaluated, further exploration of other options is necessary to enhance wash fastness. The wastewater analysis indicated that the dyeing and post-processing effluents were within acceptable wastewater standards. However, higher total dissolved solids (TDS) levels were observed due to the use of metal mordants, which should be considered in future research.

References

Ab Kadir, M. I., Wan Ahmad, W. Y., Ahmad, M. R., Abdul Jabbar, H., Ngalib, K., & Ismail, A. (2014a). Dyeing properties and absorption study of natural dyes from seaweeds, Kappaphycus alvarezii. In *Proceedings of the International Colloquium in Textile Engineering, Fashion, Apparel and Design* 2014 *(ICTEFAD 2014)*. https://doi.org/10.1007/978-981-287-011-7_18

Ab Kadir, M. I., et al. (2014b). Utilization of eco-colourant from green seaweed on textile dyeing. In *Proceedings of the International Colloquium in Textile Engineering, Fashion, Apparel and Design* 2014 *(ICTEFAD 2014)*. https://doi.org/10.1007/978-981-287-011-7_14

Ab Kadir, M. I., Ahmad, M. R., Ismail, A., & Abdul Jabbar, H. (2016). Investigations on the cytotoxicity, neurotoxicity and dyeing performances of natural dye extracted from *Caulerpa lentillifera* and *Sargassum* sp. seaweeds. *Advances in Applied Sciences, 1*(3).

Ab Kadir, M. I., Ahmad, M. R., & Ismail, A. (2017). Partial characterisation and colorimetric characteristics of sargassum sp. colorant on treated polyester fabric with dendrimer. *Key Engineering Materials, 737*, 329. https://doi.org/10.4028/www.scientific.net/KEM.737.329

Ajmal, M., Adeel, S., Azeem, M., Zuber, M., Akhtar, N., & Iqbal, N. (2014). Modulation of pomegranate peel colourant characteristics for textile dyeing using high energy radiations. *Industrial Crops and Products, 58*, 188. https://doi.org/10.1016/j.indcrop.2014.04.026

Ali, S. (2007). *Evaluation of cotton dyeing with aqueous extracts of natural dyes from indigenous plants*. University of Agriculture, Faisalabad, Pakistan. Retrieved October 2, 2024, from http://142.54.178.187:9060/xmlui/handle/123456789/10203

Azeem, M., et al. (2019). Harnessing natural colorants from algal species for fabric dyeing: A sustainable eco-friendly approach for textile processing. *Journal of Applied Phycology, 31*(6), 3941. https://doi.org/10.1007/s10811-019-01848-z

Baek, N. W., Zhang, X., Lou, J. F., & Fan, X. R. (2022). Dyeing fabrics with a colorant extracted from blue-green algae. *AATCC Journal of Research, 9*(5), 223. https://doi.org/10.1177/24723444221103673

Bechtold, T., Mahmud-Ali, A., & Mussak, R. (2007). Natural dyes for textile dyeing: A comparison of methods to assess the quality of Canadian golden rod plant material. *Dyes and Pigments, 75*(2), 287. https://doi.org/10.1016/j.dyepig.2006.06.004

Bisht, G., Srivastava, S., Kulshreshtha, R., Sourirajan, A., Baumler, D. J., & Dev, K. (2020). Applications of red pigments from psychrophilic Rhodonellum psychrophilum GL8 in health, food and antimicrobial finishes on textiles. *Process Biochemistry, 94*, 15. https://doi.org/10.1016/j.procbio.2020.03.021

Burkinshaw, S. M., & Kumar, N. (2009). The mordant dyeing of wool using tannic acid and $FeSO_4$, Part 1: Initial findings. *Dyes and Pigments, 80*(1), 53. https://doi.org/10.1016/j.dyepig.2008.05.008

Chander, M. (2022). Microbial production and applications of pigments. *International Research Journal of Engineering and Technology*. 9(9),366–375

Ciptandi, F., Susilowati, T. H., & Ramadhan, M. S. (2021). Opportunities of using Spirulina platensis as homemade natural dyes for textiles. *Open Agriculture, 6*(1), 819. https://doi.org/10.1515/opag-2020-0167

Delgado, C., Forman, G., & Apolinário, A. (2024). Blue-green horizons: Redefining alginate bioplastics with spirulina dyes. *Materials Circular Economy, 6*(1). https://doi.org/10.1007/s42824-023-00099-1

Doble, M., & Kumar, A. (2005). *Biotreatment of industrial effluents*. Elsevier. https://doi.org/10.1016/B978-0-7506-7838-4.X5000-3

El-Khatib, E. M., Ali, N. F., & El-Mohamedy, R. S. R. (2016). Enhancing dyeing of wool fibers with colorant pigment extracted from green algae. *Journal of Chemical and Pharmaceutical Research, 8*(2).

Ghaffar, A., et al. (2023). Plant extracted natural fluorescent protein C-phycocyanin doped in PVA nanofibers for advanced apparel application. *Materials Research Express, 10*(5). https://doi.org/10.1088/2053-1591/acd50f

Guesmi, A., Ladhari, N., Ben Hamadi, N., Msaddek, M., & Sakli, F. (2013). First application of chlorophyll-a as biomordant: Sonicator dyeing of wool with betanin dye. *Journal of Cleaner Production, 39*, 97. https://doi.org/10.1016/j.jclepro.2012.08.029

Hoseini, S. M., Khosravi-Darani, K., & Mozafari, M. R. (2013). Nutritional and medical applications of spirulina microalgae. *Mini-Reviews in Medicinal Chemistry, 13*(8). https://doi.org/10.2174/1389557511313080009

Hossain, S., Jalil, M. A., Kader, A., & Bin Kamal, S. A. (2022). Excellent dyeing properties of a natural dye extracted from the leaves of Phoenix dactylifera Linn on cotton and silk fabrics. *Journal of Natural Fibers, 19*(15), 9803. https://doi.org/10.1080/15440478.2021.1993410

ILO. (2021). *International Labour Organization X Exposure to hazardous chemicals at work and resulting health impacts: A global review*. Geneva.

Imchen, T., & Singh, K. S. (2023). *Marine algae colorants: Antioxidant, anti-diabetic properties and applications in food industry*. https://doi.org/10.1016/j.algal.2022.102898

Ismail, M., et al. (2016). *The effect of cationic surfactant treatment on the dyeability of cotton and silk fabrics with natural dye from brown seaweeds Sargassum sp*. International Journal of Applied and Natural Sciences, 5 (5), 19–28.

Jihad, R. (2014). Dyeing of silk using natural dyes extracted from local plants. International Journal of Scientific and Engineering Research, 5(11), 809–818.

Kadir, M. I. A., Ahmad, M. R., & Jabbar, H. A. (2023). Dyeability of surface-modified cotton and silk fabrics with PAMAM dendrimer using Sargassum sp. seaweed extract. *Trends in Sciences, 20*(9). https://doi.org/10.48048/tis.2023.5807

Kanchana, R., Fernandes, A., Bhat, B., Budkule, S., Dessai, S., & Mohan, R. (2013). Dyeing of textiles with natural dyes—An eco-friendly approach. *International Journal of ChemTech Research, 5*(5), 2102–2109.

Khan, A. A., Iqbal, N., Adeel, S., Azeem, M., Batool, F., & Bhatti, I. A. (2014). Extraction of natural dye from red calico leaves: Gamma ray assisted improvements in colour strength and fastness properties. *Dyes and Pigments, 103*, 50. https://doi.org/10.1016/j.dyepig.2013.11.024

Kimura, M. (1993). *Dyeing with phycocyanin*. JPH07166480A.

Kraseasintra, O., et al. (2022). Application of phycocyanin from Arthrospira (Spirulina) platensis as a hair dye. *Frontiers in Marine Science, 9*. https://doi.org/10.3389/fmars.2022.1024988

Licina, D., Morrison, G. C., Bekö, G., Weschler, C. J., & Nazaroff, W. W. (2019). Clothing-mediated exposures to chemicals and particles. *Environmental Science & Technology*. https://doi.org/10.1021/acs.est.9b00272

Lyon, F. (2010) *IARC monographs on the evaluation of carcinogenic risks to humans. Vol. 93. Carbon black, titanium dioxide, and talc.*

Mao, M., Han, G., Zhao, Y., Xu, X., & Zhao, Y. (2024). A review of phycocyanin: Production, extraction, stability and food applications. *International Journal of Biological Macromolecules, 280*, 135860. https://doi.org/10.1016/j.ijbiomac.2024.135860

Mir, R. A., et al. (2019). Green algae, Cladophora glomerata L.–based natural colorants: Dyeing optimization and mordanting for textile processing. *Journal of Applied Phycology, 31*(4), 2541. https://doi.org/10.1007/s10811-018-1717-6

Moldovan, S., Ferrandiz, M., Franco, E., Mira, E., Capablanca, L., & Bonet M. (2017). Printing of cotton with eco-friendly, red algal pigment from Gracilaria sp. In *IOP Conference Series: Materials Science and Engineering*. https://doi.org/10.1088/1757-899X/254/19/192011

Mona, S. F. P., & Sekarenthiran, C. S. (2019). *Extraction of algal pigments and their suitability as natural dyes*.

Mongkholrattanasit, R., Krystufek, J., Wiener, J., & Studnickov, J. (2011). Natural dye from eucalyptus leaves and application for wool fabric dyeing by using padding techniques. In *Natural dyes*. https://doi.org/10.5772/20738

Moorhead, K., & Capelli, B. (1993). Spirulina nature's superfood. *Agricultural and Food Sciences, 5*, 66–78.

Muruganandham, M., et al. (2023). An eco-friendly ultrasound approach to extracting yellow dye from Cassia alata flower petals: Characterization, dyeing, and antibacterial properties. *Ultrasonics Sonochemistry, 98*, 106519. https://doi.org/10.1016/j.ultsonch.2023.106519

Mutaf-Kılıc, T., Demir, A., Elibol, M., & Oncel, S. S. (2023). *Microalgae pigments as a sustainable approach to textile dyeing: A critical review*. https://doi.org/10.1016/j.algal.2023.103291

Pizzicato, B., Pacifico, S., Cayuela, D., Mijas, G., & Riba-Moliner, M. (2023). Advancements in sustainable natural dyes for textile applications: A review. *Molecules, 28*. https://doi.org/10.3390/molecules28165954

Prabhu, K. H., & Teli, M. D. (2014). Eco-dyeing using Tamarindus indica L. seed coat tannin as a natural mordant for textiles with antibacterial activity. *Journal of Saudi Chemical Society, 18*(6), 864. https://doi.org/10.1016/j.jscs.2011.10.014

Rahman Bhuiyan, M. A., Ali, A., Islam, A., Hannan, M. A., Fijul Kabir, S. M., & Islam, M. N. (2018). Coloration of polyester fiber with natural dye henna (Lawsonia inermis L.) without using mordant: A new approach towards a cleaner production. *Fashion and Textiles, 5*(1). https://doi.org/10.1186/s40691-017-0121-1

Rani, K. (2020). Utilizing natural colorants of brown seaweeds in bioactive fabric dyeing. *Journal of Weed Science Research, 27*(2), 403–414. https://doi.org/10.28941/pjwsr.v26i4.860

Rani, K., et al. (2020). A potential benefit of brown seaweed (Stoechospermum marginatum) using for sustainable fabric dyeing. *Journal of Research in Weed Science Journal, 3*. https://doi.org/10.26655/JRWEEDSCI.2020.2.1

Ranson, B. (n.d.). *The true cost of colour: The impact of textile dyes on water systems*. Fashion Revolution. https://www.fashionrevolution.org/the-true-cost-of-colour-the-impact-of-textile-dyes-on-water-systems/

Roda-Serrat, M. C., Christensen, K. V., El-Houri, R. B., Fretté, X., & Christensen, L. P. (2018). Fast cleavage of phycocyanobilin from phycocyanin for use in food colouring. *Food Chemistry, 240*, 655. https://doi.org/10.1016/j.foodchem.2017.07.149

Ferrándiz, M., Moldovan, S., Mira, E., Pinchetti, J.L.G., Rodríguez, T., Abreu, H., Rego, A.M., Palomo, B., Caro, P. (2016). Phycobiliproteins – new natural dyes from algae as a sustainable method. Fibres and Textiles, 23(3), 56–61.

Samanta, A. K., & Konar, A. (2011). Dyeing of textiles with natural dyes. In *Natural dyes* (p. 30). InTechOpen.

Saxena, S., & Raja, A. S. M. (2014). Natural dyes: Sources, chemistry. *Application and Sustainability Issues*. https://doi.org/10.1007/978-981-287-065-0_2

Shahid, M., Shahid-Ul-Islam, & Mohammad, F. (2013). Recent advancements in natural dye applications: A review. *Journal of Cleaner Production*. https://doi.org/10.1016/j.jclepro.2013.03.031

Swami, C., Saini, S., & Gupta, V. B. (2012). Extraction of a natural dye from Sesbania aculeata plant. *Journal of Textile and Apparel, Technology and Management, 7*(4).

Swedish Chemicals Agency. (2016). *Hazardous chemical substances in textiles—Proposals for risk management measures*. Report from a Government assignment, Stockholm. Retrieved October 2, 2024, from https://www.kemi.se/download/18.6df1d3df171c243fb23a98e7/1591454108832/report-8-16-hazardous-chemical-substances-in-textiles.pdf#page=64.08

Wicker, A. (2023). *To dye for: How toxic fashion is making us sick—And how we can fight back* (1st ed.). G.P. Putnam's Son.

Willis, S. (2023). How sustainable dyeing is changing the textile industry. Retrieved October 2, 2024, from https://www.plugandplaytechcenter.com/insights/how-sustainable-dyeing-changing-textile-industry

Wiraningtyan, A., Ruslan, R., Ayu Mutmainnah, P., & Perkasa, M. (2020). Application of natural dyes and sodium alginate from Sargassum sp. sea weed in coloring Bima woven fabric. *Oriental Journal of Chemistry, 36*(05), 964. https://doi.org/10.13005/ojc/360524

Wu, M., Zhou, Y., & Tang, R. C. (2023). Bridging phycocyanin onto silk by genipin towards durable colouristic, antioxidant and UV protective properties: A sustainable strategy for fully bio-based textile. *Chemical Engineering Journal, 477*, 146808. https://doi.org/10.1016/j.cej.2023.146808

Zarkogianni, M., Mikropoulou, E., Varella, E., & Tsatsaroni, E. (2011). Colour and fastness of natural dyes: Revival of traditional dyeing techniques. *Coloration Technology, 127*(1), 18. https://doi.org/10.1111/j.1478-4408.2010.00273.x

Zerin, I., Farzana, N., Sayem, A. S. M., Anang, D. M., & Haider, J. (2020). Potentials of natural dyes for textile applications. *Encyclopedia of Renewable and Sustainable Materials, 1–5*. https://doi.org/10.1016/B978-0-12-803581-8.11668-6

A Critical Analysis on the Coloration of Textiles Using Natural Colorants

Chandrasekaran Karuppuswamy

Abstract Natural colorants are equipped with medicinal rich qualities, sustainable, comfortable to human skin, and biodegradable. In the ancient times, the clothes were colored with natural colorants only. Later the emergence of synthetic colorants, their availability and their consistency in shade reproducibility shifted the dyers toward synthetic colorants. Even though various colors can be produced using synthetic colorants, their effect on human skin is not skin-friendly and many people report skin allergy/ailments. Nowadays people are very much conscious and expect eco-friendliness in whatever they use in their day-to-day activities and are paying more attention toward natural colored clothes. Even though the color fastness and the shade reproducibility are the challenging task in natural coloration, efforts have been taken by the natural dyers and researchers to resolve these problems. In this chapter, an attempt is made to explore the various natural resources and the possible colors produced from them. A critical analysis was carried out pertaining to the various techniques used for the natural coloration of textiles using different types of mordants and the coloration technique employed without the use of mordants.

Keywords Natural/Ayurvedic dyeing or coloration · Natural dyes or colorants · Mordants · Sustainability · Medicinal value

1 Introduction

Dyes and pigments have been in use for a long time and find wide uses in various fields especially in textile industries. This has led to widespread research activities on the emergence of colorants from natural origin. Natural colorants are possessed

C. Karuppuswamy (✉)
Department of Fashion Technology, PSG College of Technology, Coimbatore, Tamil Nadu, India
e-mail: kcs.fashion@psgtech.ac.in

S. S. Muthu (ed.), *Sustainable Coloration of Textiles*, Sustainable Textiles: Production, Processing, Manufacturing & Chemistry,
https://doi.org/10.1007/978-3-031-91217-7_5

with benefits such as they are sustainable, soft, lustrous, and calming to the human eye. Natural colorants are free from health hazards and disposal problems. A colorant should offer decent act on the substrates, economically feasible, and eco-friendly. Even though natural colorants offer many advantages such as non-toxicity, eco-friendliness, and effortless extraction, few colorants often equipped with medicinal properties, suitable for coloration of materials. Despite these advantages, they are uneconomical due to low produce and deprived effectiveness. Natural colorants now have begun as capable alternatives to the toxic synthetic colorants (Yadav et al., 2023). Now more information is available to know the chemistry of natural colorants and co-ordination between metallic mordant, natural colorants, and natural fibers, to yield higher depth of shade of natural color with repeatable shades by means of standardization of extraction, mordanting techniques, and coloration conditions and further to maximize color fastness for such natural colored textiles using natural fibers (Samanta et al., 2018). With the growing alertness about the health issues and hazards of synthetic colorants, the old age practice of textile coloration with natural sources are emerging yet again. People who have worried about the human health and the environment, making tremendous efforts to recover the traditional culture of coloration with the natural resources using plants, insects, and minerals. These natural resources are biodegradable as well as possessing medicinal properties with them (Srivastava & Singh, 2019). Between all the plant-based colorant sources, plant leaf colorant sources are very essential for textile coloration as it offers both colorant and fragrance. As natural colorant demonstrates non-toxic, non-allergic effects and results in lower pollution and unintended consequences, it emerged as a key area in the field of textile coloration investigation. Despite having an extensive range of flora and fauna, very few has been exploited for the fulfillment of textile coloration (Sujitha et al., 2019). Natural colorants are an essential group of resources which can be seen used in various types of industries including textiles (Křížová, 2015).

This chapter aims to explore the various aspects associated with the sustainable coloration of textiles using natural colorants such as its sources, extraction methods, mordants, mordanting methods, coloration techniques, need for standardization, disposal of effluents, and sustainability.

2 Sources of Natural Colorants and Their Significance

Natural colorants are attained from diverse natural origins. These colorants are normally utilized in the coloration of many fields including textiles. Owing to the imperishable and toxic nature inherent with the synthetic colorants, people transitioned to utilizing natural colorants. Textile industry is the major area for the use of natural colorants, on the other hand, it can also be utilized in other fields. There are many investigations that has been carried out by the research community affiliated to various categories of organizations with regard to the rehabilitation of natural colorants, but more investigations are needed to regularize the methods for its uses (Chungkrang et al., 2021). Colorants for textile coloration can be grouped into two main categories, natural and synthetic colorants. Since olden days, natural colorants are used for coloring food medium, leather, and natural fibers such as wool, silk, and cotton. Now, synthetic compounds are used for coloring textile materials and they lead to water pollution as well as waste disposal problems as they are imperishable and hazardous. These problems can be sorted out with the use of natural colorants. Natural dyes are colorants obtained from various natural origins. Majority of natural colorants are plant-based colorants, the major origins of which are the parts of plants. The other notable origin is from living sources such as fungi, snails, insects, and many more. Since all the natural materials are possessed with color producing ability, there are some origins excluding plants that offer some costly colors like sea snails, Cochineal insects, and many more. Currently, many commercial dyeing units have started using natural colorants to combat the ecological harm arising from synthetic colorants. The bulk of the plants utilized for the removal of colorants are categorized as therapeutic and few of them are proven with their antimicrobial effect. The microbe-fighting attributes of the plant colorants pay to the longer durability of the textile materials if they are applied to textile fabrics. Natural colorants are proved to be a cheap, non-toxic, renewable, and sustainable resources (Tamilarasi & Banuchitra, 2021). At present, the requirement for natural colorants has been rising quickly because of the growing consciousness on harmful, poisonous, and allergenic responses linked with chemical colorants. Among various plant-derived colorant origins, floral colorant sources are very much important for textile coloration as it offers both colorant and fragrance. Even though the availability of various flora and fauna, very few has been utilized for the requirement of textile coloration. Data documentation and its analysis are available for precious floral resources with regard to natural coloration. It emphasize on forming strategy for conservation of floral dye resources (Singh & Srivastava, 2015). The production of synthetic dyes need more energy and is associated with environment pollution. Major synthetic colorants particularly azo dyes are reported to be carcinogenic. Natural colorants have mostly been demonstrated as preservation and re-establishment of textiles with substitution of synthetic colorants with natural colorants. The research activities conducted with regard to standardization of natural colorants are found low. Very few thoughtful efforts have been made so far to create fresh details on the utilization of natural colorants. The standardized natural extracts are very much essential for certain fields including textiles. The advantages of natural dyes are often linked in terms of fastness to washing and light and further enhanced by correct choice of natural mordants and separation coupled with effective use of technology and sustainable process. Natural coloration of textiles by processing operations in extensive coloration unit is currently a certainty in the textile field of sustainable textiles (Prabhu & Bhute, 2012). Currently, there exist a huge demand for the utilization of natural colors around the globe due to the imperishable and toxic nature inherent with synthetic colorants. The colors produced by natural colorants are visually calming, earthbound, pleasant, and charming (Chungkrang & Bhuyan, 2020).

3 Extraction of Natural Colorants

Preventing ecological pollution is the major objective of colorant extraction from plant bases. The study reported uses three different raw materials such as onion peel, marigold, pomegranate through which dye extraction was carried out using different extraction processes. The color derived from pomegranate, marigold, and onion peels reported to offer many benefits. The main benefit reported was that waste material could be utilized to isolate the colorant. It controls waste materials to yield required colorant, a renewable and biodegradable colorant base. Obtaining raw materials reported to be at substantially low cost. A cotton fabric was dyed and the color observed reported with highly economical and eco-friendly since it is prepared from onion peels, marigold flowers, and pomegranate seeds. The study further reported that methanol and water solution used for the extraction process yielded more color strength than other extraction methods. The study further reported that a diversity of soft and light colures can be developed by joining many mordants and mordanting methods (Bhangale et al., 2023). In the study reported on the separation of colorants from ten plant materials, the attained colorants obtained variety of colors and are treated with various types of mordants. The cotton and jute yarns processed with the mordant sodium chloride reported with best results and exhibited lower fastness to washing (Kumar & Prabha, 2018). In the study conducted with microwave radiations (MW) as a eco-friendly separation tool, to demonstrate the intrinsic coloration capability of tea leaves for the coloration of cotton fabrics, for separation of natural colorant from tea leaves, MW irradiation for 1–6 min was given to color cotton fabric both in aqueous and basic media. The study reported that a rise in color yield (K/S) was observed when 6 min of microwave energy was used during the use of basic extract of tea leaves on the coloration of cotton fabric. The study further revealed that 6% of Cu and 8% of Fe offered the best fastness properties and the maximum color yields in pre-mordanting and post-mordanting methods, respectively. The study concluded that MW energy has the potential for the separation of colorant from tea leaves for coloration of cotton fabric under lowered optimal coloration conditions. Microwave radiation treatment given for 6 min in the basic medium was reported to yield high color strength. The study results further show that catechin from tea leaves separated in a basic medium by using microwave treatment of 6 min has offered enhanced results of natural brown colorant on irradiated cotton fabric. It has been further reported that superior color strengths were obtained with the use of eco-friendly chemical mordants (Adeel et al., 2021).

4 Effect of Mordants and Mordanting Methods

In the study conducted on dyeing of cotton fabric with natural dye from Haldi, Anar, and Katha extract, to know the effect of natural and chemical modern fixation of dye on cotton fabric and fastness to repeated washing, the process of mordanting was carried out for increasing color fastness of natural dye. Each mordant gave different shades and their fastness also reported to be different. The fastness of a natural dye on cotton fabric can be increased by using alum or potassium dichloride and or vinegar during pre-mordanting and mordanting process. The study concluded that chemical mordant can be substituted by the natural mordant to make the entire process more eco-friendly and easily (Bhairam, 2023). There are huge gaps exist in optimizing coloration conditions for particular fiber-mordant-colorant groupings and there is further research needed on the effects of various eco-friendly chemical mordants, which minimize environmental hazards and ensures effective dye fixation and bio mordants, which are derived from tannin based natural compounds and mordanting assistants, which are derived from gallic acid or chebulinic acid based natural compounds for the regularization of coloration process variables for getting even and consistent shades to attain with natural colors. Another aspect of using bacterial/fungal resistance and UV protection action or deodorizing action of particular natural colorants are with thorough investigation of the influence of various post-treatment compounds, bacterial resistance/UV uptake materials for enhancing its applications as renowned textiles. Research activities carried out using various natural and eco-friendly synthetic colorant fixers for enhancing wash-resistant color stability and sunlight/UV protection of natural colored textiles can be maximized by appropriate post-treatment with UV absorbers. For enhancing rubbing fastness, post-treatment with natural binders or various reactive thickeners and eco-friendly synthetic binding agents are needed, moreover efforts to maximize microbicidal and ultraviolet shielding potential of such natural colored textiles. It is further reported that it is equally imperative to see and comprehend well the precise fiber-mordant dye co-ordination and function of various prior and after treatments on stimulating color yield (K/S value), uniformity of color yield (CV% of K/S value) along with rating for microbicidal and ultraviolet protection factor for various fiber-mordant-natural colorant grouping as appropriate especially to cotton, silk, wool and jute fibers when colored with aqueous extract of whichever identified natural colorant (Samanta, 2020). The study conducted with the Syzgium cumini (SC) plant extract was dyed with Silk fabrics by various mordanting techniques with different mordants. Next to coloration, the colored Silk was put into various fastness tests. As the flower of Syzgium cumini (SC) are from a sustainable source and are plentifully obtainable, the natural colorant yielded may be recognized as a promising origin of sustainable natural colorant for textiles. The study results reported that natural colorant can be effectively obtained from the flower of Syzgium cumini (SC). The quantum yield of the coloring matter from Syzigum Cumini (SC) was reported to be 4% and offered better results. The study further reported that mordant treatment improved the color strength and also durability of the natural colorant, yielded varieties of shades; therefore, this natural colorant source was reported as a promising alternative to synthetic colorants (Santhi & Kowsalya, 2021). In the reported work, colorants were extracted from various parts of plants and coloration process of silk was carried out using these colorants by various mordanting techniques with diverse range of mordants. After the coloration process, various fastness tests were carried out and from the study results it was observed that the root of rumex abisinicuss jaq and eucalyptus leave, colorant can be effectively applied for coloration of silk in order to attain an extensive variety of soft, pastel, and light colors with the natural colorants and metallic mordants. The color fastness test results reported with excellent fastness to washing except in the case of prior and after mordanting methods; excellent fastness to rubbing except in the case of pre-, simultaneous, and post-mordanting with $K_2Cr_2O_7$. When comparing with the various fiber-mordanting systems investigated, it was reported that the application of 3% of ferrous sulphate using simultaneous mordanting for succeeding coloration on cotton and silk with extract of eucalyptus colorant, 3% of aluminum sulphate applied using simultaneous mordanting for succeeding coloration on silk with 5% extract of rumex abisinicuss jaq and 3% of stannous chloride applied with simultaneous mordanting for succeeding coloration on silk with 5% extract of rumex abisinicuss jaq reported with high K/S values when compared to other selective mordanting systems (Jihad, 2014). In the study reported on the determination of the capability of colorants separated from mango leaves in the coloration process of silk, the extraction medium was standardized by separating colorants from specified amount of crushed leaves by maintaining the pH value range from 3 to 12. The study further reported that at pH 10, maximum relative color strength of the separated colorant liquor was observed, the optimum colorant separation parameters were reported to be 98 °C (Temperature), 60 min (time), and 1:10 (M:L Ratio), respectively. The study further reported that coloration was done using the standardized colorant extract using silk fabrics treated with and without mordants and the colored silk fabrics were tested for the color yield and durability. The study further emphasized that the color values observed were affected with the addition of mordants. It was further reported that mango leaves have strong capability for dyeing of silk fabric; extracted dye liquors possessed with remarkable pH stability in acidic conditions. The order of color yield reported in single mordanting as ferrous sulphate followed by alum and tin. In the case of combined mordanting, the order reported was the combinations of ferrous sulfate and alum followed by alum, tin and tannic acid; ferrous sulfate, alum and tin followed by alum and tin. The study further emphasized that remaining color values were also observed to be affected with mordanting. Fastness against washing and light were also reported to be from good to excellent in major cases. The study report finally reaffirmed that mango leaves have great potential for use on silk fabrics (Uddin, 2015). Irrespective of the mordanting methods (pre/simultaneous/post), the color rest on natural colorant, mordant and mordanting assistants. Though, for a particular pre-mordanting method, color matching database for natural coloration of any specific textile material and match identification are possible with computerized program (Samanta & Agarwal, 2009). The study reported utilized pomegranate and turmeric extracts to color bleached cotton fabric applying several mordants, involving every mordanting techniques such as pre-mordanting, post-mordanting, and simultaneous mordanting. The natural colorants in the mentioned extracts are reported with the presence of various phytochemicals which transmit various colors to the fabrics. The study explored the light sensitivity of these colorants along with the contribution of pH of the coloration process. The study further reported that attainment of diverse ranges of colors are promising by altering the process parameters such as concentration, ratio, pH of the extracts and choice of mordants. The reported work effectively formed an eco-friendly coloration process for cotton fabric with colorants obtained from pomegranate and turmeric. The study further highlighted that the capability of the colorant combination for textile uses, with important observations exploring its adaptability and efficacy. The study further reported that obtained colorant offered a diverse range of color shades with the assistance of many binders, giving design versatility. Particularly, utilizing particular mordants considerably enhanced the color durability of the colored fabric, warranting long life and opposing to fading. Light fastness was reported as moderately well as grade 3–4. The study concluded that natural extracts exhibit in-built bacteria-resistance properties, contributing an additional advantage for textile application and also innovative approach adopted in the study may encourage additional investigation into natural colorants for various types of fabrics, endorsing eco-friendly textile manufacture (Ahire et al., 2024). In an innovative approach adopted in a study, instead of mordants, the fabric was dyed with an-ionic vegetable dye first, then the dyeing process was repeated with cat-ionic vegetable dye at correct pH, temperature and pressure and dyeing time that are specific to particular dye. Common salt was used as an electrolyte with the dyes, which have poor affinity for fiber. The study reported that results were encouraging such as the textile/fiber attains characteristic softness, brilliancy, and strength which were lacking when mordants are used. It has been further reported that the color obtained was also fast in the sense that it does not fade when exposed to sunlight and washed with soap and ordinary detergents (Bharathan & Rajagopalan, 2002).

5 Coloration of Textiles with Various Natural Dyes

Peanut pod (Arachis hypogaea) with firm uses was reported as one of the primary contributing origin of natural colorants. The key coloring constituent in peanut reported was pods. In the study conducted for the evaluation of peanut pod powder as natural textile dyestuff, the work was carried out in three steps. The colorant separation procedure adopted was an olden days practice. Coloration of cotton fabrics was carried out with the extract of peanut pod powder and the coloration method was standardized using three mordanting agents. Ultimately, colored fabrics have been put into to various lab tests. The study concluded that the colorant source, peanut pod are very much opt for cotton material with ferrous mordant and pre-mordanting with 10% and 15% of dye, reported appropriate for cotton material (Chhipa et al., 2017).

Ayurvedic coloration is a traditional method of coloration using diverse natural sources. This coloration method utilizes natural mordants for setting shades and natural gums for infusing the benefits of herbs into the fabric. The aim of ayurvedic dyeing reported is to prepare the fabric focused on the wellness of body. Rust dyeing is a sustainable dyeing process that generates unique surface patterns applying reclaimed rusty iron remnants. Being a colorant origin source, the oxidized iron reported to produce long-lasting, elegant, and intense orange tones on fabric and paper. Rust dyeing is reported to be a fabulous way to upcycle a garment. The rust natural dyeing technique is reported with a great capability to be made into alternative natural colorants since the colors formed are often graded and have different characteristics of each fabric utilized as a transmission pathway and can also penetrate on polyester fabric. In rust dyeing with a pre-mordanting of salt in various fabric types reported to produce a smoke effect and printing flawless motifs, dense color and graded (Kaur & Kaur, 2023).

A vast number of natural colorants/color are sustainable, excluding a little. A range of natural colors are not only sustainable, but also having additional worth for its therapeutic performance and skin-compatibility. Application of appropriate dual or triple combinations of alike or compatible natural colorants for the coloration of sustainable textiles in diverse range of relaxing / unique shades with sustainable mordants and finishing agents are reported to be the best wanted customer centric products for the future. The unrepeatability and low color fastness, etc., have been partially settled by several scientists' uninterrupted efforts in this venture. A textile dyeing industry person must recognize the influence of inconsistency for separation, mordanting, and coloration and utilization of optimized process for the choice of components in the system comprising of fiber, mordant and natural colorant to acquire repeatable color yield and color matching as well to adopt various sustainable ways to enhance color fastness to a maximum extent. Hence, natural coloration of textiles by processing systems in a big coloration unit is currently a certainty in the textile market of sustainable textiles (Samanta & Konar, 2011).

The dyeing and finishing processes in the textile chemical processing require substantial quantity of water, energy, and chemicals. So as to overcome ecological problems, natural colorants are attaining focus from scientists as an alternative to chemical colorants. Capable new technologies and practices have been effectively incorporated to enhance the effectiveness and eco-friendliness of natural coloration processes, but numerous constraints like the low firm fixation of colors with natural colorants, their poor association with textiles substrates, complications in the consistent shades formation and also the other factors such as economic aspects, right now inhibit industry from utilizing natural colorants on a greater extent and still need further investigations so as to maximize their utility outside specialized uses. With regard to sourcing, exciting prospects of eco-conscious advancements are developed through the adaptation of by products and waste from agro-industry as natural colorants origins, which prevents the land use for direct collection and avoids opposing with further uses like food. In the extraction of natural colorants, cutting-edge approaches, like ultrasound and microwave-aided separation or enzymatic separation are currently utilized to lower the intake of resources like water, energy, and chemicals portraying the conventional methods of separation. In this regard, the capability of supercritical fluid separation is even now partially explored in the area of natural colorants but certainly could permit water-less and more eco-friendly separation, assisting to lower the vast quantity of water required to separate natural colorants. In the pre-treatment processes, developments in the use of bio-mordants and enzymes are reported to substitute metallic salts as mordants, while new capable technologies identified are ultrasound, ultraviolet, gamma radiation technologies, electron beam irradiation technology, and plasma treatment. These technologies are expected to help tackle the low fastness properties of natural colorants via eco-friendly process, as they normally have fast, low-temperature processes, needing minimum quantities of water and chemicals. In the coloration process, promising developments and technologies are reported. However, only few information is present regarding the coloration process with ionic liquids use to natural colorants, denote a potential replacement, but hugely lowering the quantity of water utilized in the process (Pizzicato et al., 2023).

6 Disposal of Effluents Generated from Natural Dyeing Process

The waste generated from synthetic dyes in textile industry leads to environmental concerns, and termed as hazardous wastes. In contrary, the usage of natural colorants is reported to yield lesser pollutants. However, the usage of mordant for enhancing the quality of the coloration process may discharge hazardous pollutants since mordant that is commonly used comprises of heavy metals. Hence, to state that the usage of natural colorants supports the sustainable eco-friendly textile sector, it is difficult to identify substitute sources of mordant-like materials that are eco-friendly. It has been stated that tannins could substitute for dyeing process using natural colorants. Eco-friendly use of natural colorants in textile industry may be attained by means of economical production of natural colorants simultaneously safeguarding organic material diversity that is not polluting the ecology, application of mordant-like material from the renewable product like tannin, and encouraging new entrepreneur in natural colorants producer, traders, and textile industry (Elsahida et al., 2019).

7 Standardization of Natural Dyes

Accelerated research advances in synthetic chemistry assisted by the mechanization of textile manufacture set the stage for the formation of synthetic replacements to famous natural colorants moreover to many synthetic colorants in numerous hues and colors that slowly moved the natural colorants into extinction. But, ecological concerns in the manufacturing and use of synthetic colorants have regained consumer attention in natural colorants during the final decades of the twentieth century. In the olden days, colorants were obtained purely from natural origins. But natural colorants struggle from few built-in draw backs of standardized use and the standardization of the colorant itself as colorants collected from related plants or natural origins are affected and subjected to the fluctuations of climate, soil, farming techniques etc. Therefore, for the natural colorants to be truly commercialized and to take a competitive place with respect to the synthetic colorants, the standardization methods play a very significant and vital role. At present, there is no economical method to eradicate both the coloration and harmful features of the colorants dispersed into the eco-system by synthetic colorants. Natural colorants are well known as economical, safe, eco-friendly, and replenishable resource with reduced ecological impact, have captured the focus of the researchers to utilize them in diverse ways of olden and modern usage areas (Verma & Gupta, 2017).

8 Conclusion

Even though the coloration process with natural dyes involves challenges, correct extraction methods, standardization of natural dyes, and careful selection of natural mordants, mordanting methods, dyeing techniques, and mordant free natural dyeing processes will pave the way for this exciting sector to flourish in near future with a tag of sustainable coloration. There is plenty of scope for re-using the remnants (effluents) generated from natural dyeing process. These remnants can be properly dried and can be used as an organic fertilizer. More research works in this area will help to generate organic fertilizers and the farmers can use them and regain the soil fertility of their agricultural lands.

References

Adeel, S., Anjum, M. N., Ahmad, M. N., Fazal-ur-Rehman, Saif, M. J., Azeem, M., Hassan, A., Imran, M., & Amin, N. (2021). Eco-friendly dyeing of cotton fabric with waste tea leaves-based tannin natural dye. *Global NEST Journal, 23*(3), 365–369. https://doi.org/10.30955/gnj.003456

Ahire, B. B., Kasabe, S. M., Mali, A. B., & Jadhav, R. B. (2024). Development of a sustainable dyeing process for cotton fabric utilizing natural dyes from Punica granatum L. and Curcuma longa. *Current World Environment, 19*(1). https://doi.org/10.12944/CWE.19.1.12

Bhairam, R. (2023). To study the dyeability of natural dye for sustainable fabric (cotton). *International Journal of Creative Research Thoughts, 11*(12), 790–795.

Bhangale, P., et al. (2023). Extraction of natural dye. *Medicon Agriculture & Environmental Sciences, 4*(5), 16–21.

Bharathan, & Rajagopalan, M. R. (2002). Natural dyes in Gandhigram. *Journal of Extension and Research, 5*(1), 1–4.

Chhipa, M. K., Srivastav, S., & Mehta, N. (2017). Study of dyeing of cotton fabric using peanut pod natural dyes using Al_2SO_4 $CuSO_4$ and $FeSO_4$ mordanting agent. *International Journal of Environmental & Agriculture Research, 3*(2), 36–44.

Chungkrang, L., & Bhuyan, S. (2020). Natural dye sources and its applications in textiles: A brief review. *International Journal of Current Microbiology and Applied Sciences, 9*(10), 261–269. https://doi.org/10.20546/ijcmas.2020.910.034

Chungkrang, L., Bhuyan, S., & Phukan, A. R. (2021). Natural dyes: Extraction and applications. *International Journal of Current Microbiology and Applied Sciences, 10*(01), 1669–1677. https://doi.org/10.20546/ijcmas.2021.1001.195

Elsahida, K., et al. (2019). Sustainability of the use of natural dyes in the textile industry. *IOP Conference Series: Earth and Environmental Science, 399*, 012065. IOP Publishing. https://doi.org/10.1088/1755-1315/399/1/012065

Jihad, R. (2014). Dyeing of silk using natural dyes extracted from local plants. *International Journal of Scientific & Engineering Research, 5*(11), 809–818.

Kaur, H., & Kaur, H. (2023). A review on rust dyeing and Ayurveda dyeing on silk with onion peel and Harad. *National Multidisciplinary Conference on Emerging Trends, Opportunities and Challenges in Higher Education, International Journal of Scientific Research in Science and Technology, 10*(7), 141–147.

Křížová, H. (2015). Natural dyes: Their past, present, future and sustainability. In D. Křemenáková, J. Militký, & R. Mishra (Eds.), *Recent developments in fibrous material science* (pp. 59–71). o.p.s.

Kumar, V., & Prabha, R. (2018). Extraction and analysis of natural dye. *Journal of Natural Product and Plant Resources, 8*(2).

Pizzicato, B., Pacifico, S., Cayuela, D., Mijas, G., & Riba-Moliner, M. (2023). Advancements in sustainable natural dyes for textile applications: A review. *Molecules, 28*(16), 5954. https://doi.org/10.3390/molecules28165954

Prabhu, K. H., & Bhute, A. S. (2012). Plant based natural dyes and mordants: A review. *Journal of Natural Product and Resources, 2*(6), 649–664.

Samanta, P. (2020). A review on application of natural dyes on textile fabrics and its revival strategy. In A. K. Samanta, S. N. Awwad, & H. M. Algarni (Eds.), *Chemistry and technology of natural and synthetic dyes and pigments*. https://doi.org/10.5772/intechopen.90038

Samanta, A. K., & Agarwal, P. (2009). Application of natural dyes on textiles. *Indian Journal of Fibre & Textile Research, 34*, 384–399.

Samanta, A. K., & Konar, A. (2011). Dyeing of textiles with natural dyes. In E. P. A. Kumbasar (Ed.), *Natural dyes*. https://doi.org/10.5772/1817

Samanta, P., Singhee, D., & Samanta, A. K. (2018). Fundamentals of natural dyeing of textiles: Pros and cons. *Current Trends in Fashion Technology & Textile Engineering, 2*(4), 555593. https://doi.org/10.19080/CTFTTE.2018.02.555593

Santhi, M., & Kowsalya, S. (2021). Dyeing of silk with eco-friendly natural dyes obtained from flower of Syzgium cumini (sc) plant using different mordants. *Journal of Global Biosciences Peer Reviewed, Refereed, Open-Access Journal, 10*(11), 9109–9116.

Singh, R., & Srivastava, S. (2015). Exploration of flower based natural dyes—A review. *Research Journal of Recent Sciences, 4(IVC-2015)*, 6–8.

Srivastava, R., & Singh, N. (2019). Importance of natural dye over synthetic dye: A critical review. *International Journal of Home Science, 5*(2), 148–150.

Sujitha, K., Sujatha, S., & Bhuvaneswar, R. (2019). Exploration of leaves based natural dyes—A review. *International Journal for Research Trends and Innovation, 4*(3), 123–127.

Tamilarasi, A., & Banuchitra, M. (2021). Classification and types of natural dyes: A brief review. *International Journal of Creative Research Thoughts, 9*(11), 527–532.

Uddin, M. G. (2015). Extraction of eco-friendly natural dyes from mango leaves and their application on silk fabric. *Uddin Textiles and Clothing Sustainability, 1*(7). https://doi.org/10.1186/s40689-015-0007-9

Verma, S., & Gupta, G. (2017). Natural dyes and its applications: A brief review. *International Journal of Research and Analytical Reviews, 4*(1), 57–60.

Yadav, S., Tiwari, K. S., Gupta, C., Tiwari, M. K., Khan, A., & Sonkar, S. P. (2023). A brief review on natural dyes, pigments: Recent advances and future perspectives. *Results in Chemistry, 5*, 100733. https://doi.org/10.1016/j.rechem.2022.100733

Sustainable Coloration of Cotton Fabric Using Mexican Mint Leaves

M. Parthiban and D. Gopalakrishnan

Abstract For many years, natural dyes have been used to color fabrics, providing vibrant hues and acting as a sustainable alternative to synthetic dyes. Sourced from plants, minerals, and sometimes insects, these dyes offer a wide range of colors, each with its own unique characteristics and qualities. Renowned for their visual appeal and environmental benefits, natural dyes require careful attention to ensure color longevity. Factors such as the selection of the right mordant, the regulation of dye concentration, control over pH levels, and the application of after-dye treatments significantly affect the permanence of the dye. Additionally, the care of naturally dyed fabrics is important and includes gentle washing and shielding them from prolonged exposure to sunlight to extend their durability. Mordanting is essential in the natural dyeing process as it secures the color to the fibers and improves color stability. Mordants create a chemical connection between the dye molecules and the fibers, promoting enduring and vivid colors. Different mordants, including alum, iron, copper, and tannin-rich materials like oak galls and myrobalan, can be employed in natural dyeing. Each type of mordant affects the final color differently, so selecting the right one for the intended outcome is crucial. Prior to mordanting, it is important to thoroughly scour the fabric to eliminate impurities and prepare the fibers for even dye absorption. Mordanting not only enhances color fastness but also enables a broader spectrum of colors to be achieved with natural dyes. By experimenting with various mordants and dyeing methods, we can produce fabrics that showcase rich, intricate colors and distinctive visual effects.

Mexican mint leaves, also referred to as Indian Borage, are recognized for their healing qualities. When ingested, they help defend against colds and coughs, detoxify the body, and enhance digestion. Additionally, they are believed to aid in wound healing. The leaves possess a strong fragrance. Natural dye can be derived from these leaves and utilized for fabric dyeing. In this chapter, cotton fabric was treated with three distinct mordants and then dyed with the extract of Mexican mint leaves. The mordants employed were myrobalan, ferrous sulphate, and copper sulphate.

M. Parthiban (✉) · D. Gopalakrishnan
Department of Fashion Technology, PSG College of Technology, Coimbatore, India

S. S. Muthu (ed.), *Sustainable Coloration of Textiles*, Sustainable Textiles: Production, Processing, Manufacturing & Chemistry,
https://doi.org/10.1007/978-3-031-91217-7_6

When myrobalan and copper sulphate were used as the mordants, a pale yellow hue was achieved. A light brown shade was produced from the sample treated with ferrous sulphate. The colorfastness of the sample dyed with copper sulphate mordant was superior. A handkerchief was created using the dyed fabric that was mordanted with copper sulphate.

Keywords Cotton · Coloration · Dyeing · Fastness · Mexican-mint · Mordants · Testing · Sustainable

1 Introduction

In the field of sustainable textile manufacturing, the investigation of natural dyes has attracted significant interest because of their environmentally friendly qualities and their ability to lessen dependence on synthetic alternatives. This study examines the natural dyeing technique, with a particular emphasis on the use of Mexican mint leaves as a dye source for fabrics. Additionally, this research explores the effects of three different mordants—myrobalan, copper sulfate, and ferrous sulfate—on the dyeing process and the color fastness of the fabrics dyed. Natural dyes present a viable substitute for traditional synthetic dyes, as they are sourced from renewable materials and often exhibit distinct esthetic characteristics (Chungkrang et al., 2021). Mexican mint leaves are celebrated not only for their medicinal properties but also for their fragrance. By leveraging the coloring potential of these leaves, this study aims to expand the array of natural dye options available while fostering sustainable methods in the textile sector. Moreover, the use of mordants in the dyeing process is crucial for enhancing color vibrancy, increasing color fastness, and ensuring the longevity of the dyed textiles (Geetha & Sumathy, 2013). Myrobalan, copper sulfate, and ferrous sulfate have been chosen as mordants because of their widespread application in natural dyeing techniques and their differing influences on color development and fabric characteristics. Through organized experimentation, this project intends to clarify the impact of each mordant on the dyeing results, thereby offering important insights into optimizing dyeing practices for Mexican mint leaf dye (Satheesh et al., 2022).

Furthermore, this research extends beyond simply examining dyeing techniques by seeking to create a tangible product that employs the dyed fabric samples (Bhatt et al., 2013). By assessing the color fastness, washability, and overall visual appeal of the dyed textiles, the study aims to determine the most effective combinations of dye sources and mordants for practical uses. Ultimately, the results of this project strive to enhance sustainable practices in textile dyeing, paving the way for environmentally friendly production methods within the fashion and textile sector (Chandhrua et al., 2023).

2 Work Methodology

Figure 1 shows the Work Process Flow diagram.

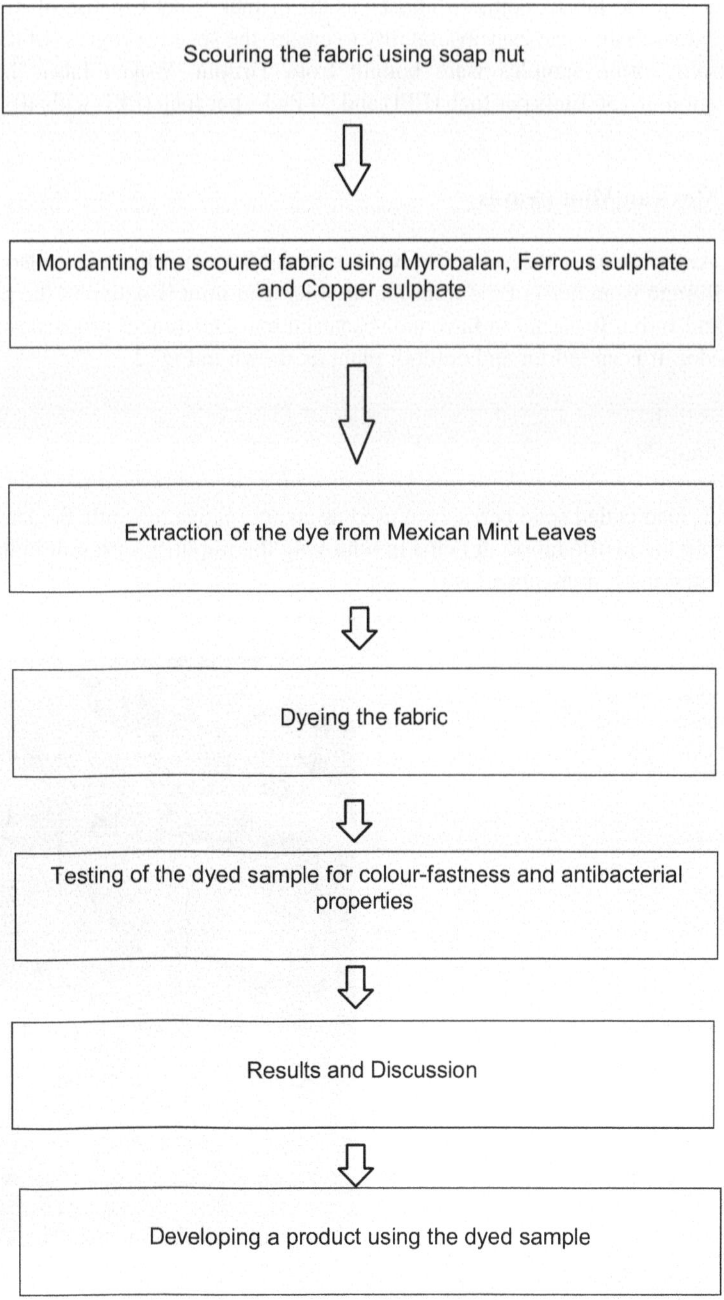

Fig. 1 Work process flow diagram

2.1 Materials

2.1.1 Fabric

When it comes to fabric, cotton is chosen as the primary fiber because of its absorbency characteristics and easy availability to ensure the speedy progress of the project. Woven cotton samples were bought from Tirupur. Woven fabric had the specification of 136 Ends per Inch (EPI) and 94 Picks per Inch (PPI) with 40s count.

2.1.2 Mexican Mint Leaves

Coleus Amboinicus is the scientific name of the Mexican Mint plant also called Indian Borage is an herb of the mint family. Mexican mint is native to the parts of Africa and India. It is said to have anti-bacterial and anti-fungal properties with a strong odor. It is an indoor and outdoor plant as shown in Fig. 2.

2.1.3 Soap-Nut

Soap-nuts also called soap berries are used as an alternative to synthetic detergents for souring the textile fabric. It helps in removing the impurities present in the fabric, thus making it more absorbent.

Fig. 2 Mexican mint leaves. (Source: https://urbanplants.co.in/products/ajwain-leaves-plant-vamu-plant-mexican-mint-plant)

2.1.4 Mordants

Natural Mordants

Myrobalan is used as one of the mordants. Myrobalan is rich in tannin and can be used for mordanting cotton fabric.

Synthetic Mordant

Ferrous sulphate and Copper sulphate were used as the other two mordants for the cotton fabric.

2.1.5 Instruments and Tools

For extracting the dye and for dyeing the sample, a large stainless steel pot is required. The overlock machine is used to stitch the handkerchief.

2.2 *Methods*

2.2.1 Scouring Using Soap-Nut

The main reason for scouring the fabric before dyeing is to remove the natural and artificial impurities present in the fabric, making it more absorbent. Synthetic detergents are normally used for the scouring process in industries. When synthetic detergent-containing wastewater is discharged into various water sources, it may harm the aquatic environment. Natural detergents like soap nuts are considered the best substitute for toxic chemicals used for scouring textile fabrics. Soap-nuts contains aponins, which are natural surfactants that acts like detergents. Thus soap-nut is used for scouring the plain cotton fabric to remove the impurities.

Scouring using soap-nuts (also known as soapberries) is a natural and eco-friendly method to remove impurities, oils, and waxes from fabrics before dyeing or washing. Soap-nut scouring is environmentally-friendly and bio-degradable, making it a sustainable alternative to synthetic detergents.

Take about 10–15 soap-nuts and remove the seeds from the soap-nut. Crush them into smaller pieces. In a large pot add the soap-nut and 4–6 cups of water. Let the soap-nut soak for a few hours so that the saponins are released into the water. Then boil the solution for about 30 min. Add the fabric to the pot turn off the heat and let the fabric soak in the solution for 1 h. Stir the solution occasionally. After 1 h take out the fabric and wash it using normal water. Then dry the sample on a flat surface.

2.2.2 Mordanting

The scoured fabric is cut into three parts so that the mordanting can be done using three different mordants. The mordants that were used for mordanting the fabric are Myrobalan, Ferrous sulphate, and Copper sulphate. Myrobalan is a natural mordant and contains tannins. Copper sulphate and Ferrous sulphate are synthetic mordants. Take the mordant for about 15% weight of the fabric and dissolve it in hot water. Add the dissolved mordant into a large pot containing water. The amount of water should be sufficient for the fabric to move freely. Add the scoured fabric into the pot after wetting it in normal water. Heat the solution to bring it to a boil and then turn off the heat. Stir the fabric occasionally. Let the fabric be in the pot for about 1–2 h. Then take the fabric out and dry it on a flat surface. The mordanting process is carried out for the three mordants.

2.2.3 Extraction of the Dye

Take Mexican mint leaves for about 50% of the weight of the fabric and add them to a large pot filled with water. Heat the solution at 120–140 °F for about 30–45 min. Turn off theheat and let the solution cool. Filter the solution so that a uniform shade can be achieved on the fabric.

2.2.4 Dyeing

Add the mordanted fabric to the dye solution and heat it. After it starts to boil turn off the heat. Let the fabric be in the solution for 1–2 h. Stir the fabric occasionally. After a few hours take out the sample and wash it with normal water to remove the excess dyes from the sample. Dry it on a flat surface. The dyeing process was carried out separately for three mordanted fabrics as represented under Figs. 2, 3, and 4, respectively (Fig. 5).

Fig. 3 Dyed fabric with myrobalan mordant

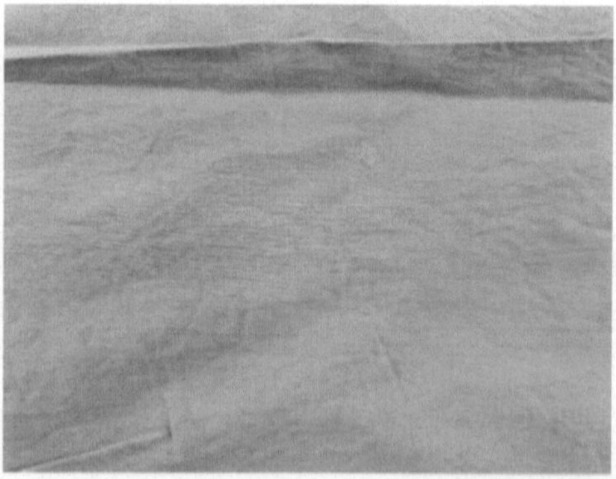

Fig. 4 Dyed fabric with ferrous sulphate mordant

Fig. 5 Dyed fabric with copper sulphate mordant

3 Results and Discussion

3.1 Testing of Fabric Toward Color Fastness to Sublimation

The dyed samples are tested for sublimation fastness to check their color strength toward heating through a sublimation tester. The dyed samples were tested under three conditions, they are dry (110 ± 2 °C), damp (150 ± 2 °C), wet (200 ± 2 °C). After testing, the samples are analyzed under a color cabinet using greyscale and are given a rating from 1 to 5. The results of the sublimation fastness test are tabulated in Table 1.

From the above-obtained result from Table 1, we can infer that the sample dyed with pre-mordanting using copper sulphate has good color fastness to sublimation compared to the other two mordants used.

Table 1 Results of color fastness to sublimation

Mordants used	Dry		Damp		Wet	
Myrobalan	Color	4	Color	4	Color	¾
	Stain	5	Stain	4	Stain	4/5
Ferrous sulphate	Color	4	Color	4	Color	4
	Stain	5	Stain	4/5	Stain	4
Copper sulphate	Color	5	Color	5	Color	4/5
	Stain	5	Stain	5	Stain	4/5

Table 2 Results of color fastness to rubbing

S. no	Mordant used	Stain grade
1	Myrobalan	4
2	Ferrous sulphate	4
3	Copper sulphate	5

3.2 Testing of Fabric Toward Color Fastness to Rubbing (Crock Meter Test)

Take a white fabric sample of 5 × 5 cm, mount it on the finger and fasten it with a clamp. Prepare the test specimen and fasten it to the baseboard of the crock meter using a clamp. Set the crock meter at ten strokes and then start the crock meter. After the test is complete, the fabric sample and rubbing cloth are taken and kept aside for assessment. Use the greyscale for assessing staining on the rubbing cloth and assign grades accordingly. The results of the color fastness to rubbing test are tabulated in Table 2.

From the above-obtained result from Table 2, we can infer that the sample dyed with pre-mordanting using copper sulphate has good color fastness to rubbing compared to the other two mordants used.

3.3 Testing of Fabric Toward Color Fastness to Washing

The treated samples were analyzed under the color cabinet for fastness toward washing and staining. The results were analyzed with the grayscale and the staining scale. The color fastness toward washing and staining is given from 1 to 5.

From the above-obtained results from Tables 3 and 4, we can infer that the sample dyed with pre-mordanting using copper sulphate has good color fastness to washing compared to the other two mordants used.

Table 3 Results of color fastness to washing

Mordant used	Color
Myrobalan	3
Ferrous sulphate	¾
Copper sulphate	5

Table 4 Results of stain due to washing

Multifiber	Myrobalan	Ferrous sulphate	Copper sulphate
Wool	5	5	5
Acrylic	5	5	5
Polyester	5	5	5
Nylon	5	5	5
Cotton	4/5	4	5
Acetate	5	5	5

Table 5 Results of the anti-microbial test

		Zone of inhibition for micro-organism (in mm)		
S. no	Organism	Fabrics 1	Fabrics 2	Gentamycin
1.	*E. coli*	No zone	No zone	29 mm
2.	*S. aureus*	No zone	No zone	28 mm

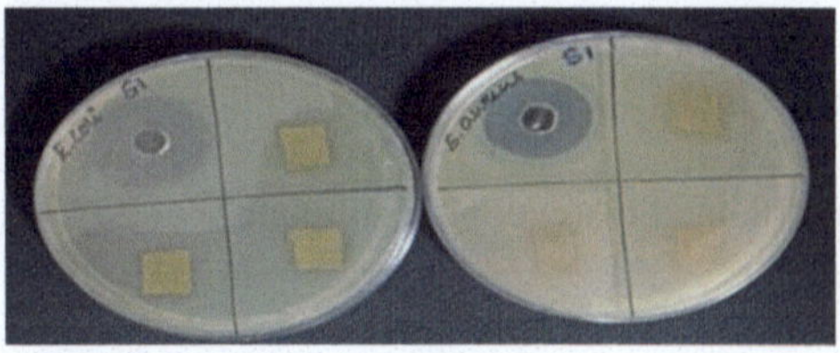

Fig. 6 Fabric S1

3.4 Determination of Anti-microbial Activity

The agar well diffusion method is widely used to evaluate the anti-microbial activity of plants or microbial extracts. Similar to the procedure used in the disk-diffusion method, the agar plate surface is inoculated by spreading a volume (500 μL) of the microbial inoculum over the entire agar surface. Then, a hole with a diameter of 6–8 mm is punched aseptically with a sterile 1000 μL tip, and 100 μL of the anti-microbial agent or extract solution at the desired concentration is introduced into the well. Then, agar plates are incubated at 37 °C under suitable conditions depending upon the test micro-organism. The anti-microbial agent diffuses in the agar medium and the zone of inhibition is checked after the incubation period. Here Gentamycin is used as an antibiotic control for Bacterial culture. The zone of inhibition sizes(in mm) average for samples determined by Agar diffusion method was represented under Table 5 and the testing samples were represented under Figs. 6 and 7.

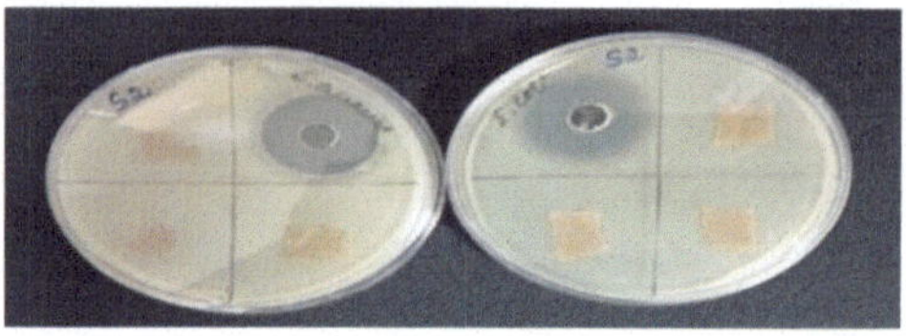

Fig. 7 Fabric S2

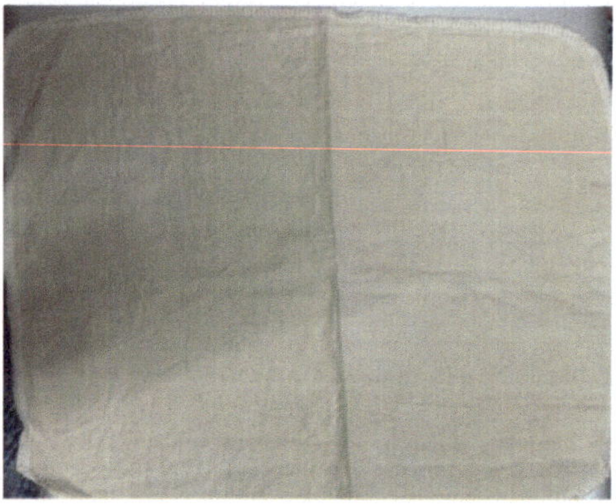

Fig. 8 Handkerchief developed from dyed fabric mordanted using copper sulphate

4 Product Development

The dyed fabric mordanted using copper sulphate showed better color fastness than the other two dyed fabrics mordanted using myrobalan and ferrous sulphate. As shown in Fig. 8, a handkerchief is developed from the dyed fabric that is pre-mordanted using copper sulphate.

5 Conclusions

In this chapter, the natural dyeing process of cotton fabric using dyes derived from Mexican leaves, utilizing three different mordants: myrobalan, ferrous sulphate, and copper sulphate had been explored. The aim of our study was to evaluate the dyeability of Mexican Mint leaves and examine the color fastness of the dyed fabrics under conditions of sublimation, washing, and crocking, along with assessing the anti-microbial properties of the dyed textiles. The cotton fabric samples underwent mordanting with the three distinct mordants: myrobalan, ferrous sulphate, and copper sulphate. Mordanting plays an essential role in natural dyeing as it improves the absorption and adherence of dyes to the fabric. Each mordant was applied to individual fabric samples to study its influence on dye uptake and color fastness. Copper sulphate can provide distinctive color shades and improve color fastness in natural dyeing processes. The dye was extracted from Mexican leaves and applied to the mordanted cotton fabric samples. Careful control of the dyeing process was maintained to ensure uniformity and consistency across all samples. The dyed fabric samples were subjected to a series of tests to assess color fastness. According to our results, the fabric mordanted with copper sulphate showed superior color fastness in comparison to the samples mordanted with myrobalan and ferrous sulphate. This indicates that copper sulphate effectively improved the fixation of the dye and the retention of color on the cotton fabric.

References

Bhatt, P., Joseph, G. S., Negi, P. S., & Varadaraj, M. C. (2013). Chemical composition and nutraceutical potential of Indian borage (Plectranthus amboinicus) stem extract. *Journal of Chemistry*. https://doi.org/10.1155/2013/320329

Chandhrua, M., Gunasekarana, P., Maruthupandia, M., Meenakshia, R., Sundarb, M., Kalanthodena, A. N., Ahmadc, N., Khanc, M. R., Santhana, A., Kumar, S. K., Rania, S. K., & Vasimalaia, N. (2023). Bio-fabricated silver nanocatalyst for photocatalytic degradation and organic transformation of toxic pollutants. *Next Materials, 1*(3), 100023.

Chungkrang, L., Bhuyan, S., & Phukan, A. R. (2021). Natural dyes: Extraction and applications. *International Journal of Current Microbiology and Applied Sciences, 10*, 1669.

Geetha, B., & Sumathy, V. J. H. (2013). Extraction of natural dyes from plants. *International Journal of Chemistry and Pharmaceutical Sciences, 1*(8), 502–509.

Satheesh, V., Kaur, J., Jarial, S., Ghosh, P., Sharma, K., Patni, M., Singh, J., & Bhadariya, V. (2022). Indian borage: A comprehensive review on the nutritional profile and diverse pharmacological significance. *The Pharma Innovation Journal, 11*(6), 42–51.

Combining Chemistry and Technology to Achieve Effective Color Fastness and Desired Color, with Foam Finishing Application

Nina Cohen and Karen Leonas

Abstract Textile production requires many steps from the fiber to final product. Imparting color through dyeing and printing is critical in achieving both functional and esthetic requirements. The coloration process, specifically dyeing, has been recognized as using great amounts of water, energy, and chemicals during the processing. Greenhouse gasses are emitted during the process contributing to Climate Change. Excessive water use is problematic in water stressed regions of the world, and the dyeing effluent, containing chemicals used in achieving the desired color, is a major contributor to water pollution. There is increased attention to the environmental impact created during the dyeing process and legislation around the world, specifically in the EU, is being implemented to address these issues. The industry is examining processing technologies that result in cleaner production systems including reducing energy, water and chemical use. There have been various processes developed in addressing these issues including foam dyeing. Foam dyeing was first introduced in the late 1970s and recent innovations have improved the process.

Foam dyeing uses significantly less water when compared to traditional dyeing methods. As less water is used, less energy is used in heating the dye bath and drying the fabric. A review of foam dyeing, the innovative changes that have been made to improve its performance, and results of research looking at the impact of dye class and the use of pretreatments on color uniformity, colorfastness, and reduction in processing time (which results in lower energy used and lower GHG emissions) are included. The study compared benchtop dyeing to scaled up processing of the fabrics and results are reported. Information on the optimum processing conditions (percent pretreatment used, dye class, processing variables) to achieve the desired performance (color, colorfastness) was determined. Various statistical methods

N. Cohen
Huntersville, NC, USA

K. Leonas (✉)
North Carolina State University, Raleigh, NC, USA
e-mail: kleonas@ncsu.edu

S. S. Muthu (ed.), *Sustainable Coloration of Textiles*, Sustainable Textiles: Production, Processing, Manufacturing & Chemistry,
https://doi.org/10.1007/978-3-031-91217-7_7

were used in the analysis including ANOVA, T-Tests, and Tukey HSD, to identify significant groups. Through testing, it was found that the dye type and level of pretreatment has an impact on some of the colorfastness and washfastness scores. Results also showed there was no statistical significance between the L*A*B readings, colorfastness, and washfastness scores between benchtop dyeing, and foam dyeing.

Keywords Foam dyeing · Sustainability · Textile wet processing · Dyeing · Low liquor · Coloration · Textile sustainability · Foam finishing

1 Introduction

Textile wet processing is an umbrella term that encompasses major categories of fiber preparation, coloration, and finishing. The textile industry is known to use large amounts of water during this processing and contributes to approximately 20% of the total industrial water pollution globally (European Parliament, 2024a, 2024b). With increased attention on current environmental issues including water scarcity, GHG emissions, and water pollution, this chapter focuses on the dyeing process and provides a case study evaluating the use of foam dyeing to reduce water use, water pollution, and energy while measuring the effectiveness of the dyeing through achieving a desired color and colorfastness.

Specifically, this chapter focuses on the use of reactive dyes on pretreated cationized cotton. Colorfastness testing, including AATCC TM8; colorfastness to crocking and AATCC TM61, 2A; colorfastness to laundering were utilized to measure the amount of color shed and color fade on multiple combinations of dye type, dye dosage, and pretreatment dosage on a 100% cotton substrate. From pre-trial lab work, specific combinations of dye and pretreatment were selected based on predetermined colorfastness scores to be recreated with the foam dyeing process. L*a*b* values were measured after foam dyeing trials, to compare the color achieved during foam dyeing to the initial lab work completed to identify optimal combinations. The results of the foam dyeing trials were statistically analyzed using ANOVA and T-Tests for variable comparison. Tukey HSD was also utilized to identify groups with significant results.

This case study was completed with support from industry partner JB Martin of Leesville, SC. They provided laboratory space and equipment, as well as the pretreatment chemistry, fabric, and auxiliary. The foam dyeing trials were completed with industry partner Gaston Systems of Stanley, NC. Finally, dyes were donated to the project by Huntsman, Cote Color, and Orco Dyes and Pigments.

2 Coloration of Textiles

Textiles are colored for many reasons and achieving the desired esthetic and functional characteristics of the finished product is of importance. Typically this coloration occurs through printing, dyeing or a combination of these processes. Coloring agents can be classified in many ways but one is based on whether they are a pigment or a dye.

2.1 Printing

Printing is characterized by the color/pattern being only on the surface (face) of the textile and dyes or pigments can be used. Commonly used methods of printing include wet printing techniques such as flatbed screen, rotary screen, and engraved copper roller printing, and dry printing techniques such as transfer printing and digital inkjet printing (Moser, 2003). Printing has the advantage of adhering to a large range of substrates, and producing very little to no waste effluent.

2.2 Dyeing

Traditionally, dyeing methods have involved the immersion of the textile into a solution that contains dyes. The solution can be immersed by either a padding or exhaust process. The solution penetrates the fabric and the goal is to achieve a textile having a uniform color, same on the face and back, and across the width of the fabric. Like printing, there are several categories of dyeing which are discussed in section. There are also speciality dyeing techniques, like tie dyeing, shibori, and others where a pattern is produced.

3 Coloring Agents

3.1 Pigments

Pigments are applied to fabric in a paste and sit on the surface of the textile. They are adhered to the surface of the textile using binders. Dyeing results in the penetration of the fabric and high quality dyeing is characterized by the textile having a uniform color, same on the face and back, and across the width of the fabric. Dyes, unlike pigments, are soluble in water and penetrate into the fibers. Rather than being bonded with a binder, the dye molecules are chemically bonded to the textile fiber.

3.2 Dyes

Dyes can be classified in many ways including whether they are natural or synthetic, by chemical class, color, and several other methods. There are a variety of natural dyes; however, they have limited use due to the availability (large amounts are needed for mass production), poor reproducibility, and the time required to dye products (may take days or weeks rather than hours) (Stewart, 2017). These dyes are most commonly derived from plants and insects and require large amounts of the raw material to produce a small quantity of dye (Stewart, 2017). There also can be inconsistencies in color produced depending on the conditions of the resources. Natural dyes are limited by their higher cost, and inconsistency in color. Because of these conditions, natural dyes are more commonly used in crafts or small production runs (Stewart, 2017).

Synthetic dyes, commonly produced from raw materials such as benzene and toluene (Abrahart, 1977), are more readily available and less expensive to produce than natural dyes. They offer a wider range of colors and are more consistent in the colors produced over time and from dye batch to dye batch. Synthetic dyes can be used on a variety of substrates and are selected based on application process, color desired, and cost.

4 Classification of Dyes

Dyes can be classified by molecular weight, source, chemical group, color index, and end use (Mock, 2002). All dyes have similar components in their molecular structure that are broadly classified into chromophores (primarily produces color) and auxochromes (assist in color production, solubility, and reactions with the fibers). However, specific dye classes have more specific molecular make-ups. For example, a reactive dye contains a chromophore, a bridging group, a solubilizing group, a reactive group, and a leaving group. Chromophores are conjugated systems of double and single carbon to carbon bonds that create color. Typically, larger conjugated systems move toward large wavelengths. A red chromophore will have a longer wavelength than blue. The bridging group is the reactive group that connects to the chromophore. The solubilizing group is what helps the dye dissolve in water. If a dye does not have a solubilizing group, then solvents like alcohols need to be used to solubilize the dye. The reactive group is what reacts with the substrate. In the case of reactive dyes, the reactive group helps to form a covalent bond with the substrate. Finally, the leaving group leaves the structure to undergo nucleophilic substitution or addition.

There are multiple ways to classify dyes; molecular weight, source, chemical group, and color index are some examples (Mock, 2002). Classifying dyes by structure is commonly used, such as reactive, direct, and vat. Dyes are selected for a variety of reasons including that is being dyed (Table 1), cost, method of application, and other considerations.

Table 1 Textile fibers and common dyes used to color textiles (K. K. Leonas, Professor, NC State University, personal communication, shared document, 2018)

Fibers	Dye by chemical structure
Cellulose	Azoic, direct, vat, sulfur, reactive
Protein	Direct (some), reactive, acid, chrome (mordant), basic,
Viscose	Direct, vat, sulfur
Nylon	Azoic, acid, chrome (mordant), basic, disperse
Acrylic	Acid, chrome (mordant),basic (some), disperse
Polyester	Azoic, vat, basic, disperse
Acetate	Azoic, disperse

Most commonly, dyes are classified based on their chemical structure. Each dye class is known for specific characteristics including colors produced, fastness characteristics, substrate compatibility, and molecular weight. There are nine classes commonly used and five commonly used on cotton. As the case study in this chapter relates to cotton, those five classes are generally described below.

4.1 *Classification by Dye Structure*

Azoic dyes produce a variety of hues in the red to orange range. These hues are typically known to perform poorly in colorfastness properties including light and washfastness, and azoic dyes filled that gap offering more acceptable performance. Azoic dyes have been referred to as ice dyes due to their reaction taking place in lower temperatures.

There are four subclasses of azo dyes based on affinity (Fu, 2016). These include low substantivity, medium substantivity, high substantivity, and very high substantivity. Wet and dry crocking results are higher with each level of increased substantivity. These dyes are limited in accessibility, and not used as commonly due to the popularity of reactive dyes.

4.2 *Direct*

Direct dyes are regarded as the first dyes to achieve acceptable washfastness properties without the use of a mordant (Mock, 2002). They are one of the most commonly used dyes due to low cost and variety of colors offered. Regardless of the acceptable colorfastness results, direct dyes still have poor colorfastness properties due to bonding to fibers through hydrogen bonds or Van Der Waals interactions. Direct dyes have a linear structure rather large in size, and this results in weaker bonds forming with the fibers. Direct dyes are commonly seen in low cost or minimally washed products.

Direct dyes are further classified into three subclass sections based on their leveling and migration patterns. They are classified as class A, class B, and class C dyes. Class A dyes have overall good migration and are self-leveling. Class A dyes are typically monoazo or diazo with two solubilizing groups. They require great amounts of salt for best exhaustion results.

Class B dyes are poorer in leveling compared to class A. They are also more sensitive. They tend to have a higher molecular weight but are similar to class A dyes in terms of migration. Class B dyes have three to four solubilizing groups.

Class C dyes are the poorest performers in regard to class A and class B dyes. They require specifically controlled temperature and salt conditions to execute an exhaust process.

4.3 Reactive

Reactive dyes are commonly used in textile coloration, because of their availability and wide color range. In 2009, it was estimated that almost 45% of all textile dyes produced annually are classified as reactive dyes as they are extensively used in dyeing cellulosics such as cotton, linen, and rayon (Tunç et al., 2009). They are easy to replicate over multiple lots of fabric and are practical in their set up and run conditions. Reactive dyes form covalent bonds with fabrics, which lead to stronger colorfastness properties.

Reactive dyes are further classified as monofunctional or bifunctional dyes based on the amount of reactive group subclasses in them. Subclasses are based on their heterocyclic ring groups, which include triazines, vinyl sulfones, quinoxaline, and pyrimidines. Within these four groups, there are eight types of reactive dyes that can be created.

Triazine reactive dyes have a base structure of a heterocyclic ring with three nitrogen atoms, and there is variation in the leaving group that is attached to the heterocyclic ring. These elemental differences affect the reactivity of each dye within the triazine subclass. Pyrimidine subclass has either two nitrogen atoms within the heterocyclic ring or one chlorine and two fluorine atoms. Their reaction is similar to triazine subclass. Vinyl sulfone (VS) subclass does not have a halogen in the electrophilic group. "An ether linkage is formed with the cellulose through a nucleophilic addition to a carbon – carbon double bond" (Broadbent, 2001, p. 335). VS subclass is ideal to use in combining functional groups. Quinoxaline subclass uses heteroatoms as the reactive structure. They are more reactive than other subclasses because they have chlorocarbonyl rather than a nitrogen bridging group.

As stated previously, reactive dyes can also be bifunctional. This takes place when two reaction groups are present in the dye molecule. Bifunctional dyes present the advantage of more opportunities for fixation. If a dye molecule hydrolyzes, it cannot fixate onto a substrate and will go to waste. Having two reactive groups reduces the amount of wasted dye and increases the temperature range a dye can operate with. Bifunctional dyes are either homobifunctional or heterobifunctional. Meaning they have identical reactive structures or two different reactive structures.

4.4 Sulfur

Sulfur dyes are typically inexpensive to use, which is favorable when dyeing dark hues, which require higher levels of dye concentration to achieve. They typically have poor to moderate colorfastness properties (Farrell, 2012). Molecularly, sulfur dyes come in a variety of shapes in sizes. They share many properties with vat dyes, including a similar application process (Rivlin, 1982).

4.5 Vat

Vat dyes are one of the oldest and most costly dyes (Fu, 2016). They do, however, have excellent colorfastness properties. Vat dyes are classified further into indigonoid (indigo) or anthraquinone derivatives. Indigo is commonly used in denim applications, as its poor washfast properties are a preference in the esthetic of denim. Sodium hydroxide and sodium hydrosulfite are common reducing agents for this step in vat dyeing (Farrell, 2012). Vat molecules are small, which allow them to penetrate into fibers much easier. The small molecule size also benefits the washfast properties.

5 Dyeing Process

Dyes come in various formulations, most commonly in powder or liquid forms. Dyes that are not as easily soluble in water, such as nonionic dyes, require additional surfactants to disperse them through the water used in their application to the textile.

The mechanism of dyeing involves three phases: (1) adsorption, (2) absorption, and (3) fixation. The first step in the dyeing process is the adsorption of the dye onto the surface of the substrate. Salt or other auxiliary chemicals may aid in the affinity of the dye to the substrate. In order for the process to work, the dye needs to be able to leave the solution it is dispersed in and move onto the substrate.

The next step in the process is absorption, where the dye begins penetration into the fiber to find a site to react with. The dye diffuses onto the surface and works its way into the middle of the fiber. Agitation is typically included into this step to increase the rate of dyeing.

The final step of the dyeing process occurs when the dye is diffused through the fiber and fixation can occur. The dye and fiber form either an ionic, Van Der Waals force, hydrogen or covalent bond causing the dye to bond to the fiber. The type of bond between the dye and fiber can have an impact on the substrate's likelihood to lose the color (colorfastness). Hydrogen bonds are typically formed between direct dyes on cellulose and are known to be a relatively weak bond so there is a greater chance for a loss of color. Ionic bonds, found when using acid dyes on wool fibers, are somewhat stronger and so colorfastness issues are less likely to occur (Broadbent, 2001). Covalent bonds, which form between a reactive dye and cellulose, are also strong. When covalent bonds are present, the substrate has better colorfastness than when other bonds are involved in the dyes fixation to the fabric. When dye is exhausted into a fiber, it will follow a specific isotherm. The three they may follow are Nernst, Langmuir, and Freundlich. Isotherms are a curve that shows the rate in which dyes move onto a fiber in the dyeing process (Perkins, 1996). The curve is dependent on the fiber type, dye type, auxiliary, and agitation (Perkins, 1996).

5.1 Current Dye Application Processes

Coloration can be applied in any stage of fabric or garment production. The stage to implement dyeing and type of dyeing method is dependent on the manufacturer and what they are making, the lot size of material, and the equipment available. Dyeing processes can be categorized as continuous or batch processes.

Continuous dyeing methods are for large, long runs of materials, where additional material can be added in to keep the process going. A continuous process runs at a typical line speed of 50–250 m/min (Raja, 2014) and can contain multiple steps in addition to dyeing like scouring, washing, and drying. Continuous dyeing processes are typically applied at the fabric stage or on warp yarns.

An example of continuous dyeing is pad dyeing. It can use room temperature conditions to dye and finish fabric. Lower temperatures along with padding minimize undesireable defects such as tailing or a variation in shade along the length of the fabric (Broadbent, 2001). If heat is needed to cause a reaction between the dye and fiber to cause the fixation of the dye to the fiber, it would be applied during a steaming or drying process. It is typical for pad processing to be used with reactive dyes on cellulose fabric. In some cases, the dye can be mixed with the finishing auxiliary and applied in one bath. In other cases, the fabric will be dyed and possibly treated with other finishes, and then go to a batching process, where the fabric is rolled up and stored to process and let reaction occur in ambient or warm conditions. In batched cases, the fabric may run through the continuous range once more to receive more wet processing treatments.

Batch dyeing is typically used for fabric runs between 50 and 100 m depending on fabric weight (Broadbent, 2001) and can be implemented in a fiber, yarn, fabric, or garment stage. In this process, dye is transferred from a large bath onto the fabric. The dye can be transferred in three ways: fabric circulating through a bath, the dyebath circulating through stationary material, and both bath and fabric circulating together. Batch dyeing may also be referred to as exhaust dyeing.

Beck dyeing is an example of batch dyeing. It is used for a wide variety of fiber types and fabric constructions. Beck dyeing may take several hours to complete, which results in the need for higher levels of water, chemicals, and energy (Cassidy & Parikshit, 2018). Flaws due to temperature variations through the dyeing process have also been identified. This can result in tailing of the fabric.

Another example of batch dyeing is Jet dyeing. The process used in Jet dyeing is similar to that of beck dyeing. What differs is the fabric transport mechanism and machine layout. Jet dyeing has several advantages over beck dyeing including using lower liquor ratios and it is a quicker process. A jet dye machine also has the capability to wash or bleach a fabric due to its shape and process.

6 Sustainability Related Issues

The textile industry has been recognized as a main contributor to sustainability, social, and environmental concerns. Water pollution, energy consumption, and greenhouse gas emissions are some of the areas textiles can have a large impact. The coloration process, specifically dyeing, has been recognized as using great amounts of water, energy, and chemicals during the processing. Energy is needed to heat and aid in the movement the dye bath to enhance agitation. Heat is also required in the post-dyeing drying process. Not often is the water utilized in the dyeing process able to be recycled or reused in any way. Wastewater is typically expensive to collect and treat. The United Nations reports that high-income countries treat approximately 70% of total generated wastewater, middle-income countries treat approximately 28–38% of total generated wastewater, and low-income countries treat approximately 8% of total generated wastewater (United Nations, 2019). In some developing nations, it is quicker and easier to dump the untreated water in the nearest river, lake, or stream.

Excessive water use is problematic in water-stressed regions of the world, and the dyeing effluent, containing chemicals used in achieving the desired color, is a major contributor to water pollution. 1.4 billion people live in areas of extreme water scarcity (UNICEF, 2021). Over 4 billion people experience water scarcity for minimally 1 month per year, and over 2 billion people live in areas where water is generally inadequate (UNICEF, 2021). Areas with poor infrastructure are more likely to experience it. Today, wet processing operations are most commonly found in India, China, and Bangladesh (Muthu, 2017). Other areas affected by water stress include Africa, South America, Asia, and sections of the US.

China, for example, produces a lot of textiles used in apparel and home products. Facilities are aware of water risk, and want more training, and financial support to implement cleaner production (McGregor, 2017). China has an uneven distribution of water resources, rapid economic development and urbanization from population growth, which all contribute to their water scarcity issue and poor management of it (Jiang, 2009).

7 Legislation Involvement

There has been an increasing amount of public awareness over water pollution. This has resulted in legislation around the world being implemented to address these issues. In the USA, the Environmental Protection Agency (EPA) has acts in place such as the Emergency Planning and Community right to know act (SARA title III) of 1986 and the 1990 Pollution Prevention act. Acts such as these have regulated how wastewater is measured, treated, and managed (EPA, 1994). Acts such as these also discourage some manufacturing in the USA, because following legislation like this leads to higher overhead costs, and possible modifications to facilities to be made to comply with regulations. Much textile production has since moved offshore to countries such as China and Bangladesh to offset the cost of wastewater treatment. Manufacturing in these developing nations comes at a lower cost than US manufacturing.

In China specifically, there is legislation around water pollution. The Water Pollution and Control Law, recently amended in 2017, focuses on protecting and improving the environment, preventing, and controlling water pollution, and protecting ecology and general public health (Ministry of Ecology and Environment The People's Republic of China, 2017). This law is the response to severe water pollution due to the rapid industrialization and urbanization of China. This law created targets for different industries to reduce the total amount of discharge into bodies of water. There is also verbage to improve wastewater treatment and improve water quality monitoring (Ministry of Ecology and Environment The People's Republic of China, 2017).

In the European Union (EU), organic matter, nitrogen, and phosphorus are found in urban wastewater, and if they are not treated, can cause eutrophication (European Parliament, 2024a). The European Union estimates that almost 50% of Europe's total emissions derive from industrial activity (European Parliament, 2024b). This includes textile manufacturing. The EU has a set of directives to address urban wastewater, industrial emissions, and overall pollution emitted.

In May 2021, the EU put in place the "Zero Pollution Action Plan" as part of the European Green Deal. This directive aims to reduce soil, air, and water pollution to a level which is no longer harmful to human health. Targets of this plan include proposals for directives that will more strictly regulate industrial emissions to make sure they comply with climate, energy, and circular economy policies.

8 Current Methods of Cleaner Production

In 2011, it was reported that the daily wet processing water consumption of an average-sized textile mill having a production of about 8000 kg of fabric per day is approximately 1.6 million liters (Kant, 2012). Textile converters will need to accommodate for the higher demand of goods which in turn will require a further increase in resources including water, chemicals, and energy. There are some methods currently in place to reduce the amount of water, energy, and greenhouse gasses emitted during textile production. To focus on water usage, lowering liquor ratios and using alternative mediums to carry the dye to the substrate have been explored.

8.1 Low Liquor Dyeing

Reducing liquor ratio is a method which meets the goal of lowering water usage while using already existing technology. A dye bath liquor ratio may range from 20:1 to 40:1 (Farrell, 2012). With some of the newest technology, low liquor ratios can go as low as 1:2 when working specifically with cellulose (Davis, 2023). This new technology needs approximately 0.6 L/kg to operate. When lowering the ratio, the amount of auxiliary needed is shifted. These imputes adjusted to the amount of reduced liquor affects the exhaustion and fixation (Irfan et al., 2018). If not carefully recalculated, the finished fabric is at risk for unlevel dyeing or a spotty appearance.

8.2 Supercritical Carbon Dioxide

Supercritical Carbon Dioxide ($SCCO_2$) utilizes pressure, CO_2, and high temperature to apply disperse dyes to a substrate. Newer technology can have a dye uptake of up to 98% (Davis, 2023). This technology is limited by its ability to only work with polyester at this time; however, other synthetic substrates are being evaluated for this application method. An advantage of $SCCO_2$ is its ability to work as a closed loop system. This means that very little material is wasted during the dyeing process, and some material may even be reused. In newer equipment, reclaimed CO_2 from existing industrial processes can be collected and utilized (Davis, 2023). The result of $SCCO_2$ is a rich color penetration that is consistent throughout the lot. More research is needed to expand $SCCO_2$ capabilities and eventually be adopted into more textile coloration facilities.

8.3 Plasma

Plasma follows a similar process to $SCCO_2$; however, its current application usages fall more into the pretreatment side of textile manufacturing (Ghoranneviss & Shahidi, 2017). Plasma does, however, have the ability to carry dye to a fabric. Rather than $SCCO_2$ use of carbon dioxide, plasma technology utilizes a constant flow of helium or oxygen into the pressure chamber (Kan et al., 2015). The gas is used to agitate the dye, which helps it mix with the substrate. The high level of pressure needed for plasma application make the usage of this technology sensitive to errors (Kan et al., 2015). Incorrect usage can end up damaging the chamber or ever the entirety of the equipment. Plasma also utilizes copious amounts of energy during processing, which is not ideal when looking to conserve resources such as that. An advantage of plasma, however, is not only can it carry dye to a fiber, but it can also absorb and enhance dyeing and coating better as well as increase bond strength (Davis, 2023).

9 Foam

Foam dyeing is a more sustainable alternative to traditional dyeing methods. It is estimated that dye, water, energy, and chemicals can be reduced up to 90% (Ondrey, 2018). Foam dyeing varies from the traditional dyeing method of using dye plus water, by using dye plus foam. Foam technology in the process of textiles, as a low add-on technology, uses foam to apply chemicals and colorants to textiles, resulting in large water and energy savings owing to the replacement of water with air (Yu et al., 2014). Since this is a method of coloration utilizing reduced liquor significantly less resources are needed. Traditional dyeing requires approximately 400 gallons of water per 100 yards of fabric, but when using foam, the amount is reduced to about 3.5 gallons of water per 100 yards of fabric (Ondrey, 2018). Foam is created when gas is trapped into a liquid, creating a new dispersion.

Foam application must be applied in a continuous setting. Padding is typical for foam application. The padding is needed to help the dispersed bubbles collapse onto the fabric, as well as push aside the excess foam. The liquor is then able to penetrate into the fibers.

Foam dyeing offers a more cost effective approach to fabric dyeing compared with other methods. Due to reduced liquor ratio, drying times are minimal for the process, resulting in reduced energy usage and cost. The amount of dye needed is also reduced, resulting in less waste. While there has been initial interest in foam application overall, adapting the technology has been a slow process. The beginning trials of foam dyeing resulted in poor quality dyeing due to non-uniformity (Booth, 1982). Advancements in technology have reduced or eliminated color quality concerns. One example of an advancement is the improvement of applying foam to one side of the fabric to applying foam to both sides of the fabric. This addressed the unlevel dyeing issue, and significantly resolved it. The introduction of the top applicator could be used to apply either the same dye mix, a resin, or a different dye mix to add more functionality to the technology (Baker et al., 1982).

Foam dyeing may also utilize cationizing pretreatment to help the substrate uptake more dye. Cotton, a natural cellulosic fiber and the substrate selected for this case study, when in water, has an anionic charge, due to the partial ionization of hydroxyl groups on the cellulose structure (Fu, 2016). When adding a pretreatment, it changes the charge from anionic to cationic. The use of cationization can also aid in reducing or eliminating the salt needed for dyeing, and further lowering the liquor ratio.

Cationizing technology is classified into monomeric reagents and polymeric reagents (Fu, 2016). Monomeric reagents are most effective with Monochlorotriazine (MCT) dyes (Fu, 2016). Polymeric reagents are best suited for Dichlorotriazine (DCT) and Fluorochloropyrimidine (FCP) reactive dyes as well as direct and acid dyes (Fu, 2016). An advantage of utilizing cationized cotton is its ability to dye within a wide range of temperatures. Cotton incorporated reported trial temperatures of 100 °F (38 °C) to 200 °F (93 °C) (Cotton Inc., 2005). They also found that depth of shade can also be positively impacted with the concentration level of the reagent.

While foam dyeing technology has been around since approximately 1970 (Booth, 1982), it has been challenging to adapt into wider scale manufacturing. Part of the poor adaptation is due to minimal research conducted on foam dyeing. It is challenging to adapt new technologies where there is not enough data supporting that fabric esthetic, performance, and quality will not be altered by a new method of textile dyeing. One study was conducted by Yu et al. in 2014, looking at the depth of color between foam and conventional dyed fabrics. The study focused on total consumption between foam dyeing and conventional exhaust dyeing, and compared the ratios of wet pick-up, fixing agent, foam stabilizer, and blow ratio used. The study measured the depth of color with the Kubelka-Munk eq. (K/S) which is defined as the absorption (K) and scattering (S) coefficients and the concentrations of the colorants on the sample with its overall reflectance (Broadbent, 2001) to discuss the dye fixation rates.

This study found that conventional dyeing required a higher wet pick-up rate of around 60–100%. The foam dyed samples achieved desired depth of shade at a 30% wet pick-up which lowered the amount of water usage and increased the intensity of the color. As for colorfastness, AATCC TM8 was utilized for colorfastness testing and found that wet and dry crocking was similar between foam dyeing and conventional dyeing. This study helped prove that if material utilizing a reduced liquor process for coloration can achieve similar results to a conventional dyeing process, then consideration can be given to replace older, resource heavy processes with foam dyeing technology. Aside from Yu et al., there are few additional studies addressing factors such as using polyoxyethylene ether as a foam controller (Chen et al., 2017), using hydrolyzed keratin as a foaming agent for foam dyeing of wool (Bhavsar et al., 2017), and foam dyeing on knit fabrics (Shang et al., 2011).

Foam dyeing is suitable for fabrics to be used in an apparel, home, or industrial application. One notable brand who has recently adapted foam dyeing into their textile processes is Wrangler (Wrangler, 2018). Since adapting foam, Wrangler has seen a net reduction in energy by 90% as well as shorter production time and chemical usage.

10 Coloration Measurement Methods

When adding color to a textile, it is imperative that the color matches the intended esthetic, maintains that color through washing, and the color does not rub off onto skin or other surfaces. There are standards in place to address these potential coloration issues, created by organizations such as American Association of Textile Chemists and Colorists (AATCC), International Standard Organization (ISO), and others. Standards are defined as "a set of characteristics or procedures that provide a basis for production decisions" (Glock & Kunz, 2005, p. 98). They define what is a baseline for quality and performance. Testing is usually performed in a laboratory setting with controlled conditions to ensure standardization no matter where the test is performed.

Measuring color is also recognized as measuring colorfastness. Colorfastness is defined as a materials' ability to retain its original hue without fading or running (Cohen et al., 2011). In ideal performance, a material will experience little to no fading or color alteration when exposed to conditions such as washing, sunlight, or rubbing.

Color can shed due to a decomposition of the dye on the fibers or a removal of dye in the washing process. Colors can also change in a variety of situations including surface rubbing, exposure to light, and during laundering. The following section will focus into more detail on colorfastness and the testing used to measure the materials.

10.1 Crocking

Crocking is known as the color transfer from one material onto another. The transfer comes from rubbing the surfaces against each other in either a dry or wet setting. A clean 100% cotton crocking square is utilized to evaluate the color transfer. The use of a crockometer is also used to make sure the pressure, and the amount of rubbing is consistent. It is typical to see more color rubbing off in darker shades as well as anything that has unfixed dye.

10.2 Lightfastness

Color has the ability to fade when exposed to UV rays and large amounts of light. There are accelerated and controlled test methods to measure the amount of color loss due to light exposure. Equipment called a "Weather-Ometer" creates a simulated sunlight exposure at a constant rate to create fading within the material. Testing can take place anywhere from 10 to 160 h. The amount of testing hours would depend on the final use of the material. Sleepwear is exposed on the lower end of around 10 h, as the end use of this apparel should have minimal sun exposure. Suiting and day wear fabrics are exposed longer at around 40 h, and drapery and canopy fabrics will be exposed in the weather-Ometer for 160 h (Cohen et al., 2011).

10.3 Laundering

Washing is expected to happen to most textiles going into apparel applications. During a washing process, color can be lost due to the soap, water, and agitation. There is specific testing in place to measure color loss due to washing. AATCC's specific laundering tests offer five variations of accelerated laundering (AATCC, 2019). Much like lightfastness testing, equipment called a "laundrometer" exists to measure colorloss due to washing in a controlled setting. This equipment uses canisters where a predetermined sample size, water, detergent, steel balls, and fiber test strip (if applicable) are mixed together and agitated uniformly over a set number of cycles.

10.4 Measuring Colorfastness Results

To measure the amount of color loss, there are three standardized tools including a greyscale, nine-step chromatic transference scale, and spectrophotometer. When testing the color of material, it is customary to look along the length and width of the substrate, as factors such as application pressure, hydrolysis, and dye bath replenishment can lead to unlevel dyeing. Evaluation of color can be either measured quantitatively with equipment, or subjective.

10.5 Subjective Methods

The grayscale and nine-step chromatic transference scale are two examples of measuring color subjectively. The greyscale is typically used for colorfastness to laundering testing. In this assessment, the original unwashed fabric would be compared to the washed fabric. The grayscale is used alongside the evaluation to match the shade variation to its closest resemblance on the grayscale. Based on the closest match, the grayscale will help to score the tested fabric between a one and five. In this scoring, one signifies an immense or extreme amount of difference in the samples while five signifies practically no difference between the original and washed samples.

The nine-step chromatic transference scale is typically utilized for colorfastness to crocking testing. This scale contains a table set up of five colors (red, green, yellow, blue, and purple) at nine different intensities of each color.

The crocking square used to test the fabric is lined up to the scale based on what level of saturation it best matches. The scale is rated one to five, where one indicates intense color rub off and five indicates practically none to no color rub off. A light box is used to provide consistent lighting while reviewing results.

10.6 Quantitative Methods

When a numerical or measurable output is needed to evaluate color, a spectrophotometer may be used. The spectrophotometer measures the intensity of electromagnetic energy at each wavelength of light in a specified range (Craic, 2019). Spectrophotometers work by utilizing a light source, a device to focus the light, a shutter to collect the light from the sample, a monochromator to separate the light into wavelengths, and a detector to measure light intensity at each wavelength. The output results are a decimal indicating how close the measured sample is to the baseline standard. The spectrophotometer outputs L*A*B values, and a Delta E (ΔE). L* represents the lightness/darkness of the sample. It is measured between 0 (black) and 100 (White). The A* value measures the red/green axis of the sample, with green starting at -128 and red ending at $+128$. The B value measures the yellow/blue axis of the sample, with blue starting at -128 and yellow ending in $+128$ (Allen, 2023). The ΔE indicates overall how close the samples are to your standard.

11 Foam Dyeing Case Study

Foam dyeing research needs to be continuously expanded to increase the knowledge base and help make the technology more applicable. Despite existing foam dyeing research, a gap was found in information available on foam dyeing regarding colorfastness properties on cationized cotton with multiple variations of dye type, and dye concentration. Typically prior to manufacturing runs, small scale trials are completed in a laboratory, to gather information on esthetic and performance of a material. Doing small scale lab work helps to predetermine what conditions to use when scaling up to a manufacturing setting. This in turn avoids wasting time, energy, and materials on an unsuccessful large scale trial. To add more information and research to foam dyeing, an experiment was designed to test predetermined dye and pretreatment combinations in a foam dyeing setting, to see if they could perform the same as their conditions in a conventional dyeing process.

12 Methodology

This methodology was created to evaluate how specific dyeing variables impact colorfastness properties when utilizing foam dyeing on cationized cotton. The variables include looking at the impact of dye type, the impact of pretreatment dosage, and the impact of dye dosage. Furthermore, this methodology aims to assess the esthetic and colorfastness performance in a foam dyeing process, which will use significantly less water.

The study utilized specific combinations of dye and pretreatment to be included in a foam dyeing application. Each combination was tested for two colorfast methods and evaluated. Lastly, the results were statistically analyzed to evaluate the variables considered in the experiment. A statistical analysis was run utilizing the dye type, and level of pretreatment as independent variables. The results measure the impact of these variables on the colorfastness results and how their properties impact the colorfastness properties. The experiment also examines how closely a foam dyed sample can match the esthetic and performance of traditionally dyed fabric.

If the foam dyed samples can match or exceed the colorfastness performance and esthetics of a traditionally dyed fabric, then a case can be better made for utilizing foam dyeing more in the industry. To test this, 25 combinations of dye type, dye concentration level, and pretreatment level were selected to be run with foam dyeing application. These combinations were selected based on previous lab work completed, identifying optimal combinations of these variables. The combinations selected showed promising results in colorfastness testing including AATCC TM8; colorfastness to crocking, and AATCC TM61-2A, Colorfastness to laundering.

12.1 Materials Utilized

This experiment was completed on 100% cotton plain weave fabric, treated with Oasis™ pretreatment prior to the foam dyeing process. The treatment was applied in increments of 20 g/L, starting at 0 g/L (untreated) to 100 g/L. Six levels of pretreatment in total were utilized for the trials.

For the foam dyeing trials, a foaming agent needed to be added to the bath in order to facilitate the baths ability to mix with and incorporate air. This was added into the calculation to create the bath. No other auxiliary chemicals were utilized in the bath formation process.

As colorfastness was a significant property to identify the ability to predict performance in foam dyeing, five dyes across the reactive subclasses were selected to be tested. Both monofunctional and bifunctional dyes were selected to test. For monofunctional dyes, an MCT, VS, and DCT dye were selected. For the bifunctional dyes, a homobifunctional FT/FT dye and a heterobifunctional VS/MCT dye

Table 2 Dyes used in foam dyeing study

Dye	Subclass	Monofunctional	Bifunctional	Dye provider	CI
Chromative Red HT-3B Powder	MCT	X		Cote color	Reactive Red 120
Chromazol Red RB Powder	VS	X		Cote color	Reactive Red 198
Novacron Scarlet EC-6G	FT/FT		X	Huntsman dyes	Reactive Red 267
Red MX-5B	DCT	X		Orco dyes and pigments	Procion Brilliant Red 5B
Red BF-6BN	VS/ MCT		X	Orco dyed and pigments	Reactive Red 195

were selected. Table 2 provides a breakdown of the materials selected to be used as well as where they came from.

Red dyes were selected strictly because of their known poor colorfastness properties. Dyes were all in a powdered form when received and were supplied by Cote Color (Spartanburg, SC), Huntsman (Woodlands, TX), and Organic Dyes and Pigments (Lincoln, RI).

It was determined that three dye concentrations would be evaluated. To decide concentration levels of dye to use, the ceiling (high) amount was determined based on the foam dyeing application limit (30 g/L). The medium and low concentrations were recommended based on what was commonly used in industry (T. Heller, Plant Manager, JB Martin, Personal communication, conversation, 2019).

Twenty-four combinations of dye type, concentration level, and pretreatment level were selected based on preliminary testing to be evaluated in foam dyeing. They were elevated for colorfastness. These combinations are given in Table 3.

The bath was calculated to factor in the wet pick-up of 30%. To run a trial in the equipment available, the total volume of the bath was determined to be 4000 L. The grams of dye needed for each combination was calculated as follows:

$$\text{Ratio of WPU} \times \text{Dye concentration g / L} \times 4\,\text{L}\,\text{H}_2\text{O} = \text{Total weight of dye}$$

To determine the quantity of water needed for each combination, the following equation was used:

$$W_{\mathrm{T}} - W_{\mathrm{FA}} - W_{\mathrm{DYE}} = W_{\mathrm{H_2O}}$$

where:

W_{T} = Total dye bath weight
W_{FA} = Weight of foaming agent
W_{D} = Weight of dye

Table 3 Dye and pretreatment combinations at each level of dye concentration

Variable combinations that were foam dyed[a]			
Dye type	High dye concentration	Medium dye concentration	Low dye concentration
Level of pretreatment			
MCT	100 g/L, 80 g/L, 60 g/L	40 g/L	20 g/L, 0 g/L
VS	100 g/L, 80 g/L, 60 g/L, 40 g/L	–	–
FT/FT	100 g/L, 80 g/L	60 g/L, 40 g/L	20 g/L
DCT	100 g/L, 80 g/L, 60 g/L	40 g/L	–
VS/MCT	100 g/L, 80 g/L, 60 g/L	40 g/L, 20 g/L	–

[a] Based on mean scores of pre-lab work colorfast results

12.2 Process

The bath was prepared by combining the predetermined amounts of dye, water, and foaming agent and then mixed for 5 min using a drill with a metal stirring rod attachment. No heat was needed during the mixing process. The prepared bath was transferred to the vat and the foam range was set to match industry standards. Once the foam was calibrated (dye evenness through the foam), it dispersed onto the substrate, then padded, steamed, and dried. The same test methods were utilized in the predetermining lab work. All samples were also measured on the spectrophotometer as well, using the previous lab work samples as the standard for each combination being looked at.

Upon completion of the dyeing, testing, and color measuring, a statistical analysis was completed between the previous lab work completed and the foam dyed samples. To analyze the results, ANOVA and T-Tests were completed to compare all AATCC scores and L*A*B readings. The analysis utilized the reactive dye subclass and level of pretreatment as the independent variables. Lastly, previous lab work results were compared to foam dyed results with ANOVA to identify any statistical differences between the two dyeing methods.

13 Review of Results

As stated previously, this experiment was created to evaluate how specific dyeing variables impact colorfastness properties when utilizing foam dyeing on cationized cotton. The study aimed to find out if predetermined dyeing combinations in a lab setting can be a good indicator for the esthetic and performance of foam dyeing. The analysis is used to evaluate the impact of dye type, level of pretreatment, and dye concentration on their impact on colorfastness properties and L*A*B values. The analysis also only utilized data which was shown to be passing (received a mean score of ≥2.75 in each colorfastness test).

The data presented is organized first by the impact of the dye type and level of pretreatment on wet crocking. It then covers the impact of dye type and level of pretreatment on wash fastness. Finally, the L*a*b* values are presented and compared between conventional and foam dyed samples.

13.1 Impact of Dye Type on Crocking

All samples were measured for wet crocking and washfastness using AATCC TM8: Colorfastness to crocking and AATCC TM61,2A: Colorfastness to laundering. Samples were tested in triplicate, and the means and standard deviation of each level of dye concentration are found in Tables 4, 5, and 6.

Table 4 Wet crocking results of foam dyed samples at high level of dye concentration

Wet crocking results at high dye concentration			
Dyestuff	Pretreatment level (g/L)	Mean[a,b]	Standard deviation
MCT	60	3.16	0.288
	80	3	0.0
	100	3.16	0.288
VS	40	3	0.0
	60	3.16	0.288
	80	3.5	0.0
	100	3	0.0
FT/FT	80	3.33	0.288
	100	3.5	0.0
DCT	60	3	0.0
	80	3	0.0
	100	3	0.0
VS/MCT	60	3.66	0.288
	80	3.66	0.288
	100	3.5	0.0

[a] Mean of 3 replications
[b] Transfer Scale 5 = no transfer of color 1 = high transfer of color

Table 5 Wet crocking results of foam dyed samples at medium level of dye concentration

Wet crocking results at medium dye concentration			
Dyestuff	Pretreatment level (g/L)	Mean[a,b]	Standard deviation
MCT	40	4	0.0
FT/FT	40	4.5	0.0
	60	4	0.5
DCT	40	4	0.0
VS/MCT	20	4.16	0.288
	40	4.16	0.288

[a] Mean of 3 replications
[b] Transfer Scale 5 = no transfer of color 1 = high transfer of color

Table 6 Wet crocking results of foam dyed samples at low level of dye concentration

Wet crocking results at low dye concentration			
Dyestuff	Pretreatment level (g/L)	Mean[a,b]	Standard deviation
MCT	0	3.5	0.0
	20	4.33	0.288
FT/FT	20	4.3	0.288

[a] Mean of 3 replications
[b] Transfer Scale 5 = no transfer of color 1 = high transfer of color

Table 7 Wet crocking by dye type results of foam dyed samples at each level of dye concentration

ANOVA of impact of dye type on wet crocking score and Tukey HSD							
ANOVA (high concentration of dye)					Tukey HSD		
Dye	Mean value[a,b]	Standard deviation	Standard error	*P*-value[c]	Dye subclass	Group A	Group B
MCT	3	0.0	0.088	0.007	MCT		X
VS	3.125	0.25	0.068		VS	X	X
FT/FT	3.5	0.0	0.108		FT/FT	X	
DCT	3	0.0	0.088		DCT		X
VS/MCT	3.5	0.0	0.088		VS/MCT	X	
ANOVA (medium concentration of dye)					Tukey HSD		
MCT	4	0.0	0.0	0.001	MCT		X
FT/FT	4.5	0.0	0.0		FT/FT	X	
DCT	4	0.0	0.0		DCT		X
VS/MCT	4	0.0	0.0		VS/MCT		X
T-test (low concentration of dye)							
MCT	4	0.547	0.233	0.17	MCT	–	–
FT/FT	4.5	0.0	0.00		FT/FT	–	–

[a] 1 = poor, 5 = good
[b] Mean of all mean passing scores
[c] $\alpha = 0.05$ for those tested

ANOVA testing of each combined mean was completed on each level of dye concentration to determine if the dye type has an impact on the colorfastness results. Alpha for each analysis was set to 0.05. The purpose of this testing was to address if there would be a significant difference between the dye type and colorfastness score. Wet Crocking by Dye type Results of Foam Dyed Samples at Each Level of Dye Concentration is tabulated in Table 7.

The results indicated that at the highest level of dye concentration, the dye type has a significant impact on the wet crocking scores. The bifunctional dyes had significantly different scores from the monofunctional dyes. In the medium level of dye concentration, when comparing the dye types, there was a significant difference in wet crocking scores. At this level of concentration, the bifunctional FT/FT dye was significantly different than the others evaluated. At the low level of dye concentration, no statistical significance by dye type was found.

Overall, the bifunctional dyes had the highest colorfastness scores for wet crocking in the high, medium, and low levels of dye concentration. It is thought that the time each dye type needs to penetrate the fiber could be what led to these results. It may take some molecules, such as those with non-linear structures, additional time to find bonding sights on the fiber. The contact time between dye and fiber is relatively quick, as once the foam is dispersed, it sits on the surface for a few seconds until it goes through the padder to remove excess foam. Agitation in traditional processes aid in helping the come in contact with available bonding sites, but there is no agitation in foam dyeing (Da Silva, 1983). It is also likely that the bifunctional dyes had increased opportunity to bond with the fiber over the monofunctional dyes.

Dye molecule shape and size also has a large impact on the dyes ability to penetrate into a substrate at different rates. A dye with a linear structure, such as the VS/MCT, can align with the cellulosic polymer chain in the cotton fiber, and form a bond more quickly than other shaped molecules. Bifunctional dyes tend to have larger molecule sizes which can lead to difficulty in finding available bond sights. They do, however, have a larger surface area which is advantageous in covering more area to find sights to bond with faster. and having two different subclasses in one dye increases opportunity to find more bonding sights.

At the lowest level of dye concentration, there was no significant difference in wet crocking scores by dye type. Both monofunctional and bifunctional dyes were not impacted by the level of dye concentration.

13.2 Impact of Dye Type on Washing

All samples were measured for colorfastness to washing utilizing AATCC TM61,2A: Colorfastness to laundering. The purpose for evaluating colorfastness after washing is to understand if the use of water, detergent, and agitation had an impact on the colorfastness at each level of dye concentration. A breakdown of the results by level of dye concentration is given in Tables 8, 9, and 10.

The data above analyzed the washfastness results at each level of dye concentration. The alpha was set to $\alpha = 0.05$. The data looks to address if there is a significant difference between dye type and colorfastness score. ANOVA analysis was completed on all results and they are represented in Table 11.

The results of the ANOVA analysis indicate that the dye type makes no significant difference on the washfastness results at all levels of dye concentration. This means that both the monofunctional and bifunctional dyes achieved similar washfastness results. The type of dye also has no significant impact on the color fastness scores at each level of dye concentration.

Table 8 Washfastness results of foam dyed samples at high level of dye concentration

Washfast results at high dye concentration			
Dyestuff	Pretreatment level (g/L)	Mean[a,b]	Standard deviation
MCT	60	4	0.0
	80	4	0.0
	100	4.33	0.288
VS	40	4.33	0.288
	60	4.66	0.288
	80	5	0.0
	100	4	0.0
FT/FT	80	4.5	0.0
	100	5	0.0
DCT	60	3.83	0.288
	80	2.83	0.288
	100	4.66	0.288
VS/MCT	60	4.66	0.288
	80	5	0.0
	100	4.66	0.288

[a] Mean of 3 replications
[b] Transfer Scale 5 = no transfer of color 1 = high transfer of color

Table 9 Washfastness results of foam dyed samples at medium level of dye concentration

Washfast results at medium dye concentration			
Dyestuff	Pretreatment level (g/L)	Mean[a,b]	Standard deviation
MCT	40	4.33	0.288
FT/FT	40	4.83	0.288
	60	4.83	0.288
DCT	40	4.16	0.288
VS/MCT	20	3.33	0.288
	40	4.66	0.288

[a] Mean of 3 replications
[b] Transfer Scale 5 = no transfer of color 1 = high transfer of color

Table 10 Washfastness results of foam dyed samples at low level of dye concentration

Washfast results at low dye concentration			
Dyestuff	Pretreatment level (g/L)	Mean[a,b]	Standard deviation
MCT	0	3	0
	20	4.83	0.288
FT/FT	20	4.66	0.288

[a] Mean of 3 replications
[b] Transfer Scale 5 = no transfer of color 1 = high transfer of color

Table 11 Colorfastness to laundering by dye type results at dye level of dye concentration

ANOVA of impact of dye type on washfastness and Tukey HSD				
ANOVA (high concentration of dye)				
Dye	Mean value[a,b]	Standard deviation	Standard error	*P*-value[c]
MCT	4.167	0.288	0.415	0.357
VS	4.5	0.408	0.321	
FT/FT	4.75	0.353	0.508	
DCT	3.667	1.04	0.415	
VS/MCT	4.667	0.288	0.415	
ANOVA (medium concentration of dye)				
MCT	4.5	0.0	0.5	0.455
FT/FT	5	0.0	0.353	
DCT	4.5	0.0	0.5	
VS/MCT	4	0.707	0.353	
T-test (low concentration of dye)				
MCT	4	1.41	1	0.821
FT/FT	4.5	0.0	1.414	

[a] 1 = poor, 5 = good
[b] Mean of all mean passing scores
[c] $\alpha = 0.05$ for those tested

13.3 Impact of Pretreatment on Wet Crocking

The level of pretreatment was also assessed as an independent variable to understand its impact on both AATCC TM8; colorfastness to wet crocking and AATCC TM61,2A; colorfastness to laundering results. The samples were tested in triplicate and each ANOVA alpha was set to $\alpha = 0.05$. This data looks to identify if there is a significant difference between level of pretreatment and colorfastness score. Table 12 provides a breakdown of all the data tested.

The results indicate that there is no statistical significance on the wet crocking results between each level of pretreatment at all levels of dye concentration. This result is similar to how the preliminary lab work performed. There is also no *P*-value for the results at the low level of dye concentration because the results are identical.

13.4 Impact of Pretreatment on Washfastness Scores

The data was also analyzed for AATCC TM61, 2A; Colorfastness to Laundering. The data is analyzed by the high, medium, and low dye concentration to determine if the level of pretreatment has an impact on washfastness results. If the P-Value was significant, the Tukey HSD analysis is also included. This data looks to assess that the level of pretreatment has no significant impact on washfastness scores. The ANOVA analysis results are given in Table 13.

Table 12 Wet crocking by level of pretreatment results of foam dyed samples at each level of dye concentration

Level of pretreatment	Mean value[a,b]	Standard deviation	Standard error	*P*-value[c]
Impact of level of pretreatment on wet crocking score (high dye concentration)				
40 g/L	3	0.0	0.154	0.447
60 g/L	3.125	0.25	0.071	
80 g/L	3.3	0.273	0.061	
100 g/L	3.2	0.273	0.061	
Impact of level of pretreatment on wet crocking score (medium dye concentration)				
20 g/L	4	0.0	0.25	0.421
40 g/L	4.125	0.25	0.125	
60 g/L	4.5	0.0	0.25	
Impact of level of pretreatment on wet crocking score (low dye concentration)				
0 g/L	3.5	0.0	0.0	–
20 g/L	4.5	0.0	0.0	

[a] 1 = poor, 5 = good
[b] Mean of all mean passing scores
[c] $\alpha = 0.05$

Table 13 Washfast by level of pretreatment results of foam dyed samples at each level of dye concentration

Level of pretreatment	Mean value[a,b]	Standard deviation	Standard error	*P*-value[c]
Impact of level of pretreatment on colorfastness to laundering score (high dye concentration)				
40 g/L	4.5	0.0	0.726	0.905
60 g/L	4.25	0.288	0.337	
80 g/L	4.2	1.03	0.287	
100 g/L	4.5	0.353	0.287	
Impact of level of pretreatment on laundering score (medium dye concentration)				
20 g/L	3.5	0.0	0.25	0.044
40 g/L	4.625	0.25	0.125	
60 g/L	5	0.125	0.25	
Impact of level of pretreatment on laundering score (low dye concentration)				
0 g/L	3	–	0.353	0.154
20 g/L	4.75	0.353	0.25	

[a] 1 = poor, 5 = good
[b] Mean of all mean passing scores
[c] $\alpha = 0.05$

The results indicate that at the level of pretreatment does not make a significant difference on the washfast scores at the highest level of dye concentration. The lowest level of dye concentration also showed no significant difference between the level of pretreatment and washfast scores. The high and low levels of dye concentration indicate that the level of pretreatment does not have an impact on washfastness scoring.

At the medium level of dye concentration, there was a significant difference between the levels of pretreatment and their impact on the washfast score. This indicates that a specific amount of pretreatment is needed to have an acceptable washfastness score. In this level of dye concentration, the 20 and 60 g/L are significantly different from each other. This indicated that 20 g/L is probably too low for the dosage of dye being used.

Generally, using a lower level of the pretreatment chemical is ideal for both economic and environmental savings. The use of pretreat still, however, helps to improve the affinity of dye to substrate, thus reducing the amount of dye wastage overall. Without pretreatment, the dyes would have a more challenging time finding sights on the substrate to bond with. More auxiliary chemicals may also need to be used to aid in the dyeing. Overall, the use of pretreatment reduces the amount of other auxiliary needed.

13.5 L*a*b* Results

This section of the analysis was completed to compare the color of the foam dyed samples to conventionally dyed samples utilizing the same dye type and levels of dye concentration. To do this, each sample was read on a spectrophotometer. The conventional dyed samples were set as the standard, and the foam dyed samples were measured against the standard, to see how closely they matched. Pooled T-tests were utilized to analyze the data. The results are indicated in Table 14.

There is no significant difference in the L*a*b* results between the foam dyed samples and their conventionally dyed counterparts. This is true for all levels of dye concentration considered. These results support that there is no significant difference in the L*a*b* values between conventional and foam dyed samples. This entails that the dyeing method can be used interchangeably.

This research aimed to find out how the properties of each dye impacted the colorfastness. Overall, some of the variations in the colorfastness had to do with the

Table 14 L*a*b* results of foam dyed samples at each level of dye concentration

Type of dyeing	Mean value[a]	Standard deviation	Standard error	*P*-value[b]
T-Test of L*a*b* results (high dye concentration)				
Conventional	37.652	20.08	3.104	0.684
Foam	35.860	21.54	3.104	
T-test of L*a*b* results (medium dye concentration)				
Conventional	41.001	22.206	5.101	0.993
Foam	40.938	24.682	5.51	
T-test of L*a*b* results (low dye concentration)				
Conventional	40.945	27.357	10.162	0.819
Foam	37.617	33.319	10.162	

[a] Mean of all passing scores
[b] $\alpha = 0.05$

specific subclass of the dye being used. The triazine based dyes have one or two leaving groups to react with the substrate, while vinylsulfone based dyes create an ether linkage to react with the substrate. Dyes with more leaving groups have a better chance of bonding with a substrate than less reactive subclasses. In this study, the VS dyes had acceptable colorfastness scores at the highest level of dye concentration, using a pretreatment level of 40 g/L and above. MCT, DCT, and VS/MCT achieved good colorfastness scores at the high level of dye concentration with pretreatment levels of 60 g/L and higher. The FT/FT dye was the least reactive, and needed a pretreatment of 80 g/L to achieve acceptable colorfastness scores. Overall, the heterobifunctional dye performed better than the homobifunctional dye.

The level of pretreatment was also assessed to understand its impact on colorfastness at the three levels of dye concentration. By changing the charge of the fabric, the affinity is increased, and dye can bond easier. On average, the lowest levels of pretreatment used were not acceptable for the highest level of dye concentration used. The highest levels of pretreatment were too high for the lowest level of dye concentration, and also resulted in failed fabric. Pretreatment levels from 40 g/L up to 100 g/L were suitable for utilizing the highest level of dye concentration. For the medium level of dye concentration, 20–40 g/L was ideal for achieving the best colorfastness results. For the lowest level of dye concentration, 20 g/L was ideal.

Less reactive dyes also require a higher level of pretreatment for better chance of reaction. FT/FT dyes were one of the least reactive dyes utilized in this study, and did not perform well with lower levels of pretreatment. To reduce the amount of pretreatment utilized, a more reactive dye like the VS subclass is recommended.

Lastly, this research sought to investigate if foam dyed samples can achieve the same color measurements as conventionally dyed samples. The research showed no significant difference between the samples considered, indicated that the colors were very similar. The same dye and the same level of pretreatment should be used when moving from one method of dyeing to another. For esthetic purposes, it is important to know that color is not compromised between coloration methods.

14 Conclusion

As textile wet processing is resource intensive, it is important to adapt methods of coloration that utilize reduced water, energy, and chemicals. Textile converters need to implement new ways of applying color to textile with a method that uses less resources than conventional dyeing methods. Reduced liquor methods of coloration reduce or eliminate effluent and overall water. The use of a pretreated material can also reduce or eliminate the need for salt and other auxiliary chemicals that could be harmful to waterways they are dumped in.

Foam dyeing has been studied for a long period and has the ability to reduce water usage by up to 90%. It has the capability to dye 100 yards of fabric with 3.6 gallons of water. Foam dyeing reduces water and chemical use, without compromising the esthetic and colorfastness properties of the material. It does not increase processing time and can ultimately reduce a manufacturing facility's effluent. The information available can help textile converters make more informed decisions in adapting foam dyeing into their textile processing.

This study evaluated the impact of independent variables dye type, dye concentration, and level of pretreatment of the colorfastness scores of foam dyed samples with cationized cotton. This study utilized previous lab work to create 24 combinations of the variables to use in a foam dyeing setting. The samples were measured with AATCC TM8; colorfastness to wet crocking and AATCC TM61,2A; colorfastness to laundering. The results of those two tests were statistically analyzed to understand the impact of the independent variables on the colorfastness results. Finally, the color was measured using spectrophotometry to determine how different the foam dyed samples were to their counterpart.

The result indicated that the dye type impacts the wet crocking scores at the high and medium level of dye concentration. There was no impact on the low level of dye concentration. The dye type does not impact the wastfastness scores at any level of dye concentration. The level of pretreatment does not impact the wet crocking scores. It also does not impact the washfast scores at the high and low levels of dye concentration. There is an impact on the washfast score at the medium level of dye concentration. The method of dyeing also has no impact on the measured L*a*b* values. This means that the color is similar between conventional dyed and foam dyed samples.

Some combinations of dye type, dye concentration, and level of pretreatment are optimal for the highest colorfastness results. For the highest level of dye concentration, MCT, DCT, and VS/MCT achieved acceptable colorfastness scores on fabric with pretreatment 60 g/L and higher. VS specifically was able to achieve good colorfastness results at 40 g/L. At the medium level of dye concentration, MCT, FT/FT, DCT, and VS/MCT achieved good colorfastness scores on fabric with pretreatment at 40 g/L and higher. Finally, at the lowest level of dye concentration, MCT and FT/FT dyes only achieved acceptable colorfastness scores on fabric with 20 g/L of pretreatment. Bifunctional dyes generally performed better than the monofunctional dyes. Ultimately the lowest levels of pretreatment required are most economical for a tactile converter.

References

AATCC. (2019). *AATCC technical manual* (34th ed.). Association by Howes Pub. Co.

Abrahart, E. N. (1977). *Dyes and their intermediates*. Chemical Publishing.

Allen, D. (2023, May 9). *Understanding the CIELAB (L*A*B*) scale*. UMF. https://www.umf-corp.com/post/understanding-the-cielab-l-a-b-scale

Baker, K., Bryant, G., Camp, J., Jr., & Kelsey, W. (1982). Foam finishing technology. *Textile Research Journal, 52*, 395–403. https://doi.org/10.1177/004051758205200607

Bhavsar, P. S., Zoccola, M., Patrucco, A., Montarsolo, A., Mossotti, R., Giansetti, M., & Tonin, C. (2017). Superheated water hydrolyzed keratin: A new application as a foaming agent in foam dyeing of cotton and wool fabrics. *ACS Sustainable Chemistry & Engineering, 10*(5), 9150–9159.

Booth, B. (1982). Foam dyeing: Advantages and disadvantages. *Carpet & Rug Industry, 10*(5), 27–29.

Broadbent, A. (2001). *Basic principles of textile coloration*. Society of Dyers and Colourists.

Cassidy, T., & Parikshit, G. (2018). *Textile and clothing design technology: Fabric dyeing and printing*. CRC Press. Retrieved from 113 https://app.knovel.com/hotlink/pdf/id:kt011N3AX2/textile-clothing-design/fabric-dyeingprinting

Chen, S., Wang, C., Fei, L., & Liu, H. (2017). A novel strategy for realizing environmentally friendly pigment foam dyeing using polyoxyethylene ether surfactant $C_{14}EO_5$ as a foam controller. *Coloration Technology, 133*(3), 253–261.

Cohen, A., Johnson, I., & Pizzuto, J. J. (2011). *JJ Pizzuto's fabric science* (10th ed.). Bloomsbury Academic.

Cotton Inc. (2005). *Technical bulletin: Dyeing cationic pretreated cotton* (pp. 2–10). Cotton Incorporated.

Craie Technologies. (2019). *What is a spectrophotometer?*

Da Silva, J. (1983). *The behavior of nonionic dyes and surfactants in foam dyeing systems* (Doctoral dissertation, Clemson University: ProQuest).

Davis, R. (2023, November 30). *ITMA 2023: Transforming the world of textiles*. Textile World. https://www.textileworld.com/textile-world/features/2023/11/sustainable-dyeing-finishing-innovations/

EPA. (1994). *Statistics and facts*. Environmental Protection Agency. https://www.epa.gov/watersense/statistics-and-facts

European Parliament. (2024a, March 21). *The impact of textile production and waste on the environment (infographics)*. European Parliament. https://www.europarl.europa.eu/topics/en/article/20201208STO93327/the-impact-of-textile-production-and-waste-on-the-environment-infographics

European Parliament. (2024b). *Water protection and management*. Fact sheets on the European Union: European Parliament. https://www.europarl.europa.eu/factsheets/en/sheet/74/water-protection-and-management

Farrell, M. J. (2012). *Sustainable cotton dyeing* (Doctoral dissertation). https://search.proquest.com/docview/921659607

Fu, S. (2016). *Studies on dyeing cationized cotton* (Doctoral dissertation). https://doi.org/10.1111/rda.13387

Ghoranneviss, M., & Shahidi, S. (2017). Color intensity and wash fastness properties of dyed cotton fabric after plasma treatment. *The Journal of the Textile Institute, 108*(3), 445–448. https://doi.org/10.1080/00405000.2016.1171026

Glock, R. E., & Kunz, G. (2005). *Apparel manufacturing: Sewn product analysis* (4th ed.). Pearson/Prentice Hall.

Irfan, M., Zhang, H., Syed, U., & Hou, A. (2018). Low liquor dyeing of cotton fabrics with reactive dye by an eco-friendly technique. *Journal of Cleaner Production, 197*. https://doi.org/10.1016/j.jclepro.2018.06.300

Jiang, Y. (2009). China's water scarcity. *Journal of Environmental Management, 90*(11). https://doi.org/10.1016/j.jenvman.2009.04.016

Kan, C., Lam, Y. L., & Li, M. Y. (2015). The effect of plasma treatment on the dyeing properties of silk fabric. *Society of Dyers and Colourists*. https://doi.org/10.1111/cote.12189

Kant, R. (2012). Textile dyeing industry an environmental hazard. *Natural Science (Irvine, Calif.), 4*(1), 22–26. https://doi.org/10.4236/ns.2012.41004

Leonas, K. K. (2018). (rep.). *Personal communication*.

McGregor, D. (2017). *Insights form China's textile manufacturers: Gaps to overcome for clean and circular fashion* (pp. 14–42). China Water Risk.

Mock, G. N. (2002). *Fundamentals of dyeing and printing*. North Carolina State University.

Moser, L. S. (2003). ITMA 2003 review: Textile printing. *Journal of Textile and Apparel, Technology and Management, 3*(3), 1–15.

Muthu, S. S. (2017). *Textiles and clothing sustainability sustainable textile chemical processes* (1st ed.). Springer.
Ondrey, G. (2018). *Foam dye process cuts water and chemicals*. Access Intelligence LLC.
Perkins, W. S. (1996). *Textile coloring and finishing*. Carolina Academic Press.
Raja, D. (2014). *Continuous dyeing machines* (Doctoral dissertation, Mehran University of Engineering and Technology).
Ministry of Ecology and Environment The People's Republic of China. (2017). *Water pollution prevention and control law of the People's Republic of China (amended in 2017)*. https://english.mee.gov.cn/Resources/laws/environmental_laws/202012/t20201211_812662.shtml
Rivlin, J. (1982). *Introduction to the dyeing of textile fibers*. Philadelphia College of Textiles and Science.
Shang, S., Hu, E., Poon, P., Jiang, S., Kan, C., & Koo, R. (2011). Foam dyeing for developing the wash-out effect on cotton knitted fabrics with pigment. *RJTA, 15*(1), 44–51.
Stewart, C. W. (2017). *A comparative study of the environmental impacts of standardized dyeing systems using natural and synthetic dyes on knitted cotton fabric* (Doctoral dissertation). http://www.pqdtcn.com/thesisDetails/A683C4630BBEC553520BA79D405D62EE
Tunç, Ö., Tanaci, H., & Aksu, Z. (2009). Potential use of cotton plant wastes for the removal of Remazol Black B reactive dye. *Journal of Hazardous Materials, 163*(1), 187198. https://doi.org/10.1016/j.jhazmat.2008.06.078
UNICEF. (2021, March 30). *Reimagining WASH: Water security for all*. Unicef.org. Retrieved September 20, 2024, from https://www.unicef.org/media/95241/file/water-security-for-all.pdf
United Nations. (2019). *Water scarcity: Facts and figures*. UN. https://www.unwater.org/water-facts/scarcity
Wrangler, L. (2018). Foam dye set to revolutionize industry. *Wrangler*. https://www.wrangler.com/sustainability/industry/foam-dyed-denim.html
Yu, H., Wang, Y., Zhong, Y., Mao, Z., & Tan, S. (2014). Foam properties and application in dyeing cotton fabrics with reactive dyes. *Coloration Technology, 130*, 266–272. https://doi.org/10.1111/cote.12088

Natural Dyes and Pigments for Sustainable Coloration of Textiles

K. Amutha

Abstract Colorants are substances used to impart specific color to textiles, paints, ceramics, plastics, and so on. Two types of colorants are used for textile applications: dyes and pigments. Both dyes and pigments could be derived from natural sources. The limitations of the natural colorants are the availability of resources and the poor color fastness of the colored textiles. Traditionally, natural colorants have been derived from specific plant, animal, and mineral sources. Recent research studies have identified dyes from various waste sources which is a circular economical approach. Additionally, the waste or biomass is huge in quantity that ensures resource adequacy. This chapter discusses natural dyes and pigments from various sources and their application on textiles.

Keywords Natural dyes · Natural pigments · Textile dyeing · Textile printing · Mordant · Sustainability

1 Introduction

Colors exist by nature and could be seen all around us. Color is a complex thing to comprehend. Color means energy (wavelength) in physics, a chemical substance in chemistry, an element in fashion design, an artistic material in domestic and industrial applications, and so on. Color is an important criterion in art, science, culture, human emotions, materials, and almost everything in and around earth. Colors symbolize many different emotions that vary across cultures. Colors could be sensed by the human eyes only in the presence of light and they are invisible in darkness. Textiles are one among the wide applications of colors in the form of dyes and pigments.

K. Amutha (✉)
Department of Textiles and Apparel Design, Bharathiar University, Coimbatore, India
e-mail: amuthatad@buc.edu.in

S. S. Muthu (ed.), *Sustainable Coloration of Textiles*, Sustainable Textiles: Production, Processing, Manufacturing & Chemistry,
https://doi.org/10.1007/978-3-031-91217-7_8

Textiles are considered to be an important commodity for human life. It is an essential item for all people ranging from a common man to a celebrity. It is used in various forms like clothing, home textiles, disposables, and many more. Textiles are esthetic materials in which color plays a crucial role. Dyes and pigments are the two different coloring substances used in the coloration of textiles. Dyeing and Printing are the two widely adopted techniques for coloration of textiles (Amutha & Grace Annapoorani, 2023). It is quite interesting that natural dyes and pigments could be applied to textiles by either dyeing or printing.

Dyeing is an ancient art. Several cultures had established dyeing technologies even before 3000 BC. Finely ground colored minerals were dispersed in water and used in paints 30,000 years ago, but their fastness properties were poor. Natural dyes were extracted from plant and animal sources with water, sometimes under conditions like fermentation; an example is indigo which is considered as the world's oldest dye. Fabric was dyed by soaking it in the aqueous extract and drying. Most of the colors obtained were dull and in limited range and the dyed materials exhibited poor fastness to washing and sunlight (Broadbent, 2001).

The practice of employing natural dyes for the purpose of coloration is regarded as one of the most ancient traditions in the annals of human civilization. This creative pursuit enabled primordial humans to articulate their identities and their surroundings by utilizing natural dyes as their primary cognitive instrument. Historically, natural colorants were utilized for the dyeing of textiles until the advent of the nineteenth century, when synthetic alternatives emerged subsequent to the revelation of mauve pigments. Natural dyes, which include colorants, pigments, and other materials derived from botanical, zoological, or geological sources, represent sustainable and renewable resources that exert negligible environmental repercussions. These dyes have been harnessed since antiquity for a multitude of applications, encompassing the coloration of textiles, food items, natural protein-based fibers (such as wool, silk, and cotton), leather, food enhancers, and cosmetic products.

The term "natural dye" conventionally signifies substances obtained from botanical sources; however, it also includes colorants extracted from specific insects such as cochineal, kermes, and lac. These dyes are composed of complex mixtures of constituents drawn from natural sources including flora, fauna, or minerals. An extensive variety of herbs, shrubs, trees, insects, animals, microorganisms, and minerals have been recognized as potential sources for the extraction of coloring agents. Natural dyes possess the capability to generate a diverse palette of colors, which encompasses red, yellow, brown, blue, black, green, and orange.

It is essential to standardize the dyeing techniques to make the natural dyes commercially successful. The traditional natural dyeing process need to be revived both in its original form and improved form with the currently available technology. Rich natural resources in India open scope for the application of natural dyes on textiles (Samanta & Agarwal, 2009).

Traditionally, natural dyes were extracted with water (aqueous extraction) but the recent research studies on natural dyes employ organic solvents with different polarity for effective extraction of colorants. Additionally, Soxhlet extraction, acidic or alkaline extraction, supercritical fluid extraction, solid phase extraction, microwave assisted extraction, ultrasonic extraction, enzymatic extraction, pressurized liquid extraction methods are used (Ebrahim et al., 2022).

Traditional dyeing methods involve heating/boiling the dye liquor and treating the fabric in it while modern techniques employ ultrasonic, microwave, supercritical carbon dioxide assisted dyeing for improved dye uptake.

In a study (Naveed et al., 2020), natural dye was extracted from pomegranate rind by solvent, ultrasonic and aqueous methods and used to dye Tencel fabric with three mordants: alum, ferrous sulfate, and stannous chloride. It was found that solvent extraction was the best while ultrasonic was better than aqueous extract. Stannous chloride gave best results among the mordants used. The dyed fabrics exhibited excellent color fastness to washing, rubbing, perspiration, and light.

Pre-treatment of textiles with enzyme improves their characteristics considerably. Natural dyes extracted from *Acacia catechu* and *Tectona grandis* were used to dye cotton fabric by ultrasonic method. The cotton fabric was pre-treated with enzymes such as Protease, Amylase, Lipase, and Diasterase. The enzyme pre-treatment increased the dye fixation and also improved the wash and light fastness of the dyed fabrics (Vankar & Shanker, 2008).

Purple dye has been extracted from *Justicia spicigera Schltdl.* of Acanthaceae family which was used as a color source since pre-Hispanic period in the Mayan area of Mexico and Central America. The spectroscopic, chromatographic, FTIR, and GC/MS confirmed the presence of anthocyanins in the dye (Espinosa-Morales et al., 2012).

This chapter discusses the nature of the natural colorants in terms of dyes and pigments and the differences between them, different sources of natural colorants, the affinity of natural colorants, the chromophore and auxochrome in dyes, mordants, natural dyes for dyeing and printing of textiles, advantages and disadvantages of natural colorants, color index and CAS number of natural colorants, sustainable aspects of natural dyes, and unsustainable aspects of natural dyeing.

2 Nature of Natural Colorants

Chemically, natural colorants exist in two different forms: dyes and pigments. Both dyes and pigments are commonly applied as a coloring material, but they are distinguished on the basis of their solubility either in water or organic solvents (Choi, 2013). Dyes are bound chemically to the material while pigments are bound physically. Dyes and pigments have some similarities: (1) both could be used as colorants exhibiting a wide range of shades or hues and they exhibit good color fastness properties; (2) they could be derived from organic and inorganic compounds; (3) they could be applied to textiles by dyeing, printing, coating, and dispersion methods; (4) dyes and pigments are used for the coloration of paints, plastics, ceramics, food, cosmetics, paper, and so on. The differences between dyes and pigments are given in Table 1.

Table 1 Differences between dyes and pigments

Property	Dyes	Pigments
Solubility	Mostly soluble in water and other solvents	Insoluble in water and most solvents
Affinity to textile fibers	Direct	Indirect
Type of binding agent	Mordant	Binder
Auxochrome group	Present	Absent
Light fastness	Good to poor	Better than dyes
Cost	Expensive	Cheap
Chemical composition	Organic	Inorganic
Reaction with the textile fiber	Reacts and forms bond	Adsorbs on the surface with the help of binder
Applications	Wide	Limited
Particle size	Small	Big
Availability	Abundant	Limited

Source: www.primachemicals.com

2.1 Dye

Oxford dictionary defines dye as *a substance that is used to change the color of things such as cloth or hair*. In chemistry, a dye is a compound or mixture of compounds that impart specific color to a substrate material like textiles, food, plastics, and many more.

The Ecological and Toxicological Association of Dyes and Organic Pigment Manufacturers (ETAD) define dyes as: Dyes are intensely colored or fluorescent organic substances only, which impart color to a substrate by selective absorption of light. They are soluble and/or go through an application process which, at least temporarily, destroys any crystal structure by absorption, solution, and mechanical retention, or by ionic or covalent chemical bonds (www.etad.com).

2.2 Pigment

Oxford dictionary defines pigment as *a colored powder that is mixed with a liquid to produce paint*, etc. or *a substance that exists naturally in people, animals and plants and gives their skin, leaves*, etc. *a particular color.* In chemistry, a pigment is a colored powder that is used to change the appearance of a material by absorbing or scattering light. A pigment in biology is a colored substance produced by living organisms that gives color to their tissues.

The CPMA—Color Pigment Manufacturers Association, Inc. defines pigments as: Pigments are colored, black, white or fluorescent particulate organic or inorganic solids which usually are insoluble in, and essentially, physically, and chemically unaffected by, the vehicle or substrate in which they are incorporated. They alter appearance by selective absorption and/or by scattering of light. Pigments are usually dispersed in vehicles or substrates for application, as for instance in the manufacture or inks, paints, plastics, or other polymeric materials. Pigments retain a crystal or particulate structure throughout the coloration process (www.pigments.org).

Indigo, extracted from leaves of the plant *Indigofera tinctoria*, and Tyrian purple from Mediterranean Sea snails of the genera Murex and Purpura, are water insoluble pigments called vat dyes. These do not require mordants (Broadbent, 2001).

3 Sources of Natural Colorants

Natural colorants could be obtained from plants, animals, minerals, and microorganisms (Tables 2 and 3). A plant has different parts like root, stem, leaf, flower, and fruit while in case of a tree the stem part grows into hard wood and bark. Many plants yield natural colorants but specifically a part of it yields particular colorant. The natural dyes extracted from various plant parts are explored in Table 2.

In addition to the sources of natural colorants mentioned in Tables 2 and 3, waste materials could be sustainable sources of natural colorants. A huge amount of

Table 2 Natural dyes from plants

Part of the plant	Dye source	Color	Reference
Rhizome	Turmeric	Yellow	Mamta vashishtha et al. (2017)
Root	Madder	Pink to red	Yang et al. (2020)
	Osage-orange	Yellow	Mansour (2013)
Bark	Madder	Brick red	Yusuf et al. (2013)
	Mangrove (*Rhizophora apiculata*) bark	Reddish brown	Punrattanasin et al. (2013)
	Satin wood, Bhirra	Yellow	N Rungruangkitkrai et al. (2020)
	Osage-orange	Yellow	Mansour (2013)
	Copaifera langsdorffii Desf.	Brown	Silva et al. (2022)
	Goat willow	Pink	Raslamol et al. (2024)
	Gumbenfong	Yellow	
Wood	Sappan	Pink to red	Bukhari et al. (2023)
	Young fustic	Yellowish orange	Maria Perla Colombini et al. (2007)
	Osage-orange	Yellow	Mansour (2013)
	Red sandalwood	Red	Venkatasubramanian Sivakumar et al. (2017)
	Gumbenfong	Yellow	

(continued)

Table 2 (continued)

Part of the plant	Dye source	Color	Reference
Leaf	Custard apple (*Annona squamosa*)	Light brown to black	Prathyusha and Rajitha (2022)
	Mango leaves	Brown	Uddin (2015)
	Papaya leaf	Green	Geetha and Sumathy (2013)
	Inthanin bok leaves	Brick red	Phitchaphorn Khammeel et al. (2021)
	Lumb	Yellow	Raslamol K et al. (2024)
	Wallflower	Yellow	
	Young fustic	Yellowish orange	Raslamol K et al. (2024)
	Turmeric	Yellow	Amutha and Annapoorani (2019)
	Cassava	Green	Amutha and Grace Annapoorani (2023)
Flower	Chamomile	Yellow	Hartl and Vogl (2003)
	African marigold	Yellow	Jothi (2008)
	Bottlebrush	Purple	Baishya et al. (2012)
	Bougainvillea	Pink	Geetha and Sumathy (2013)
	African tulip	Red	Lokesh and Swamy (2013)
	Broom	Yellow	Mustafa Karaboyacı et al. (2014)
	Harshinghar, Night Jasmine	Bright orange	Dedhia and Srivastava (2018)
	Ixora coccinea	Red	Ghongade et al. (2021)
	Gul Mohar	Deep crimson	Lutamyo Nambela and Haule (2024)
Fruit	Alnus	Red	Raslamol K et al. (2024)
	Aringudi	Yellowish	Harshal Patil and Dr. Ashok Athalye et al. (2024)
	Galab	Brown	
	Terminalia arjuna	Yellow	Amutha and Sudhapriya (2020)
	Thespesia populnea	Yellow	Mohini et al. (2018)
	Spathodea campanulata	Light to dark brown	Amutha and Ramya (2021)
Seed	Annatto	Red-orange	Riyanto (2019)
	Arecanut	Brown	Kushalatha et al. (2024)
Whole plant	Indigo	Blue	Mahltig (2024)

organic waste generated by agriculture, forestry, and food processing industries are sources of natural colorants (Singhee, 2020). Such dyes/pigments obtained from waste sources are given in Table 4. This approach not only provides a zero cost source of natural colorants but also aids in effective waste management.

Table 3 Natural dyes from animal, mineral, and microorganism

Origin	Dye source	Color	Reference
Animal	Lac	Red	Huang et al. (2024)
	Cochineal	Deep red	Čorak et al. (2022)
	Kermes	Red	Alaskar and Hassabo (2021)
	Shellfish	Tryian purple	Casselman and Terada (2012)
	Carmine	Red	Vankar and Shukla (2019)
	Levantine sea snails (mollusks)	Royal Purple	Verma and Gupta et al. (2017)
Mineral	Iron	Black	–
	Malachite	Green	–
	Chrome	Yellow	Hartford (1977)
	Manganese	Black	Chungkrang et al. (2020)
Microorganism	Staphylococcus sp.	Golden	Pabba et al. (2015)
	Pseudomonas	Green	
	Serratia	Red	Metwally et al. (2021)
	Thermomyces sp.	Yellow pigment	Poorniammal et al. (2013)
	Aspergillus sp.	Green	Sarkar et al. (2017)
	Penicillum sp.	Red	

Table 4 Natural dyes from agro and food processing waste

Dye source	Color	Coloring compound	Reference
Food and beverage industry waste Raspberries Sour cherries Black elder Blackcurrant Grapes	Beige; Gray Beige Rose; red-brown Beige; gray Beige; gray; mauve; rose	–	Bechtold et al. (2006)
Tamarind seed coat	Yellow	–	Tepparin et al. (2012)
Saffron flower waste	Greenish yellow to yellow in acidic pH Reddish brown in alkaline pH	Anthocyanidin and kaempferol	Asm et al. (2012)
Tamarind fruit pods	Soft, light color	–	Umar and Nor (2013)
Banana leaves • Alkaline extraction • Acetone extraction	–	• Luteolin and Apigenin • Chlorophyll a, b and β-carotene	Saleh et al. (2013)
Olive solid waste (pomace)	–	Phenolic compounds	Elksibi et al. (2014)

(continued)

Table 4 (continued)

Dye source	Color	Coloring compound	Reference
Grape pomace	Dark brown	Anthocyanins	Baaka et al. (2017)
			Mansour et al. (2017)
Pomegranate rind	Yellow	Flavonoids and tannins	Adeel et al. (2009)
			Mahmud-Ali et al. (2012)
			Sinnur et al. (2018)
Onion peel	Yellow	Quercetin, proto-catechuic acid	Singh et al. (2019)
Walnut shells	Gray brown	Juglone	Ghaheh et al. (2012)
			Mirjalili and Karmi 2013b)
			Singhee et al. (2016)
Almond shells	Reddish brown	Catechin and epicatechin	İşmal et al. (2014)
			Deepali and Shraddha (2017)
Fallen leaves of *Madhuca longifolia*	Dark brown	Quercetin, Myricetin and β-carotene	Swamy et al. (2016)
Peanut pods	Light green	–	Chhipa et al. (2017)
Waste barks of Turkish red pine	Beige, brown, and brownish yellow	Tannic acid and quercetin	Avinc et al. (2013)
Eucalyptus liquid residue from lumber steaming	–	–	Rossi et al. (2017)
Groundnut/Peanut skin	Peach	Flavonoids	Ritu Pandey et al. (2018)
Spent coffee grounds	Light to dark brown	Chlorogenic acid, caffeine and melanoidins	Bae and Hong (2019)
Waste leaves from green tea	Dark brown	–	Sukemi et al. (2019)
Waste tea leaves	Brown	Catechin	Adeel et al. (2021)
Moroccan *Crocus sativus* L. (Saffron) flower waste	Brown to green	–	Lachguer et al. (2021)
Agricultural waste—seed skin of *Eriobotrya japonic*a L. fruit	Brown	Catechin	Zhang et al. (2021)
Agro-waste: *Ipomoea batatas* leaves	Light to dark green	Chlorophyll	Fang et al. (2022)
Spent coffee grounds	Yellow	Oxyaromatic acids and flavonoids	Amutha et al. (2022)
Roasted peanut skin	Pink	Xanthone and flavonoids	
Agro-waste: Cassava leaves	Green	–	Amutha and Grace Annapoorani (2023)
Sugar beet leaves	Range of shades with different mordants	–	Batool et al. (2024)

The handling of natural dye and its dyeing process requires scientific intervention in four main steps:

1. Extraction.
2. Isolation of dye molecule.
3. Dyeing.
4. Dye fixing.

Applications of natural dyed textiles can be extended to diverse fields, such as sportswear and medicine, owing to their various functional properties such as antimicrobial and UV protection.

4 Affinity of Natural Dyes

Natural dyes work best with natural fibers such as cotton, linen, wool, silk, jute, ramie and sisal. Among these, wool takes up dyes most easily followed by cotton, linen, silk and then the coarse fibers such as sisal and jute. Nearly all of them require some sort of a mordant. The trick is to choose the right dye from the right source that gives not only beautiful tones, but colorfast shades as well. Some of the natural dyes are fugitive and need a mordant for the enhancement of their fastness properties.

4.1 Chromophore and Auxochrome

Chromophores and auxochromes are different molecules that work together to form a specific color:

Chromophore: A group of atoms or electrons in a dye molecule that's responsible for its color. A compound with a chromophore is called a chromogen.

Auxochrome: A group of atoms or electrons attached to a chromophore that enhances the color of the chromogen. Auxochromes don't produce color on their own.

Examples of chromophores and auxochromes:

Chromophores: Carbon–carbon double bond (C=C), carbon–nitrogen double bond (C=N), carbonyl (C=O), azo (N=N), nitro (NO_2), methine (CH), and quinoid rings.

Auxochromes: Carboxyl (COOH), amine (NH_3), sulfonate (SO_3H), and hydroxyl (OH).

Auxochromes modify a chromophore's ability to absorb light by altering the wavelength or intensity of the absorption. They can increase the wavelength at which light is absorbed, which intensifies the absorption.

Table 5 Natural pigments and the color

Pigment	Color
Chlorophyll	Green
Carotenoids	Orange, yellow
Anthocyanins and flavonoids	Red, blue, purple
Indigo and indigoidin	Blue
Betacyanins (Betalain)	Reddish to violet
Betaxanthins (Betalain)	Yellow to orange

4.2 *Natural Pigments*

Many plants contain pigments, which are responsible for their natural hues. Natural pigments give animals and insects their hues as well. Carotenoids are found in the feathers, scales, and fur of many birds, fish, and reptiles. Microorganisms that create pigments, such as bacteria and fungus, are also responsible for their natural hues. Carotenoids are pigments found in orange, yellow, and red those are generated by a variety of microorganisms, including bacteria and algae. Certain species of algae and cyanobacteria create phycobiliproteins, which come in red, blue, and purple hues. Certain microbes also create pigments such as riboflavin, flavins, and niacin, which provide yellow, orange, and red colors, respectively. These pigments can be used to impart natural color to fermentation operations such as yogurt and cheese manufacture. Some pigments found in nature and their colors are given in Table 5.

5 Natural Dyes for Dyeing of Textiles

Table 6 lists a wide range of natural dyes extracted from different plant sources for the coloration of textiles.

5.1 *Mordant*

The direct affinity of the natural dyes to the fiber substrate, especially cotton, is poor in most cases and hence an additional substance called 'mordant' is employed to fix the dye to the fiber. Mordant is a term derived from Latin (mordere) and means "to bite." A mordant acts like a linking agent between the dye and the fiber. Common metal mordants used with natural dyes are potassium aluminum sulfate (alum), stannous chloride, ferrous sulfate, copper sulfate, potassium dichromate, etc. Natural mordants obtained from plant sources include Myrobalan, Indian gooseberries (amla), tannins from various sources, gallnut, pomegranate rind, etc. (Singh & Singh, 2018). Classification of mordants is shown in Fig. 1.

Table 6 Various natural dyes used to dye different textile fibers

Natural dye	Mordants	Fiber dyed	Result	Reference
Henna and onion skin	Nil	Cotton, Ployester, and P/C blend	Good wash and light fastness, no fabric strength deterioration	Mohamed (2010)
Turmeric & Cassumunar—rhizome	Nil	PLA and PET	Fluorescent and antimicrobial properties	Sriumaoum et al. (2012)
Turmeric—rhizome	Ferric sulfate, cupric sulfate, and potassium aluminum sulfate	Polyamide (nylon 6,6)	Good fastness properties & excellent antibacterial activity	Mirjalili and Karimi (2013a, 2013b)
Clitoria and Marigold flowers, pomegranate rind	Lemon, ferrous sulfate, and copper sulfate	Cotton, synthetic	Good fastness to washing, light, and rubbing	Kanchana et al. (2013)
Turmeric—curcumin	Tartaric acid, alum, copper sulfate	Cotton, polyester	Average color fastness to washing, light and rubbing	Hasan et al. (2014)
Cranberry (*Carissa carandas*)	Vinegar	Cotton, silk, and cotton/polyester	Average color fastness to washing	Aliño (2014)
Curcumin and saffron	Ferrous sulfate	Polyester, nylon	UV/O_3 treatment improved the wash and light fastness	Elnagar et al. (2014)
Hibiscus vittifolius. L—flower	Copper sulfate, ferrous sulfate, aluminum sulfate, potassium dichromate, stannous chloride	Cotton	Good fastness to washing, light, and rubbing	Thiyagarajan et al. (2015)
Arecanut—seed	Aluminum sulfate	Silk	Good fastness to washing, perspiration, and rubbing	Le and Nguyen (2021)

5.2 Mordanting

The process of application of a mordant on fabric is called mordanting which could be implemented in three different ways:

1. Pre-mordanting: The textile substrate is first treated with the mordant and then with the dye.
2. Meta-mordanting: The mordant is added with the dye liquor and the textile substrate is dyed.
3. Post-mordanting: The textile substrate is first dyed and then treated with the mordant.

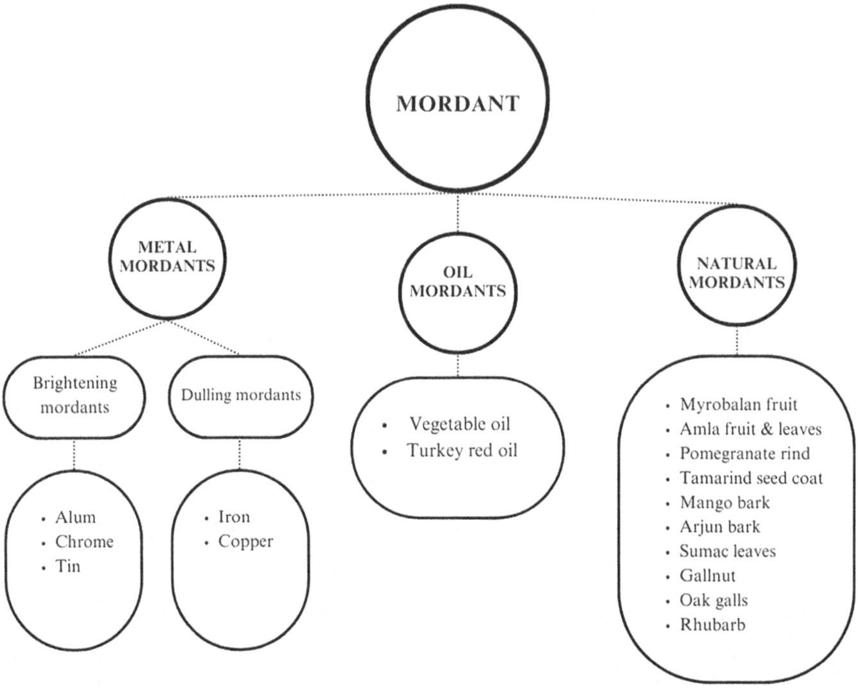

Fig. 1 Classification of mordants. (Source: Vankar, 2000)

The mordanting process has a significant effect in improving the color fastness to laundering and light (Purwar, 2016). The only eco-friendly metal mordant is alum (potassium aluminum sulfate—the most common type of alum) which is widely used in natural coloration of textiles.

The oil extraction from Antarctic krill (*Euphausia superba*) generates residue like any other oil extraction process and protein has been recovered from it for use as a natural mordant for cotton fabric with natural tannin. The color fastness properties rated from good to excellent. Hence this is a sustainable alternative to the metal mordants used in natural dyeing of textiles (Pisitsak et al., 2018).

Natural mordants are superior to metal mordants in case with certain natural dyes. In a study Arecanut natural dye was used to dye cotton fabric with natural mordants like pomegranate rind, myrobalan, and lodh. It was observed that premordanting is more efficient than the other two mordanting techniques and the mordanting improved the color fastness of the dyed fabrics. Lodh acted only as mordant and did not influence or alter the shade of Arecanut dye while pomegranate rind and myrobalan altered the actual shade of the dye. The study has been claimed to be sustainable by the authors (Jain & Vasantha, 2016).

6 Natural Dyes for Printing of Textiles

Natural dyes and pigments are widely used in printing of textiles prominently by screen printing method as listed in Table 7.

Table 7 Various natural dyes used in printing of textile substrates

Natural dye	Mordants/ additives	Textile substrate	Printing technique	Result	Reference
Alkanet powder	Aluminum ammonium sulfate Stannous chloride Magnesium sulfate Ferric chloride Tannic acid Ferrous sulfate	Wool Silk Cotton Flax	Screen printing	Violet to dark brown	Rekaby et al. (2009)
Rhubarb powder				Dark yellow, yellowish Brown and brown	
Turmeric, madder, rhubarb	Alum, tartaric acid	Cotton/ wool (50/50)	Screen printing	–	Maamoun et al. (2014)
Catechu, turmeric, Marigold	Alum, chitosan	Wool	Screen printing	–	Teli et al. (2014)
Madder	Alum, copper sulfate, ferric sulfate, tin chloride	Cotton, Wool, Silk	Screen printing	–	Özgüney et al. (2015)
Butea monosperma flowers	Copper sulfate and ferrous sulfate	Silk	Screen printing	Green shades	Babel and Gupta (2016)
Custard apple leaf extract	Stannous chloride Alum Copper sulfate Ferrous sulfate	Cotton	Screen printing	Lightest brown Light brown Dark brown Black	Prathyusha and Rajitha (2022)
Spent coffee grounds	Myrobalan	Silk and cotton	Screen printing	Brown	Ramya K Amutha K
Sweet potato leaves				Green	
Terminalia bellerica	Alum, chalk and iron	Silk and cotton	Screen printing	Green	Tri and Nooryan (2023)
Ceriops candolleana				Purple and gray	
Mahogany				Purple and gray	
Japanese knotweed rhizome	–	Cotton and polyester	Screen printing	Light to dark yellow shades	Klančnik (2021)

(continued)

Table 7 (continued)

Natural dye	Mordants/ additives	Textile substrate	Printing technique	Result	Reference
Thyme	Alum	Cotton	Screen printing	Golden shades	Ayvaz and Teker (2024)
Japanese pagoda tree, madder, Clove, log wood, gallnut, alder	Alum Copper sulfate Ferrous sulfate Stannous chloride	100% silk, satin	Screen printing	–	Nam (2010)
Terminalia arjuna	Aluminum phosphate and ferrous sulfate	Khadi silk fabric	Screen printing	Light brown shades	Sethi and Rains (2018)
Seeds of annatto, roots of manjistha and bark of ratnajot	Myrobolan powder Potash alum	Gray jute fabric	Screen printing	–	Chattopadhyay and Pan (2019)
Myrobalan fruit, pomegranate peel and catechu	–	100% cotton	Screen printing	Dark brown shades	Rana et al. (2022)
Cochineal	Stannous chloride	Cotton fabric	Screen printing	Purple color	Dasgupta and Ghosh (2018)
Indigo dyes	Ferrous sulfate and copper sulfate	Plain white cotton fabric	Screen printing	Black	Babel et al. (2015)
Pomegranate peel	Alum and tannic acid	Cotton wool silk	Screen printing	Light red	Ragheb et al. (2019)

6.1 Advantages of Natural Colorants

- The sources are renewable.
- The colorants are biodegradable.
- The process is sustainable.

6.2 Disadvantages of Natural Colorants

- Huge quantity of raw material yields only little colorant.
- Only poor to moderate color fastness could be achieved, especially poor light fastness caused by photo-oxidation of the chromophore.
- Extraction and dye application processes are difficult to standardize.
- Shades obtained are difficult to reproduce exactly unlike synthetic dyes.
- Mordants contain heavy metals that are hazardous to the environment.
- Both the extraction and dyeing processes consume high energy.
- High cost and hence low affordability.
- Low rate of dye exhaustion.

6.3 Color Index

The first edition of Color Index™ was published by the Society of Dyers and Colourists (SDC) in 1924 as a definitive guide for essential colorants (colour-index.com). Then the revised third edition was published by the SDC and the American Association of Textile Chemists and Colourists (AATCC) jointly. The Color Index™ is classified as follows:

1. Color Index™ Generic Name (CIGN).
2. Color Index™ Constitution Number (CICN).

A CIGN describes a commercial product by its recognized usage class, its hue and a serial number, e.g., CI Acid Blue 52, CI Acid Orange 7, CI Acid Red 1, CI Direct Red 28, etc.

6.4 CAS number of Natural Dyes

A CAS (Chemical Abstracts Service) Registry Number (CAS RN) is a unique identification number assigned to every chemical substance. The color index and the respective CAS RN with other details are given in Table 8 (www.artiscreation.com).

Table 8 CAS RN of some natural dyes

Color index	Color description	Source	Chemical composition	CAS number
Natural yellow 3	Lemon yellow	Reseda	Apigenin Luteolin	520-36-5 491-70-3
Natural yellow 3	Bright reddish yellow	Turmeric	Curcumin	91884-86-5
Natural yellow 8	Olive green to intense bright yellow	Wood of the dyer's mulberry	2′,3,4,5,7-pentahydroxyflavone	480-16-0
Natural red 1	Sophoretin	Laked extract of Toon or Indian mahogany tree	Sophoretin Meletin	117-39-5
Natural red 3	Yellow red	Kermes	Kermesic acid	476-35-7
Natural red 4	Dark deep to mid-red	Cochineal	Cochineal	1343-78-8
		Coccus cacti extract	Carmine	1390-65-4
		Cochinilla	Carminic acid	1260-17-9
Natural red 6	Deep bluish red	Chayroot	Alizarin	72-48-0
Natural red 8	Deep violet to bluish red to cardinal red to reddish orange	Natural rose madder	Xanthopurpurin	478-06-8

(continued)

Table 8 (continued)

Color index	Color description	Source	Chemical composition	CAS number
Natural red 25	Deep red	Lac insect	Laccaic acid	60687-93-6
Natural red 28	Reddish brown to purple	Archil	Orcein	1400-62-0
Natural violet 1	Deep purple	Murex shellfish	6,6′-dibromindigo	9201-53-7
Natural blue 2	Dark blueish purple	Indigo carmine	Natural sodium indigotin disulfonate	860-22-0
Natural green 2	Deep dull yellow-green	Laked extract of ripe buckthorn berries	Chlorophyll	1406-65-1
Natural Brown 3	Brown	Catechin	Catechinic acid	154-23-4
Natural Brown 6	Yellow brown	Terminalia chebula	Chebulinic acid	114921-10-7
Natural Brown 7	Dark Brown	Black walnut	Naphtoquinones	481-39-0
Natural Brown 8	Dull dark earth brown	Coal residues	Lignite	72669-22-8
Natural Brown 9	Dark reddish brown Black	Cuttlefish	Sepia	8030-69-1
Natural black 1	Black to blue-black	Logwood Lake	Hematoxylin (leuco form) Hematein (oxidized form)	8005-33-2
Natural white 1	Iridescent white to silver to yellow off white	Guanine	Guanin, Iminoxanthin, 2-Amino-6-oxypurin	73-40-5

6.5 Additional Benefits of Natural Dyes

Apart from the color imparted to the textile substrate, some natural dyes impart additional benefits like UV protection, antimicrobial activity, fire resistance, and, many more which are given in Table 9.

6.6 Sustainable Aspects of Natural Dyes

- Biodegradable: As natural dyes are easily broken down by the microbes in the soil, the biodegradability of natural dyes is superior to the synthetic dyes. No toxic substances are released during the biodegradation.

Table 9 Additional benefits of natural dyes

Natural dye	Mordants	Textile substrate	Property	Reference
Annatto seeds Eclipta leaves	Alum	Bamboo knit fabric (single Jersey, rib and interlock)	UV protection (expressed as UPF)	Sri Vidhya and Bhanu Rekha (2012)
Arjuna Terminalia	Aluminum phosphate and ferrous sulfate	Khadi silk	Presence of polyphenols, Arjuna dye offers natural UV-blocking properties	Sethi and Rains (2018)
Thyme Sage	Alum	Cotton and silk	Thyme contains thymol and carvacrol, known for their potent antimicrobial properties. The presence of phenolic compounds in sage can offer some degree of UV-blocking properties	Ayvaz and Teker (2024)
Manjistha Rantanjot Annatto	Myrobolan and potash alum	Gray jute fabric	Manjistha, Rantanjot contains bioactive compounds shows antibacterial activity Annatto contains carotenoids, which prevents bacterial growth	Chattopadhyay and Pan (2019)
Myrobalan peel	–	Cotton	The tannins in myrobalan provide natural UV-blocking properties	Ragheb et al. (2019)
Pomegranate wood (catechu)		Cotton	The high antioxidant content in pomegranate provides some UV-blocking properties	Bipul Rana et al. (2022)
Japanese pagoda tree, madder, clove, log wood, gallnut, alder	$AlK(SO_4)_3$, $CuSO_4$, $SnCl_2$, $FeSO_4$	100% silk satin	Japanese pagoda contains flavonoids e provide UV-blocking properties Manjistha, Rantanjot contains bioactive compounds shows antibacterial activity Clove contains eugenol, a compound with natural antimicrobial properties Gallnuts are rich in tannins which provide antibacterial properties	Nam (2010)

- Non-toxic: If dyed using natural mordants natural dyes are safe to the soil and aqueous environment. Among all the metal mordants used in natural dyeing alum is claimed to be eco-friendly and safe.
- Hypoallergenic: Natural dyes do not cause any skin allergies while some synthetic dyes pose allergies like contact redness, itching, blisters, swelling, rashes, dermatitis or eczema, and many more. Natural dyed textiles could be used as healing element in the name of Ayurvastra which uses herbal materials to dye textiles.
- Renewable: Natural dye sources are renewable be it the plant, animal or mineral. Some agricultural waste and food industry waste could be effective sources of natural dyes.
- Low environmental impact: In comparison to synthetic dyes natural dyes exhibit low environmental impact, particularly in the absence of heavy-metal-mordants during the dyeing process.
- Usable wastewater (effluent): The effluent generated could be used to water plants or agricultural crops directly.
- Additional qualities: Many natural dyes possess qualities like antimicrobial activity, UV protection, flame retardancy, and many more. in addition to imparting color to the textile substrate.
- Multiple applications: Natural dyes can be used to dye textiles, food, medicines, toys, and more.

6.7 Unsustainable Aspects of Natural Dyeing

Natural dyes may be safe for the environment but the process of natural dyeing cannot be claimed sustainable at all times. This is because, the additives used in natural dyeing process and the dyeing process may not be sustainable. Hence the superior qualities of natural dyes are not carried forward to the finished product.

7 Conclusion

This chapter discussed a wide array of aspects related to natural colorants and their sustainability. In spite of numerous researches on natural colorants their application to textiles is very limited. Natural colorants are economical for food coloration while it is vice versa for textile coloration. Measures could be taken by industry, government, and consumers to promote the use of natural colorants for textile applications.

References

Adeel, S., Ali, S., Bhatti, I. A., & Zsila, F. (2009). Dyeing of cotton fabric using pomegranate (Punica granatum) aqueous extract. *Asian Journal of Chemistry, 21*(5), 3493.

Adeel, S., Anjum, M. N., Ahmad, M. N., Rehman, F., Saif, M. J., Azeem, M., Hassan, A., Imran, M., & Amin, N. (2021). Eco-friendly dyeing of cotton fabric with waste tea leaves-based tannin natural dye. *Global NEST Journal, 23*(3), 365–369.

Alaskar, A. A., & Hassabo, A. G. (2021). Recent use of natural animal dyes in various field. *Journal of Textiles, Coloration and Polymer Science, 18*(2), 191–210.

Aliño, M. F. A. (2014). Carissa carandas fruit extract as a natural fabric dye. *Journal of Nature Studies, 13*(2), 1–12.

Amutha, K., & Annapoorani, S. G. (2019). Natural dye extraction from agro-waste and its application on textiles. *Asian Dyers, 16*(1), 35–39.

Amutha, K., Annapoorani, S. G., Sakthivel, P., & Sudhapriya, N. (2022). Ecofriendly dyeing of textiles with natural dyes extracted from commercial food processing waste materials. *Journal of Natural Fibers, 19*(15), 10394–10411.

Amutha, K., & Grace Annapoorani, S. (2023). Colouration of textiles with natural dyes from Cassava (*Manihot esculenta*) leaves. *Asian Dyer, 10*(2), 39–43.

Amutha, K., & Ramya, K. (2021). Natural dye from Spathodea campanulata fruit for textile applications. *Asian Dyer*, 58–61.

Amutha, K. S. G. A., & Sudhapriya, N. (2020). Dyeing of textiles with natural dyes extracted from Terminalia arjuna and Thespesia populnea fruits. *Industrial Crops and Products, 148*, 112303.

Asm, R., Pareek, P. K., Shakyawar, D. B., & Sofi, A. (2012). Extraction of natural dye from saffron flower waste and its application on pashmina fabric. *Advances in Applied Science Research, 3*(1), 156–161.

Avinc, O., Celik, A., Gedik, G., & Yavas, A. (2013). Natural dye extraction from waste barks of Turkish red pine (Pinus brutia Ten.) timber and eco-friendly natural dyeing of various textile fibers. *Fibers and Polymers*, 14, 866–873.

Ayvaz, K. M., & Teker, M. S. (2024). Usability of essential oil extracted plant pulp in screen printing in the context of sustainability. *Sanat ve Yorum, 44*, 58–67.

Baaka, N., Haddar, W., Ben Ticha, M., Amorim, M. T. P., & M'Henni, M. F. (2017). Sustainability issues of ultrasonic wool dyeing with grape pomace colourant. Natural Product Research, 31(14), 1655–1662.

Babel, S., & Gupta, R. (2016). Screen printing on silk fabric using natural dye and natural thickening agent. *Journal of Textile Science & Engineering, 6*(230), 2.

Babel, S., Gupta, R., & Mogra, D. (2015). Eco friendly screen printing using natural and indigenous thickener-mango kernel and cassia tora gum. *Journal of Environmental Science, Computer Science Engineering Technology, 4*(2), 417–423.

Bae, J., & Hong, K. H. (2019). Optimized dyeing process for enhancing the functionalities of spent coffee dyed wool fabrics using a facile extraction process. *Polymers, 11*(4), 574.

Baishya, D., Talukdar, J., & Sandhya, S. (2012). Cotton dying with natural dye extracted from flower of bottlebrush (Callistemon citrinus). *Universal Journal of Environmental Research & Technology, 2*(5), 377–382.

Batool, F., Iqbal, N., Adeel, S., Azeem, M., Hussaan, M., & Mia, R. (2024). Sugar beet (Beta vulgaris L.) leaves as natural colorant for cotton dyeing using an ecofriendly approach toward industrial progress. *Science Progress, 107*(3). https://doi.org/10.1177/00368504241271737

Bechtold, T., Mussak, R., Mahmud-Ali, A., Ganglberger, E., & Geissler, S. (2006). Extraction of natural dyes for textile dyeing from coloured plant wastes released from the food and beverage industry. *Journal of the Science of Food and Agriculture, 86*(2), 233–242.

Broadbent, A. D. (2001). Basic principles of textile coloration. *Society of Dyers and Colorists*.

Bukhari, M. N., Wani, M. A., Fatima, M., Bukhari, J. S. S., Shabbir, M., Rather, L. J., & Mohammad, F. (2023). Dyeing of wool with sappan wood natural dye using metal salts for enhancement in color and fastness properties. *Journal of Natural Fibers, 20*(2), 2208890.

Casselman, K. D., & Terada, T. (2012). The politics of purple: Dyes from shellfish and lichens. In *Textile Society of America Symposium Proceedings* (Vol. 666, p. 1).

Chattopadhyay, S. N., & Pan, N. C. (2019). Ecofriendly printing of jute fabric with natural dyes and thickener. *Journal of Natural Fibers, 16*, 1077.

Chhipa, M. K., Srivastav, S., & Mehta, N. (2017). Study of dyeing of cotton fabric using peanut pod natural dyes using Al_2SO_4 $CuSO_4$ and $FeSO_4$ mordanting agent. *International Journal of Environmental & Agriculture Research, 3*(2), 36–44.

Choi, J. H. (2013). Dye. In R. Luo (Ed.), *Encyclopedia of color science and technology*. Springer. https://doi.org/10.1007/978-3-642-27851-8_182-1

Chungkrang, L., Bhuyan, S., & Phukan, A. R. (2020). Natural dye sources and its applications in textiles: A brief review. *International Journal of Current Microbiology and Applied Sciences, 9*(10), 261–269.

Colombini, M. P., Andreotti, A., Baraldi, C., Degano, I., & Łucejko, J. J. (2007). Colour fading in textiles: A model study on the decomposition of natural dyes. *Microchemical Journal, 85*(1), 174–182.

Čorak, I., Brlek, I., Sutlović, A., & Tarbuk, A. (2022). Natural dyeing of modified cotton fabric with cochineal dye. *Molecules, 27*(3), 1100.

Dasgupta, S., & Ghosh, S. (2018). *Evaluation of screen printing using cochineal natural dye*.

Dedhia, E., & Srivastava, A. (2018). A study on dyeing of cotton fabric with Nyctanthes Arbortristis, Linn. (Coral jasmine) flower. *Journal of Asian Regional Association for Home Economics, 25*(1).

Deepali, S., & Shraddha, N. (2017). A study on dyeing of silk fabric with almond shells (P. Amygdalus L.) waste. *International Journal of Research, 4*(5), 921–935.

Ebrahim, S. A., Mosaad, M. M., Othman, H., & Hassabo, A. G. (2022). A valuable observation of eco-friendly natural dyes for valuable utilisation in the textile industry. *Journal of Textiles, Coloration and Polymer Science, 19*(1), 25–37.

Elksibi, I., Haddar, W., Ticha, M. B., & Mhenni, M. F. (2014). Development and optimisation of a non conventional extraction process of natural dye from olive solid waste using response surface methodology (RSM). *Food Chemistry, 161*, 345–352.

Elnagar, K., Abou Elmaaty, T., & Raouf, S. (2014). Dyeing of polyester and polyamide synthetic fabrics with natural dyes using ecofriendly technique. *Journal of Textiles, 2014*(1), 363079.

Espinosa-Morales, Y., Reyes, J., Hermosín, B., & Azamar-Barrios, J. A. (2012). Characterization of a natural dye by spectroscopic and chromatographic techniques. *MRS Online Proceedings Library (OPL), 1374*, 49–59.

Fang, J., Meng, C., & Zhang, G. (2022). Agricultural waste of Ipomoea batatas leaves as a source of natural dye for green coloration and bio-functional finishing for textile fabrics. *Industrial Crops and Products, 177*, 114440.

Geetha, B., & Sumathy, V. J. H. (2013). Extraction of natural dyes from plants. *International Journal of Chemistry and Pharmaceutical Sciences, 1*(8), 502–509.

Ghaheh, F. S., Nateri, A. S., Mortazavi, S. M., Abedi, D., & Mokhtari, J. (2012). The effect of mordant salts on antibacterial activity of wool fabric dyed with pomegranate and walnut shell extracts. *Coloration Technology, 128*(6), 473–478.

Ghongade, K., Bodake, V., Badadare, S., Magdum, M., Gawande, N., Kate, S., & Waghmare, K. (2021). Formulation and evaluation of some cosmetic preparations using novel natural colorant from Ixora coccinea. *Asian Journal of Research in Pharmaceutical Sciences, 11*(1), 22–28.

Hartl, A., & Vogl, C. R. (2003). The potential use of organically grown dye plants in the organic textile industry: Experiences and results on cultivation and yields of dyer's chamomile (Anthemis tinctoria L.), dyer's knotweed (Polygonum tinctorium Ait.), and weld (Reseda luteola L.). *Journal of sustainable Agriculture, 23*(2), 17–40.

Hartford, W. H. (1977). Color and the chromium minerals. *Rocks & Minerals, 52*(4), 169–175.

Hasan, M. M., Hossain, M. B., Azim, A. Y. M. A., Ghosh, N. C., & Reza, M. S. (2014). Application of purified curcumin as natural dye on cotton and polyester. *International Journal of Engineering & Technology, 14*(5), 17–23.

http://www.etad.com

http://www.pigments.org/cms
https://colour-index.com/
https://www.artiscreation.com/Color_index_names.html
https://www.primachemicals.com/dyes-vs-pigments-key-differences/

Huang, Q., Wang, Z., Zhao, L., Li, X., Cai, H., Yang, S., & Xing, J. (2024). Environmental dyeing and functionalization of silk fabrics with natural dye extracted from lac. *Molecules, 29*(10), 2358.

İşmal, Ö. E., Yıldırım, L., & Özdoğan, E. (2014). Use of almond shell extracts plus biomordants as effective textile dye. *Journal of Cleaner Production, 70*, 61–67.

Jain, H., & Vasantha, M. (2016). Eco friendly dyeing with natural dye-areca nut; enhancing colour fastness with natural mordants (myrobalan, lodhra and pomegranate) and increasing the anti-bacterial activity. *Archives of Applied Science Research, 8*(8), 1–7.

Jothi, D. (2008). Extraction of natural dyes from African marigold flower (Tagetes ereecta L) for textile coloration. *Autex Research Journal, 8*(2), 49–53.

Kanchana, R., Fernandes, A., Bhat, B., Budkule, S., Dessai, S., & Mohan, R. (2013). Dyeing of textiles with natural dyes-an eco-friendly approach. *International Journal of ChemTech Research, 5*(5), 2102–2109.

Karaboyacı, M., & Uğur, Ş. S. (2014). Ecological wool dyeing with pulps of lavender, broom, and red wine. *The Journal of the Textile Institute, 105*(8), 821–827.

Khammee, P., Unpaprom, Y., Thurakitseree, T., Dussadee, N., Kojinok, S., & Ramaraj, R. (2021). Natural dyes extracted from Inthanin bok leaves as light-harvesting units for dye-sensitized solar cells. *Applied Nanoscience*, 1–13.

Klančnik, M. (2021). Screen printing with natural dye extract from Japanese knotweed rhizome. *Fibers and Polymers, 22*(9), 2498–2506.

Kushalatha, M., Suneetha, P., & Umesh, H. R. (2024). Extraction of natural dye from areca precipitate using diverse techniques and investigation of its industrial uses. *International Journal of Fundamental and Applied Sciences (IJFAS), 13*(1), 1–9.

Lachguer, K., El Ouali, M., Essaket, I., El Merzougui, S., Cherkaoui, O., & Serghini, M. A. (2021). Eco-friendly dyeing of wool with natural dye extracted from Moroccan Crocus sativus L. flower waste. *Fibers and Polymers, 22*, 3368–3377.

Le, H. T., & Nguyen, L. A. T. (2021). Study on the extraction of colorant from areca nut and employing it in dyeing silk fabric. *Ergonomics International Journal, 5*(4), 000273.

Lokesh, P., & Swamy, M. K. (2013). *Extraction of natural dyes from Spathodea campanulata and its application on silk fabrics and cotton*. Der Chemica Sinica, (1), 111–115.

Maamoun, D., Osman, H., & Nassar, S. H. (2014). Cotton/wool printing with natural dyes nano-particles. *Journal of International Environmental Application Science, 9*(1), 90.

Mahltig, B. (2024). Overview on natural dyes and their IR-spectra–part II: Indigo containing plant dyes. *Communications in Development and Assembling of Textile Products, 5*(1), 66–80.

Mahmud-Ali, A., Fitz-Binder, C., & Bechtold, T. (2012). Aluminium based dye lakes from plant extracts for textile coloration. *Dyes and Pigments, 94*(3), 533–540.

Mansour, H. (2013). Textile dyeing: Environmental friendly Osage Orange. *Eco-Friendly Textile Dyeing and Finishing, 207*.

Mansour, R., Ezzili, B., & Farouk, M. (2017). The use of response surface method to optimize the extraction of natural dye from winery waste in textile dyeing. *The Journal of the Textile Institute, 108*(4), 528–537.

Metwally, R. A., El Sikaily, A., El-Sersy, N. A., Ghozlan, H. A., & Sabry, S. A. (2021). Antimicrobial activity of textile fabrics dyed with prodigiosin pigment extracted from marine Serratia rubidaea RAM_Alex bacteria. *The Egyptian Journal of Aquatic Research, 47*(3), 301–305.

Mirjalili, M., & Karimi, L. (2013a). Antibacterial dyeing of polyamide using turmeric as a natural dye. *Autex Research Journal, 13*(2), 51–56.

Mirjalili, M., & Karimi, L. (2013b). Extraction and characterization of natural dye from green walnut shells and its use in dyeing polyamide: Focus on antibacterial properties. *Journal of Chemistry, 2013*(1), 375352.

Mohamed, A. S. (2010). An economical dyeing process for cotton, polyester and cotton/polyester blended fabrics. *Journal of textile and apparel, technology and management, 6*(4).

Mohini, K., Tejashree, L., & Vijay, N. (2018). Dataset on analysis of dyeing property of natural dye from *Thespesia populnea* bark on different fabrics. *Data in Brief, 16*, 401–410.

Nam, S. W. (2010). Screen printing of silk fabric by natural dyes. In *Proceedings of the Korean Society of Dyers and Finishers Conference* (pp. 72–73). The Korean Society of Dyers and Finishers.

Nambela, L., & Haule, L. V. (2024). Extraction of natural dyes from Delonix regia flowers for cotton fibres. *Tanzania Journal of Science, 50*(1), 126–133.

Naveed, T., Rehman, F., Sanbhal, N., Ali, B. A., Yueqi, Z., Farooq, O., & Wang, W. (2020). Novel natural dye extraction methods and mordants for textile applications. *Surface Review and Letters, 27*(04), 1950135.

Özgüney, A. T., Seçim, P., Demir, A., Gülümser, T., & Özdoğan, E. (2015). Ecological printing of madder over various natural fibres. *Textile and Apparel, 25*(2), 166–171.

Pabba, S. K., Krishna, G., Prakasham, R. S., & Charya, M. A. (2015). Antibacterial activity of textile fabrics treated with red pigments from marine bacteria. *Journal of Marine Biosciences, 1*, 11–19.

Patil, H., Yadav, S., & Athalye, A. (2024). Valorisation of Napier grass through fibre extraction and coloration by natural dye from Rubia tinctorum. *Coloration Technology, 140*(6), 858–867.

Pisitsak, P., Tungsombatvisit, N., & Singhanu, K. (2018). Utilization of waste protein from Antarctic krill oil production and natural dye to impart durable UV-properties to cotton textiles. *Journal of Cleaner Production, 174*, 1215–1223.

Poorniammal, R., Parthiban, M., Gunasekaran, S., Murugesan, R., & Thilagavathi, G. (2013). *Natural dye production from Thermomyces sp fungi for textile application*. NISCAIR-CSIR, India, 276–279.

Prathyusha, K., & Rajitha, I. (2022). Eco-friendly printing of cotton fabric using natural dye from Annona squamosa L. *Journal of Krishi Vigyan, 11*(1), 72–76.

Punrattanasin, N., Nakpathom, M., Somboon, B., Narumol, N., Rungruangkitkrai, N., & Mongkholrattanasit, R. (2013). Silk fabric dyeing with natural dye from mangrove bark (Rhizophora apiculata Blume) extract. *Industrial Crops and Products, 49*, 122–129.

Purwar, S. (2016). Application of natural dye on synthetic fabrics: A review. *International Journal of Home Science, 2*(2), 283–287.

Ragheb, A., Mosaad, M., Mahmoud, S., & Abd Thaloth, J. I. (2019). The impact of nanotechnologies on developing the printing of natural fabrics with pomegranate peel. *Egyptian Journal of Chemistry, 62*(7), 1249–1261.

Rana, M. B., Hasan, M. Z., Islam, M. A., Rokonuzzaman, M., Khandaker, S., & Bashar, M. M. (2022). Printing effects of different natural colourants on cotton fabric. *Textile & Leather Review, 5*, 280–295.

Raslamol, K., Jobi, S., Joy, S., Sreya, M. G., Mariya, L., & Janilan, P. (2024). Natural colorants: Eco friendly alternatives to synthetic colorants. *International Journal of Pharmaceutical Science Invention, 13*(3), 60–70.

Rekaby, M., Salem, A. A., & Nassar, S. H. (2009). Eco-friendly printing of natural fabrics using natural dyes from alkanet and rhubarb. *The Journal of The Textile Institute, 100*(6), 486–495.

Ritu Pandey, R. P., Shweta Patel, S. P., Pintu Pandit, P. P., Shanmugam Nachimuthu, S. N., & Seiko Jose, S. J. (2018). *Colouration of textiles using roasted peanut skin-an agro processing residue*.

Riyanto, C. A. (2019, April). Characterization of annatto (Bixa orellana) peels activated carbon and its application as adsorbent for natural dyes from annatto seeds. *IOP Conference Series: Materials Science and Engineering, 509*(1), 012029.

Rossi, T., Silva, P. M. D. S., De Moura, L. F., Araújo, M. D. C., Brito, J. O., & Freeman, H. S. (2017). Waste from eucalyptus wood steaming as a natural dye source for textile fibers. *Journal of Cleaner Production, 143*, 303–310.

Rungruangkitkrai, N., Mongkholrattanasit, R., Phoophat, P., Chartvivatpornchai, N., Sirimungkararat, S., Wongkasem, K., et al. (2020, February). UV-protection property of Eri silk fabric dyed with natural dyes for eco-friendly textiles. In *IOP Conference Series: Materials Science and Engineering* (Vol. 773, No. 1, p. 012027). IOP Publishing.

Saleh, S. M., Abd-El-Hady, Y. A., & El-Badry, K. (2013). Eco-friendly dyeing of cotton fabric with natural colorants extracted from banana leaves. *Reactions, 5*, 7.

Samanta, A. K., & Agarwal, P. (2009). *Application of natural dyes on textiles*. Indian Journal of Fibre and Textile Research, 34, 384–389.

Sarkar, S., Saha, P., Sultana, N., & Akter, S. (2017). Exploring textile dye from microorganisms, an eco-friendly alternative. *Microbiology Research Journal International, 18*(3), 1–9.

Sethi, M., & Rains, S. (2018). *Screen printing of silk fabric with natural dye Arjuna terminelia using cassia tora as eco friendly thickener. International Journal of Basic and Applied Agricultural Research*, 16(2), 76–179.

Silva, P. M. D. S., França, V. H., Queiroz, R. S. D., Lima, F. S. D., Freeman, H. S., Costa, S. A. D., & Costa, S. M. D. (2022). Copaifera langsdorffii Desf. Bark extract: Optimisation of dyeing conditions to wool and colour fastness properties. *Natural Product Research, 36*(14), 3744–3749.

Singh, S. H. W. E. T. A., & Singh, D. R. (2018). Application of natural mordants on textile. *International Journal of Applied Home Science, 5*(1), 252–260.

Singh, N., Singh, V., & Punia, P. (2019). Antibacterial finish on silk fabric with pomegranate and onion peel extracts. *Journal of Pharmacognosy and Phytochemistry, 8*(5S), 129–131.

Singhee, D. (2020). Review on natural dyes for textiles from wastes. In *Chemistry and technology of natural and synthetic dyes and pigments* (pp. 99–122). IntechOpen.

Singhee, D., Dhanania, Y., & Samanta, A. K. (2016). Dyeing of silk fabric with onion peel waste using unconventional methods. *Indian Journal of Natural Fibers, 3*(1), 45–57.

Sinnur, H. D., Samanta, A. K., Verma, D. K., & Kaware, R. (2018). Studies on coloration and UV protective action of anar peel (pomegranate rind) as an effective natural colorant for cotton khadi fabric. *Journal of The Institution of Engineers (India): Series E, 99*, 9–26.

Sivakumar, V., Rani, K., & Kumari, M. (2017). Efficient extraction of natural dye from red sandal wood (Pterocarpus Sandalinus) using ultrasound. *International Wood Products Journal, 8*(1), 6–9.

Sri Vidhya, M., & Bhanu Rekha, V. (2012). Effect of knitted bamboo structures dyed with natural colorants on ultraviolet radiation protection. *Journal of Textile Science and Engineering, 2*, 2–4.

Sriumaoum, V., Suesat, J., & Suwanruji, P. (2012). Dyeing and spectroscopic properties of natural dyes on poly (lactic acid) and poly (ethylene terephthalate) fabrics. *International Journal of Bioscience, Biochemistry and Bioinformatics, 2*(3), 155–158.

Sukemi, Pratumyot, K., Srisuwannaket, C., Niamnont, N., & Mingvanish, W. (2019). Dyeing of cotton with the natural dye extracted from waste leaves of green tea (Camellia sinensis var. assamica). *Coloration Technology, 135*(2), 121–126.

Swamy, V. N., Gowda, K. N., & Sudhakar, R. (2016). Dyeing of silk using Madhuca longifolia as natural dye source. *Indian Journal of Fibre & Textile Research (IJFTR), 40*(4), 419–424.

Teli, M. D., Valia, S. P., Maurya, S., & Shitole, P. (2014, August). Sustainability based upcycling and value addition of textile apparels. In *Proceedings of the International Conference on Multidisciplinary Innovation for Sustainability and Growth, Kuala Lumpur, Malaysia* (pp. 27–28).

Tepparin, S., Sae-be, P., Suesat, J., Chumrum, S., & Hongmeng, W. (2012). Dyeing of cotton, Bombyx mori and eri silk fabrics with the natural dye extracted from tamarind seed. *International Journal of Bioscience, Biochemistry and Bioinformatics, 2*(3), 159.

Thiyagarajan, S., Balakrishnan, K., & Tamilarasi, S. (2015). A study of extraction and dyeing behavior of natural dye obtained from cotton a study. *IOSR Journal of Applied Chemistry, 8*(5), 85–89.

Tri, Y., & Nooryan, B. (2023). The development and analysis of eco-print and screen printing combination using natural dyes. *Fibres Text, 30*, 51–55.

Uddin, M. G. (2015). Extraction of eco-friendly natural dyes from mango leaves and their application on silk fabric. *Textiles and Clothing Sustainability, 1*, 1–8.

Umar, I. A., & Nor, M. N. M. (2013). Fastness properties of colorant extracted from tamarind fruits pods to dye cotton and silk fabrics. *Journal of Natural Science Research, 3*(4), 60–67.

Vankar, P. S. (2000). Chemistry of natural dyes. *Resonance, 5*(10), 73–80.

Vankar, P. S., & Shanker, R. (2008). Ecofriendly ultrasonic natural dyeing of cotton fabric with enzyme pretreatments. *Desalination, 230*(1–3), 62–69.

Vankar, P. S., & Shukla, D. (2019). *New trends in natural dyes for textiles*. Woodhead Publishing.

Verma, S., & Gupta, G. (2017). Natural dyes and its applications: A brief review. *International Journal of Research and Analytical Reviews, 4*(4), 57–60.Yang, R., Zhang, Y., Ranjitkar, S., Li, M., Guo, Y., Yan, X., & Yang, L. (2020). *Traditional knowledge of plant dyeing, sustainable practices and potential for improved techniques by Monpa in Tibet, China*.

Yusuf, M., Shahid, M., Khan, S. A., Khan, M. I., Islam, S. U., Mohammad, F., & Khan, M. A. (2013). Eco-dyeing of wool using aqueous extract of the roots of Indian madder (Rubia cordifolia) as natural dye. *Journal of Natural Fibers, 10*(1), 14–28.

Zhang, Y., Zhou, Q., Rather, L. J., & Li, Q. (2021). Agricultural waste of Eriobotrya japonica L. (Loquat) seeds and flora leaves as source of natural dye and bio-mordant for coloration and bio-functional finishing of wool textile. *Industrial Crops and Products, 169*, 113633.

Natural Dyeing of Cotton Fibres by Radiation Technologies

Cansu Var, Buket Arık, and Sema Palamutcu

Abstract Cotton, a cellulose-based fibre, is the most commonly used natural fibre in the textile industry. Its well-known chemical structure with numerous hydroxyl groups enable cotton fibre to be highly absorbent and easily dyeable with several types of synthetic dyes, where large quantities of dyestuff and auxiliaries consumed with resulting toxic effluent. The environment and health issues, arising from synthetic dyestuff utilization during the colouration process of textiles, recently proximite to natural dyes. Even though natural dyes stand out in their environmentally friendly characteristics, they have significant challenges of poor adsorption and poor fastness properties in comparison with synthetic dyes on cellulose-based textiles.

Radiation technologies, such as plasma treatment, microwave technology, ultraviolet irradiation, ultrasound technology, and infrared radiation, have been proven to possess not only the potential to enhance the colour and fastness properties of cellulose-based textiles, but also the tendency to reduce required energy, water, and chemical consumption during process. In this regard, combining these technologies with natural dyestuff use would be a promising alternative method to the conventional textile dyeing process, paving the way for the transition to sustainable textile dyeing processes. In the literature, it is revealed that irradiation technologies have been used not only on the fabric surface but also in the powder and extract of natural dyestuff in cotton dyeing process. Several types of natural dyes, such as curcumin, harmala seeds, cocklebur, *Acacia catechu*, *Acacia cyanophylla*, marigold flower, haar singhar flower, green tea, coffee extract, pomegranate rinds, and insect-based dye, cochineal have been utilized to dye cotton fabrics with irradiation technologies. On the other hand, various process-related parameters are of great importance in the efficiency of radiation technologies in the natural dyeing of cotton. This chapter delves into the principles of radiation technologies for textile surface treatment,

C. Var (✉) · B. Arık · S. Palamutcu
Engineering Faculty, Textile Engineering Department, Pamukkale University, Denizli, Türkiye
e-mail: cvar@pau.edu.tr

S. S. Muthu (ed.), *Sustainable Coloration of Textiles*, Sustainable Textiles: Production, Processing, Manufacturing & Chemistry,
https://doi.org/10.1007/978-3-031-91217-7_9

with particular attention to cotton fabric applications. Additionally, a comparative study on the effects of the proses-related parameters on colour and fastness properties as well as other attributes such as antibacterial activity and UV protection is presented.

Keywords Irradiation technologies · Natural dye · Cotton · Colour properties · Fastness properties

1 Introduction

Cotton, mainly composed of cellulose, is the most commonly used natural fibre in the textile industry. Numerous –OH groups enable cotton to be dyed with several types of synthetic dyes including direct, sulphur, vat, and reactive dyes (Haji & Naebe, 2020). Among the synthetic dyes, reactive dyes are the most popular group, because they provide a diverse range of hues and high washing fastness (Pruś et al., 2022). However, synthetic dyes pose many risks regarding environmental and health issues because of the requirement of strong acids, alkalis, solvents, high temperatures, and heavy metals during synthetic dyestuff utilization. The textile industry consumes annually over 400,000 tonnes of reactive dyes during the dyeing and printing process. The highest volumes of these dyes are utilized in the exhaustion procedure of dyeing. The use of reactive dyes brings about the use of great amounts of salts including sodium chloride and sodium sulphate to achieve dye adsorption. This means large amounts of these chemicals are released into wastewater, raising the salinity of water in the environment (Pruś et al., 2022). Due to the toxicity threat of synthetic dyes, chemical waste management is another important environmental concern. In certain situations, colouration chemicals can release carcinogenic and ecotoxic metals during landfill or incineration. Also, the process needs a high amount of water to remove the hydrolysed dyes, which consequence a high volume of wastewater and require a high amount of energy (Pei et al., 2021; Var & Palamutcu, 2024a). Furthermore, some synthetic dyes such as azo-dyes possess even carcinogenic characteristics (Hossain et al., 2021). With the rising environmental awareness, interest and researches in dyeing cotton with natural dyestuffs have significantly increased. Several environmentally friendly characteristics of natural dyestuffs, such as low toxicity and non-allergic effects, shine out natural dyes which are extracted from natural sources such as plants, insects, minerals, and fungi (Ibrahim et al., 2010). Moreover, antibacterial and antioxidant characteristics as well as UV protection potential are additional benefits of some natural dyes (Hamida et al., 2024). Notwithstanding their promising eco-friendly characteristics, utilization of natural dyestuff in the dyeing process of cotton contains certain challenges arising from their low-affinity tendency with cotton fabrics. When in water-based solutions, both plant-based natural dyes and cotton cellulose are ionized with negative charges. The negative charges of cotton and dye molecules repel each other, leading to poor washing fastness, light fastness, and restricted colour intensity (Ji et al., 2023). When compared to dye exhaustion on silk and wool fibres, natural dyes have a quite low propensity of dye exhaustion on cotton fibre (Ben Ticha et al., 2016). To enhance the uptake ratio of natural dye molecules, the treatment of cotton fabrics with radiation technologies, such as plasma treatment, microwave, UV irradiation, ultrasound technology, and infrared radiation, is not only a promising alternative method to the conventional textile dyeing process but also an effective way to shift environmentally harmful dyeing process to environmentally friendly process. Radiation technologies also reduce production cost, energy, water, and chemical consumption where also shorten processing times and reduce pollutant emissions. In the natural dyeing of cotton fabrics using radiation technologies, there exist various process-related parameters that affect the efficiency of the process regarding both the colour and fastness properties of the final product. In the current literature, studies are focused on the effect of gas type and treatment time parameters in plasma technology utilization in the natural dyeing of cotton, while the effect of irradiation time and microwave power was examined in studies on microwave technology. While the effect of irradiation time was most emphasized in studies involving the use of ultraviolet irradiation, the effect of the power level, ultrasonic frequency, and treatment time was emphasized in studies on ultrasonic irradiation.

In this chapter, the principles of radiation technologies for textile surface treatment, such as plasma treatment, microwave technology, ultraviolet irradiation, ultrasound technology, and infrared radiation, are explained and evaluated in terms of sustainability. Then, the applications in which radiation technologies are used in cotton fibre treatment are mentioned. In addition, an overview is given of natural dyes, which are commonly used in combination with radiation technologies for dyeing cotton fibres. Finally, studies investigating process-related parameters and their effects on the colour, fastness, and functional properties (such as antibacterial activity and UV protection) of naturally dyed cotton fabrics, are supported by a comparative analysis of relevant literature.

2 Radiation Technologies for Textile Surface Treatment

In recent studies, researchers have attention to the potential use of radiation technologies like plasma, microwave, ultraviolet, ultrasound, and infrared irradiation technologies to develop more sustainable, economic and ecological pre-treatment, dyeing, finishing, and even sizing process methods as alternative to conventional methods (Khattab et al., 2020; Panda et al., 2021; Var & Palamutcu, 2024b). These technologies promise cost saving with low energy, water and chemical consumption, decreased processing times and amounts of pollutant (Panda et al., 2021; Zhang et al., 2022). The main usage areas of these technologies in textile treatment processes are pre-treatment, dyeing, printing, and functional finishing (Zhang et al., 2022).

2.1 Plasma Treatment

Plasma, which is a partially ionized gas, is known as the fourth state of the matter and has highly conductive properties due to the various free-charge carriers (Haji & Naebe, 2020; Hassabo & El-Sayed, 2021; Panda et al., 2021; Tudoran et al., 2020). Plasma can be initiated by air or by gases like argon, helium, hydrogen, nitrogen, oxygen, fluorine, and carbon dioxide. Ionization of the gases can be induced by the ways of glow discharge, corona discharge, dielectric barrier discharge, and plasma jet (Haji & Naebe, 2020; Hassabo & El-Sayed, 2021; Yilma et al., 2020). Plasma functionalizes the textile surfaces easily and quickly without affecting the bulk property and conserves water, energy and chemical since the pşasma treatment is accomplished in dry condition without any residual effluents (Ayesh et al., 2022; Mostafa et al., 2024; Panda et al., 2021; Tudoran et al., 2020; Ennaceur et al., 2022; Hamida et al., 2024; Lara et al., 2022; Zhang et al., 2022).

Plasma application systems are divided into two types as "thermal plasma" generated by direct and alternate current or by microwave source at high pressure (>10 kPa) and "cold plasma" with a higher electron temperature than ion, working between 0.1 and 100 Pa. In textile applications, cold plasmas have commonly used at vacuum condition (lower pressure) then the atmosphere pressure) due to the degradation of textile materials at high temperatures (Haji & Naebe, 2020; Hassabo & El-Sayed, 2021; Yilma et al., 2020). In addition, they have properties like high concentration of reactive species, superior chemical selectivity, high controllability, consistency of a large surface area and reproducibility of results (Mostafa et al., 2024; Raslan et al., 2020).

Plasma technology imparts hydrophilic and adhesive characteristics to textile fibre surfaces by polymerization, cleaning, etching, grafting, and activation reactions with the help of excited and energetic species like ions, free radicals, electrons, and metastables (Correia et al., 2021; Haji & Naebe, 2020; Panda et al., 2021; Zhang et al., 2022). These species are generated when an electromagnetic field interacts with gas at the available pressure and results etching effect on the fibre surface (Hassabo & El-Sayed, 2021; Zhang et al., 2022). Due to the etching, micro-cracks and fragmentation occur on the surface of the fibre changing the surface through porous and rough structure (Correia et al., 2021; Panda et al., 2021). In this way, plasma treatment improves the wettability, dyeing affinity, and reactivity of textile fibres in an environmental friendly approach (Hamida et al., 2024; Khattab et al., 2020). The important parameters that affect the efficiency of the plasma treatments are the nature of the gas, application pressure, flow rate, power, frequency, and treatment time (Haji & Naebe, 2020; Yilma et al., 2020; Zhang et al., 2022).

2.2 Microwave Technology

Microwave (MW) technology is clean, eco-friendly, and efficient heating technology that converts energy from microwave to heat in the form of dielectric materials (Irfan et al., 2024; Zhang et al., 2022). During microwave irradiation, dielectric materials like water, salt, and ethanol heat themselves by dipole oscillations and mass is transferred from the substrate to the solvent (Zhang et al., 2022).

Microwave irradiation is used in the low-frequency range between 300 GHz and 300 MHz (Bokhari et al., 2023; Panda et al., 2021). Dipole rotation and ionic conduction take place to heat polar substances and then orientation polarization occurs. In orientation polarization, dipoles in dielectric materials (e.g., water) are reoriented under an alternate electric field. Consequently, intermolecular friction and movement of molecules generate heat and provide quick heating of the materials (Panda et al., 2021).

In textile dyeing, microwave irradiation induces dye exhaustion by increased dyeing efficiency and reduced dyeing time, lower amount of chemical, and energy consumption. Microwave technology can be used in textile processes as an alternative method to conventional heating since it ensures quick, efficient, and uniform heating (Bokhari et al., 2023; Bydoon & Saad, 2023; Khattab et al., 2020; Zhang et al., 2022). In addition, it uses less energy than conventional conductive heating with effective cost and ecological advantages (Bokhari et al., 2023; Panda et al., 2021).

In addition, microwave technology can be used for the isolation of plant pigments by bursting the cell wall of plant materials, causing the disruption of hydrogen bonds, inducing the migration of ions by the dipole rotations of molecules, and interacting the colourant part with solvent through mass transfer kinetics. This method gives the benefits of less energy, time, and solvent use with high production yield (Adeel et al., 2023; Hussaan et al., 2023). Adeel et al. (2023) used microwave radiations as extraction tool for chlorogenic acid from the powder of coffee beans (Adeel et al., 2023), Hussaan et al. (2023) used microwave radiations to extract cocklebur leaves (Hussaan et al., 2023) and Trinh Thi Kim and Le Song Thanh (2023) used to extract curcumin colourant of turmeric and verified its potential utilization areas (Trinh Thi Kim & Le Song Thanh, 2023).

2.3 Ultraviolet Irradiation

Ultraviolet (UV) light (100–400 nm) is divided into three groups according to the wavelength ranges as UVA (315–400 nm), UVB (280–315 nm), and UVC (100–280 nm). Under specific conditions, UV light forms carboxyl, aldehyde, hydroxyl, carbonyl, and other reactive groups and these groups initiate the photochemical oxidation of the fibre surface (Hong, 2024; Zhang et al., 2022). Thus, the affinity that changes the dyeing performance of the fibres has been improved and the fibre surface transforms from hydrophobic character to hydrophilic character (Hong, 2024; Zhang et al., 2022). The important parameters that change the efficiency of the UV irradiation process are exposure time, wavelength, intensity of incident light, and other related process factors (Hong, 2024).

UV irradiation gives the advantages of rapid, easy, low cost, and environmentally friendly application in textile treatment processes and contributes the sustainable textile production (Hong, 2024; Panda et al., 2021). On the other hand, strength loss of cellulosic fabrics because of the radical attack on thecellulose chain is an important disadvantage of UV application in textile treatment (Panda et al., 2021, 2023).

In textile functionalization, UV irradiation can be applied during dyeing and finishing processes via pre- or post-treatment to enhance fibre swelling in water, improve pilling resistance and dye take up (Hong, 2024; Panda et al., 2023). For instance, in the printing process, UV irradiation provides increased wettability of hydrophobic fibres and decreased pilling ability (Lara et al., 2022).

2.4 *Ultrasound Technology*

Ultrasound technology has generally been used in textile processes to improve dyestuff dispersion and cleaning treatment by cavitation phenomenon (Zhang et al., 2022). Cavitation, that is the generation, growth, and destruction of the gas-filled microbubbles or cavities in a liquid, provides chemical and physical effects of ultrasonic energy in liquid–liquid or solid–liquid systems (Khattab et al., 2020; Panda et al., 2021). For instance, in a fibre/dye bath system, cavitation causes micro-flow towards the fibre surface and promotes mass transfer in that direction (Zhang et al., 2022).

Ultrasonic waves can be generated to the limit of 5 MHz in gases and 500 MHz in liquids which are very high rates compared to the audible frequency for humans that is in the range of 20 Hz to 20 kHz (Panda et al., 2021). This energy concentrates the mass transfer rate in the inter- and intra-fibre pores of the textile substrate in textile treatment (Hassan & Saifullah, 2023; Panda et al., 2021).

The application of ultrasonic irradiation in textile processes gives advantages like presenting an acceleration effect under low-temperature and low-chemical concentration and being an eco-friendly energy source with low cost and minor fibre damage (Khattab et al., 2020; Zhang et al., 2022). Ultrasound technology significantly reduces treatment time, temperature, and amount of chemical consumption, enables low-liquor ratio dyeing of textiles and plays a positive role in terms of sustainability of textile materials (Hassan & Saifullah, 2023; Nasreen et al., 2023). Concisely, ultrasound technology can be applied in all wet textile processes like pre-treatment, dyeing and finishing.

2.5 *Infrared Radiation*

Infrared radiation (IR) which is type of electromagnetic radiation is another kind of radiation technologies used to enhance dye fixation in textile dyeing and develop efficient heating in textile finishing processes. Infrared radiation has longer wavelengths than visible light but shorter wavelengths than microwaves and presents an eco-friendly nature since it reduces pollution originating from residues of dyes and other chemicals in the wastewater. Infrared heating enables quick and even heating, reduced energy consumption, and low production costs similar to those advantages of microwave technology utilization (Bydoon & Saad, 2023).

3 Radiation Technologies for Cotton Fibre Treatment

Plasma treatment can be effectively used in pre-treatment processes and provides better efficiency than conventional methods (Panda et al., 2021; Yilma et al., 2020). Particularly, it is useful alternative method comparing to the desizing, alkaline scouring, and bleaching processes that need intensive usage of water and energy with resulting environmental pollution (Ayesh et al., 2022; Correia et al., 2021; Lara et al., 2022; Sandanuwan et al., 2021). In finishing, it can be used to impart wrinkle-resistance, self-cleaning, antibacterial, ultraviolet-protection, and flame-retardancy properties to cotton fibre (Hassabo & El-Sayed, 2021; Tudoran et al., 2020; Yilma et al., 2020). In dyeing, it is known that plasma treatment can enhance colour strength and fastness properties (Lara et al., 2022; Yilma et al., 2020; Zhang et al., 2022). Meng et al. (2021) employed plasma-induced graft polymerization using a cationic modifier on cotton, where improved dyeability with reactive dyes under salt-free conditions and satisfactory washing and rubbing fastness results are obtained (Meng et al., 2021). Mostafa et al. (2024) rendered cotton fabrics dyeable with acid dye, which in fact does not have any direct affinity to fix on cotton and enhanced the light, rubbing, and perspiration fastness properties against untreated cotton fabric by low-temperature nitrogen plasma treatment (Mostafa et al., 2024). Moreover, plasma treatment has been proven to be effective in dyeability of cellulosic fibres with natural dyes that have low affinity and require additional treatment for attachment. The combination of plasma treatment and natural dyeing is an important green and eco-friendly method in terms of the sustainable colouration of textiles (Haji & Naebe, 2020). Ennaceur et al. (2022) applied air atmospheric plasma to lower the concentrations of metallic mordants and enhance dyeability of cotton with pomegranate peel extract and obtained promising results with good to very good fastness values and reduced toxic metallic mordant traces in wastewater (Ennaceur et al., 2022). Li et al. (2023) used plasma modification with soy protein coating and natural dye and obtained increased natural dye uptake, strong antibacterial activity and antioxidant properties and improved UV protection (Li et al., 2023). Hamida et al. (2024) studied the effect of plasma pre-treatment on the dyeability of cotton fabric with synthetic and natural dye *Acacia catechu* and the effect of cationic polymers to obtain a cleaner salt-free process. The study verified the improvement of the dye uptake without the involvement of salt or harmful additives, along with significant antibacterial activity through the cationic polymers (Hamida et al., 2024).

Microwave technology reduces the time period, energy and solvent consumption of the pre-treatment processes like desizing, scouring, and bleaching (Lara et al., 2022; Panda et al., 2021). On the other hand, since microwave irradiation affects the microscopic structure and changes the orientation of the fibre beside the advantages already described, it can also be used in textile dyeing (Lara et al., 2022). Bokhari et al. (2023) studied clean and sustainable cotton dyeing method with the use of microwave irradiation and observed a significant effect on enhancing the dyeability of fabrics as well as reducing the time and energy consumption (Bokhari et al., 2023). Irfan et al. (2024) performed exhaust microwave dyeing technique and conventional cotton reactive dyeing for comparison and revealed that the microwave technique provided better results by consuming less salt with a lower liquor volume and a shorter processing time (Irfan et al., 2024). Particularly, it is used to maximize the potential of natural dyes for dyeing cotton fabrics due to the decreased crystallinity of the treated cotton. The researchers also point out the improved colour intensity and fastness properties that can be attributed to the interaction of microwave irradiation with the colourants (Zhang et al., 2022). Adeel et al. (2023) investigated the colouring efficiency of coffee-based natural brown colourant for cotton dyeing under microwave and obtained good colour characteristics (Adeel et al., 2023). Bydoon and Saad (2023) dyed cotton fabric by using natural plants such as marigold flowers, onion outer skins, fenugreek seeds, and green coffee and by assisting with microwave heating and emphasized the superior properties of microwave dyeing against traditional dyeing methods (Bydoon & Saad, 2023). Hussaan et al. (2023) extracted natural colourants from cocklebur leaves by applying a microwave extraction process and dyed cotton fabric with and without microwave irradiation. It is observed that microwave treatment was quite effective for colourant extraction and fabric modification (Hussaan et al., 2023). Trinh Thi Kim and Le Song Thanh (2023) investigated the method of dyeing cotton fabric with fresh turmeric extract using microwave irradiation and demonstrated the ability to form a bond between colourant molecules and cotton owing to the microwave technology (Trinh Thi Kim & Le Song Thanh, 2023).

UV irradiation assures sustainable low-temperature pre-treatment processes with water, time, and energy-saving results compared to conventional treatment methods (Panda et al., 2023). For instance, UV irradiation works as a catalyst during hydrogen peroxide bleaching and increases the bleaching effect (Panda et al., 2021). In dyeing, it enhances colour strength and provides acceptable colour fastness to light, washing, and rubbing (Panda et al., 2023; Zhang et al., 2022). In case of natural dyeing, UV irradiation improves the reactivity of organic substrates as well as the increase in the dye uptake and durability of natural dyes (Hong, 2024). Hong (2024) pre-treated the cotton and wool fabrics with UV radiation and dyed these pre-treated fabrics with the waste coffee extract derived from agricultural waste. The results of the study showed that UV irradiation enhanced the reactivity and dye affinity of the fabrics but decreased the fabric tensile strength (Hong, 2024).

Ultrasound-assisted desizing and washing treatments reduce the processing times without any damage to fabric mechanical properties. By this method, scouring and bleaching can be achieved at a low temperature and short time (Hassan & Saifullah, 2023; Panda et al., 2021). It is also effective in the mercerization of cotton since ultrasound considerably helps to reduce the alkali consumption and increase the conversion of cellulose I to cellulose II (Hassan & Saifullah, 2023). The application of ultrasonic technology in dyeing cotton fibres results in a significant increase in the dyeing rate especially for natural dyes, improved wash, light, perspiration and rubbing fastness values, minimum fibre damage, reduced energy consumption, and less number of chemical contaminants in the waste water (Hassan & Saifullah, 2023; Hoque et al., 2022; Zhang et al., 2022). Furthermore, ultrasound technology has been studied for the extraction process of natural dye sources to eliminate chemical use and related environmental damage in recent studies. Banna et al. (2023) applied ultrasonic-assisted sustainable extraction of dyes from the leaf of *Dillenia indica* tree and ultrasonic dyeing of organic cotton with this extract. They reported promising results for the sustainable dyeing of cotton fabrics (Banna et al., 2023). Nasreen et al. (2023) investigated the green application of ultrasonic waves to extract yellow colourant from *Haar singhar* flower and the use of the extract in cotton colouring and they found that ultrasonic radiations had a high potential for isolating colourants and dyeing cotton under low temperature, time and volume conditions (Nasreen et al., 2023). Atakan et al. (2023) employed ultrasound-assisted extraction for *Rosa canina* with different solvents such as distilled water, methanol, and a water/methanol mixture (50/50% v/v). Then they applied cotton fabric by using bio-mordant chitosan and the results showed that using water as a solvent was more effective than other solvents, ultrasound-assisted extraction made notable contributions to greener textile industry and treated cotton fabrics presented high dye uptake and additional functional properties like UV protection and antibacterial activity (Atakan et al., 2023).

Moreover, radiation technologies can be combined to obtain multifunctional effects or compared to each other in terms of their effect on colour strength. Ashraf et al. (2019) used three different radiation technologies of microwave, ultraviolet and ultrasonic radiations and compared the colour strength values of reactive dyed cotton fabrics with statistical tests and reported that the highest shade depth was achieved in case of microwave treatment (Ashraf et al., 2019). Ayesh et al. (2022) studied the impact of atmospheric plasma and UV treatment on the chemical and physical properties of cotton fabrics and found that the use of UV irradiation during the plasma treatment destroyed or caused volatilisation of some plasma-generated species, it was revealed that UV presence was not essential simultaneously with plasma (Ayesh et al., 2022). Bydoon and Saad (2023) conducted natural dyeing process using conventional, microwave, and infrared radiation methods and observed the microwave dyeing process indicated the highest colour strength results (Bydoon & Saad, 2023). Li et al. (2023) studied plasma pre-treatment followed by ultrasonic soy coating and achieved improved colour strength, UV protection functionality, and soft handle on cotton fabric (Li et al., 2023).

4 Natural Dyes for Cotton Fibres and Their Application

Natural colourants include all dyes and pigments derived from natural resources of plants, animals, insects, minerals, soil, and even microorganisms (fungi, bacteria, yeasts, and algae). Plant-based natural dyes like indigo, madder, and turmeric can be derived from parts of plants, such as roots, leaves, and flowers, while protein-based dyes, like cuttlefish ink, are derived from animals. Cochineal and lac are the two most known examples of insect-based dyes. On the other hand, iron oxide and copper sulfate are the most common mineral-based dyes used especially in textile printing. Tyrian purple is a prominent example of fungal-based dyes derived from fungi or metabolic by-products. (Pranta & Rahaman, 2024). Having been employed to dye textiles since prehistoric times, natural dye sources also held cultural and symbolic meanings (Alegbe & Uthman, 2024). Besides their colourant properties, some of the sources of natural dyes also stand out as functional benefits such as antimicrobial and deodorizing properties, moth-resistant and insect-repellent characteristics, UV protection potential, and even food and cosmetic colouration (Arik et al., 2020; Arık et al., 2020; Yusuf et al., 2017b). Some sources are also recognized for their curative properties in various traditional medicinal purposes. Especially, plant- and tree-based dyes and pigments offer a vast range of therapeutic benefits among natural dyes and pigments (Islam et al., 2024; Saxena & Raja, 2014). Also, some manufacturers are promoting naturally dyed textiles as well-being textiles (Saxena & Raja, 2014). The natural dyes preferred in studies examining the use of radiation technologies in the natural dyeing of cotton can be listed as follows: Curcumin, Harmala seeds, cocklebur, *Acacia catechu*, *Acacia cyanophylla*, Marigold flower, Haar singhar flower, green tea, coffee extract, pomegranate rinds, and an insect-based dye, cochineal.

4.1 Curcumin

Curcumin, the most popular yellow natural dye, is an active ingredient in turmeric (*Curcuma longa* L.). It marks the first recognized diarylheptanoid, a class of compounds characterized by a 7-carbon chain with aromatic rings on either side of unit chemical structure. It possesses a unique conjugated structure containing two methoxylated phenols and the enol form of β-diketone. It exists in a keto-enol tautomerism with equilibrium strongly favouring the enol form. This enol configuration allows curcumin to form inter as well as intramolecular hydrogen bonds. Studies have shown that curcumin also exhibits various functional characteristics including antimicrobial properties, which is attributed the existence of methoxy and hydroxyl groups (Han & Yang, 2005; Lykidou et al., 2021).

4.2 Harmal Seeds

Harmal seeds, also known as wild rue, belong to the Zygophyllaceae family and are commonly distributed in regions including Asia, Australia, and North America. Its seeds which were utilized in ancient Ayurvedic medicine are known for various pharmaceutical characteristics. Harmal seeds containing alkaloids, including harmine, harmaline, harmalol, and harmol, are a valuable source of natural colourants. Its extract gives a yellowish-red colour and has the potential of being an alternative to Turkish red dye that is attained from madder (Tanrıkulu, 2024). Such a yellowish-red dye is mostly utilized to dye carpets and wool yarns. As well, the seeds have also been used for their fluorescent qualities, combined with hypothermic and hallucinogenic effects, which enable them to be used in making ink, tattoos, and rug–wool dyeing (Adeel et al., 2018, 2022b).

4.3 Cocklebur

Cocklebur which belongs to the Zygophyllaceae family mostly grows as a weed and is predominantly found in Brazil, China, North America, North Africa, Pakistan, India, and Malaysia. Leaves of Cocklebur contain tannin, xanthatin, xanthumin, phenols, xanthanol, lactones, terpenoids, alkaloids, steroids, and flavonoids. Since ancient times, this plant has been known as a traditional herbal remedy plant and used to treat various diseases thanks to its ingredients (Hussaan et al., 2023).

4.4 **Acacia catechu**

Acacia catechu, mostly grown in the Indian sub-Himalayan zone, stands out in its colouring characteristics and pharmacological as well medicinal attributes. Since early times, the powdered heartwood of *A. catechu* has been utilized for textile dyeing to add brown tint shades to textiles (Yusuf et al., 2017a). Catechin is the primary colouring component of catechu. Containing tannin phytoconstituents, *Catechu* has been also investigated as a bio-mordant (Samant et al., 2022).

4.5 **Acacia cyanophylla**

Acacia cyanophylla, a shrub species, belongs to the Fabaceae (legume) family. The chemical structure of the plant consists mainly of phenolic compounds (flavonoids such as quercetin and kaempferol), dense tannins, and polysaccharides (Dhouibi et al., 2021). There is growing scientific interest in the potential health benefits of its flowers. Studies have demonstrated that its extracts have antioxidant properties and display anti-acetylcholinesterase activity (Fajraoui et al., 2019).

4.6 Marigold Flower

Marigold flower is well-known for its bright colour range spanning from yellow to red tones. Marigold flower which is utilized commonly for decoration, medicine, food products, and cosmetics is an abundantly available crop in tropical regions. Marigold flower is a valuable natural dye source with lutein as its main colourant compound. Studies demonstrated that lutein is a water-soluble pigment acting like a dyestuff (Ali & Noor, 2016; Baig et al., 2021).

4.7 Haar Singhar Flower

Haar singhar, also known as Coral Jasmine, contains several biologically active compounds, with nyctanthin identified as the primary pigment for fabric dyeing (Adeel et al., 2022a). It gives yellow colour on fabrics including cotton, wool, and silk fibre-based surfaces (Yameen et al., 2022).

4.8 Green Tea

Leaves of green tea, rich in natural pigments and antioxidants, are considered a promising source for the natural dyeing of cotton with additional advantages beyond colouration. Among the pigments, catechins–epicatechin, epigallocatechin, epicatechin gallate, and epigallocatechin gallate are known for their UV-protective characteristic. These natural compounds not only serve as natural colourants but also contribute potential antimicrobial and antioxidant properties to the dyed fabric. Moreover, using waste green tea leaves, a by-product of the beverage industry contributes to both sustainable practices in natural dyeing and a reduction of agricultural waste (Sukemi et al., 2019).

4.9 Coffee Extract

Coffee is attracting significant attention as an environmentally friendly option for natural dyeing processes. Coffee grounds are used to attain various shades of brown, especially in cotton and woollen fabrics. Thanks to its chemical constituent containing alkaloids, quinolinic acid, nicotinic acid, tannic acid, pyrogallic acid, and trigonelline, coffee may also offer a wide range of health benefits. The primary phytochemicals in coffee include caffein, chlorogenic acid, polyphenols, flavonoids, alkaloids, with chlorogenic acid, as the main functional component responsible for imparting the brown colour to the fabric (Adeel et al., 2023). As well as this, using coffee waste which is abundantly available in the coffee industry offers both economic and environmentally sustainable solutions for the dyeing process of textiles, encouraging waste reduction.

4.10 Pomegranate Rinds

Pomegranate belonging to the *Punicacea* family is commonly cultivated in warm regions of the world. Pomegranate rinds, rich in polyphenols such as flavonoids and hydrolysable tannins, have been noted not only as a natural dye but also as a natural binder/mordant because of tannin content. Its rinds have also significant pharmaceutical and antimicrobial characteristics beyond colouring properties. The rinds are also a rich source of vitamin C and serve as effective moisturizers, providing various other medicinal benefits (Ajmal et al., 2014; Otaviano et al., 2023; Sinnur et al., 2018).

4.11 Cochineal

Cochineal dye is sourced from female *Dactylopius coccus* Costa insects, which belong to the Coccoidea suborder. Cochineal is commonly utilized as a colourant in different fields including food additives, cosmetics, pharmaceuticals, and textiles. In textile dyeing, it is frequently utilized to dye animal-based fibres. Even though it can be challenging to dye cotton with cochineal dye with anionic characteristics due to the surface electronegativity of cotton fibre, efforts are being made to dye cotton with current technologies (Ke et al., 2021).

5 Process-Related Parameters and Their Effect on Colour, Fastness, and Other Properties of Dyed Cotton Fabrics

Researchers have focused on different parameters in studies examining the use of radiation technologies in the natural dyeing of cotton fabrics. In the literature, irradiation technologies have been applied not only on the fabric surface but also to the powder form and extracts of natural dyes in the natural dye process of cotton fabrics. In this section, a comparative analysis is given on how alterations in process parameters influence the colour and fastness properties of fabrics and other properties, such as antibacterial or UV protection.

5.1 Plasma Treatment

Plasma treatment is proven to have a great effect on dyeing of cotton fabrics with natural dyes with or without mordant (Haji, 2019). Dyeing performance of fabrics with plasma treatment depends on various parameters, such as plasma power dosage, gas type, pressure, flow rate, frequency, and treatment time (Ayesh et al., 2022; Haji & Naebe, 2020). The type of plasma gases and treatment time are among the most researched plasma-related process parameters for the natural dyeing of cotton fabrics.

5.1.1 Type of Plasma Gases

The type of gas which is selected to generate plasma is one of the most important parameters that affect the efficiency of the treatment. It is possible to observe different surface modifications depending on the type of gases utilized during plasma treatment. In relation to the gas type, the introduction of surface cross-linking, production of reactive free radicals, and alteration in surface tension can be accomplished (Hamida et al., 2024). Improvement of dyeability of cotton fabrics by plasma is regarding to increasing the surface wettability. The wettability is directly affected by the surface energy. Higher surface energy brings about greater hydrophilicity, whereas lower surface energy brings about hydrophobicity. (Naebe et al., 2021). Due to high numbers of –OH functional groups, cellulose carries negative charge when in water. When cellulose substrates are treated with oxygen plasma, their negative potential is enhanced, which leads to an increase in the repellence of negatively charged dye molecules. On the other hand, ammonia plasma treatment is proven to improve the adsorption of natural dye molecules onto cotton surfaces (Gorjanc et al., 2018).

Gorjanc et al. (2018) examine the effect of oxygen or ammonia gases on the dyeing of cotton fabric with curcumin and green tea dye. In curcumin dye use, the colour yield was enhanced in ammonia plasma-treated fabrics, while the yield decreased in oxygen plasma-treated fabrics. Furthermore, the same effect was observed when green tea extract was used as a natural dye. Moreover, treatment with ammonia plasma for 30 s and 300 s increased the UV protection factor to 77.19 and 95.37, respectively (Gorjanc et al., 2018). Hamida et al. (2024) examine the effects of ambient air and nitrogen gases on the dyeing performance of cotton fabric with *Acacia catechu*. Also, in the study, chitosan and quaternized poly[bis(2-chloroethyl) ether-alt-1,3-bis[3-(dimethylamino) propyl] urea (PQ-2) were selected as cationic polymers for substitution the salt. The nitrogen plasma-treated cationized fabrics displayed better performance regarding dye uptake. Nitro compounds addition to the cotton fabric surface decreased the negative charge, which supported the influence of PQ-2. The fabrics dyed with *Acacia catechu* exhibited a suitable-rated fastness for both air and nitrogen plasma treatment (Hamida et al., 2024).

5.1.2 Plasma Treatment Time

The treatment time is one of the key parameters in determining the extent of surface modification on the fabric. Short treatment times may cause limited dye uptake due to poor surface reactivity, while longer treatment times may cause unnecessary etching effect on the fabric surface. Over-etching may adversely affect the mechanical characteristics and also quality of the fabric.

Haji (2017) examined the effect of different treatment times ranging from 0 to 5 min on the dyeing of cotton fabric with pomegranate rinds. To enhance the dyeability of cotton fabrics, chitosan treatment was applied on the surface of the fabrics. It was observed that increasing the treatment time to 2.5 min did not make any significant difference in colour strength but increasing the treatment time to 5 min led to an increase in the *K*/*S* value. Even though the fastness properties of untreated fabrics which were dyed with natural dye were acceptable, plasma and chitosan-treated dyed fabrics displayed improved washing and light fastness properties. Oxygen plasma-treated fabric with chitosan coating had the highest washing fastness, attributed to the ionic interaction between chitosan with the amine groups and the dye molecules with the hydroxyl groups. Both dyed and non-dyed fabrics with chitosan coating exhibited high antibacterial activity against both Gram-negative and Gram-positive bacteria (Haji, 2017).

5.2 *Microwave Technology*

In the literature, studies are conducted to examine the effects of microwave (MW) rays not only on the textile surface but also on the extraction of biomolecules. At this point, it has been revealed that the effect of MW rays on the dyeing of cotton fabrics depends on various key parameters such as irradiation time and microwave power (Adeel et al., 2023).

5.2.1 Irradiation Time

The exposure time of the dye powder, extract, and fabric to microwave radiation directly affects the dyeing process of the fabrics. Microwave irradiation ruptures plant cell walls, creating a mass transfer effect from the substance to the solvent. Namely, microwave heat breaks the cell walls and promotes the colourants to pass into the solvent quickly and effectively. This allows to extraction of the colourant without any deterioration (Adeel et al., 2018). If microwave irradiation is applied for too short time, the colourant molecules will not be completely extracted from the cell wall and dissolved in the solution, which may lead to poor colouration during the dyeing process. If irradiation is applied for too long time, impurities other than the colour substance may evolve into the solvent or denaturation of colourant molecules may occur, which can lead to poor colourant yields and undesired colour shade. Also, it is possible to obtain desorption of molecules and weak colour tones under excessive irradiation, even though MW rays increase dye uptake by making physical modifications on the surface of the cotton fabric (Adeel et al., 2018, 2023; Hussaan et al., 2023; Naveed et al., 2022).

Adeel et al. (2018) treated harmala seeds powder and extracts as well as cotton fabrics with different irradiation times ranging from 1 to 6 min. In the study, extraction of the colourant from irradiated and non-irradiated powder was employed in four different media such as aqueous, alkaline, acidic, and organic (methanol). When the powder was irradiated for 5 min and the extraction was employed with methanol, maximum colour strength was observed on the non-irradiated fabric. When the acidified methanol extract was irradiated for 4 min, darker shade and high colour strength for non-irradiated fabrics were observed (Adeel et al., 2018). Naveed et al. (2022) examined the effect of different irradiation times ranging from 1 to 5 min on the dyeing of irradiated cotton fabric with irradiated pomegranate rind/turmeric rhizome blending powder. When 3 min irradiated mixture was used, good colour strength was attained onto 3 min irradiated fabric (Naveed et al., 2022). In another study, Adeel et al. (2023) examined the effect of different irradiation times ranging from 1 to 6 min on the dyeing of cotton fabric with coffee extract. 1 min aqueous extract brought about high colour yield onto 1 min irradiated fabric when compared to the non-irradiated fabric which was dyed with non-irradiated extract. Additionally, excellent light and washing and good rubbing fastness ratings were obtained when MW treatment of the extract and fabric was followed by the application of chemical and plant-based phenolics from turmeric and pomegranate as bio-stabilizers (Adeel et al., 2023). Hussaan et al. (2023) examined the effect of different irradiation times ranging from 1 to 5 min on the dyeing of irradiated cotton fabric with irradiated natural dyes of cocklebur leaves. It was concluded that 4 min irradiated alkaline extraction of colourants displayed significantly remarkable colour strength onto irradiated fabrics in comparison to aqueous medium (Hussaan et al., 2023).

5.2.2 Microwave Power

Microwave power determines the energy which is delivered to the system. With increasing microwave power, more dyes can be extracted from natural sources. On the other hand, microwave dyeing relies on dielectric and thermal properties. High-frequency microwaves produce frictional heat by increasing the vibrational energy of water and dye molecules. This heating makes it easier for the dye to penetrate the fabric. Ionic resistance provides additional heating by accelerating ions in solution and colliding with dye and fibre molecules. High power levels lead to quick heating and also accelerated dyeing process. It is also noted that higher MW power may cause fabric deterioration and uneven colour distribution (Elshemy, 2011).

Trinh Thi Kim and Le Song Thanh (2023) examined the effect of microwave power (low, medium, and high) on the dyeing performance of cotton fabrics with fresh turmeric extract. The highest *K*/*S* value (21.2) was obtained under high power conditions (Trinh Thi Kim & Le Song Thanh, 2023).

5.3 Ultraviolet Irradiation

Ultraviolet (UV) irradiation, another innovative technique, provides more effective dye fixation, dye uptake, and deeper shades. It can also enhance the wettability of fabrics. Although the outcome of the process depends on several parameters, such as irradiation time, wavelength, and intensity of incident light (Hong, 2024; Rehman et al., 2017), the irradiation time parameter was mostly examined in studies investigating the use of UV irradiation in the natural dyeing of cotton fabrics.

5.3.1 Irradiation Time

Short irradiation times are inadequate to activate the carboxyl and OH groups of cotton structure, while long irradiation times may deteriorate or dislocate the fibre chemical structure. If the colourant is not strongly bonded with the fabric due to inappropriate conditions, dye removal may occur during the washing process.

Adeel et al. (2014) examined the effect of irradiation time on the dyeing performance of cotton fabrics with extract of *Acacia nilotica* bark. Both cotton and bark powder were irradiated for 30, 45, 60, 90, or 120 min. 90-min irradiation modified the fabric surface. Dyeing the fabric for 50 min at 55 °C and pH 6 conditions with aqueous extract utilizing 8 g/100 mL of salt provided the best colour strength. Also, using iron brought about enhanced colour fastness properties from moderate to good (Adeel et al., 2014). Rehman et al. (2017) examined the effect of irradiation time on dyeing performance of cotton fabrics with Marigold flowers. Both powder of Marigold flowers and cotton fabrics were exposed to UV irradiation for 30, 45, 60, 90, and 120 min. 90-min irradiation brought about the modification of the fabric surface and also oxidation which assisted in firm bonding. 90-min irradiated fabric provided good colour strength when alkaline extract was used (Rehman et al., 2017). Hong (2024) applied UV irradiation pre-treatment to cotton fabrics for 5 or 15 h and examined the dyeing process of the fabrics with coffee extract as a natural dye. For *K*/*S* values under both irradiation times, no significant difference was obtained. 15-h treated cotton fabric had a grade of 4.5 after one cycle of washing. Also, 15-h treated fabric had a grade of 4 in terms of colour fastness to light. Moreover, as UV exposure time increased, the reduction rate of *S. aureus* bacteria on cotton fabrics also increased significantly. *S. aureus* reduction was found to be 13.0%, 78.2% and 93.2% for non-irradiated, 5-h irradiated, and 15-h irradiated cotton fabric, respectively. Similarly, *K. pneumoniae* reduction was found to be %12.1, %13.2, and %48.6 for non-irradiated, 5-h irradiated, and 15-h irradiated cotton fabric, respectively (Hong, 2024).

5.4 *Ultrasound Technology*

Ultrasonic irradiation, an innovative method, has great potential to enhance the dyeability of cotton fabrics with natural dyes. Several parameters affecting the efficiency of ultrasonic-assisted natural dyeing include power level, frequency, and irradiation time.

5.4.1 Power Level

The power determines the cavitation threshold. When a certain power is exceeded, cavitation occurs. On the other hand, below the threshold, waves only propagate (Guesmi et al., 2013). High power levels may lead to more effective dye–fibre interactions. It may be attributed that ultrasonic power increases the mobility of molecules in the aqueous medium, which facilitates dyes to dissolve more quickly, gases to be removed from the liquid, and penetration of dyes into the fibres. However, applying excessively high power may degrade natural dye or damage the cotton fibres.

Guesmi et al. (2013) examined the ultrasonic dyeing process of cotton fabric with sosalipurposide dye, which was isolated from *Acacia cyanophylla* yellow flowers, under different ultrasonic frequencies (25 or 40 kHz) and different power levels (100, 200, 300, 400, or 500 W). In the study, cationization of the fabric was achieved by a sonication reaction of cellulose with bromoacetyl bromide, followed by substitution of the terminal bromo groups with triethylamine under different ultrasonic frequencies (25 or 40 kHz) and different power levels (100–500 W). As the power level increased, an improvement in colour strength was observed (Guesmi et al., 2013). Ahmed et al. (2017) examined the influence of ultrasonic power levels (100–500 W) on the extraction of cochineal natural dye to use for the natural dyeing of cotton and cotton/polyester blends. Extract absorption increased with increasing power level and maximum extraction was achieved at 500 W (Ahmed et al., 2017). Ma et al. (2020) examined the optimization of dyeing conditions for cotton fabric with turmeric dye. Different power levels (100, 200, and 400 W), different dyeing temperatures (30, 40, and 50 °C) and different dyeing times (20, 30, and 40 min) were selected. The best dyeing performance was attained with a neutral pH of the bath, the dyeing temperature of 40 °C and the dyeing time of 30 min, and ultrasonic power of 200 W, resulting in a *K*/*S* value of 2.53. In the study, also the dyed fabrics exhibited good dry and wet rubbing fastness (Ma et al., 2020).

5.4.2 Ultrasonic Frequency

Ultrasound causes the ejection of entrapped gas from cotton fibre into liquid where it is removed by the cavitation process. Such a mechanism facilitates dye–fibre interaction and better dye absorption (Guesmi et al., 2013).

Colour and Fastness Properties

Guesmi et al. (2013) examined the ultrasonic dyeing process of cotton fabric with sosalipurposide dye, which was isolated from *Acacia cyanophylla* yellow flowers, under different ultrasonic frequencies (25 or 40 kHz) and different power levels (100, 200, 300, 400, or 500 W). In the study, cationization of the fabric was achieved by a sonication reaction of cellulose with bromoacetyl bromide, followed by substitution of the terminal bromo groups with triethylamine under different ultrasonic frequencies (25 or 40 kHz) and different power levels (100–500 W). The *K*/*S* value at 25 kHz ultrasound was higher than the value at 40-kHz ultrasound. Both the rubbing and light fastness of the dyed fabrics were around five, where the washing and water fastness were about 4–5 (Guesmi et al., 2013).

5.4.3 Treatment Time

Ultrasonic irradiation of cotton fabric in a wet state improves its dyeability with natural dyes. Natural colourants are frequently firmly embedded within the cell wall of plants, so an effective extraction method effectively facilitates the release and transfer of these colourants into the solvent (Sadeghi-Kiakhani et al., 2021). Treatment time is a critical parameter in extraction of natural dye since it directly affects the efficiency of releasing and transferring colourants from the cell wall. While short irradiation time does not promote rupture of the cell wall, extended irradiation time may stimulate the separation of other components affecting dye shade stability (Nasreen et al., 2023).

Ahmed et al. (2017) examined the influence of different treatment times (20–120 min) on the extraction of cochineal natural dye to use for the natural dyeing of cotton and cotton/polyester blends. The absorption value was 1.98 mg/g after 60 min when ultrasonic extraction was used, whereas the value was only 1.63 mg/g after 100 min when conventional heating was used (Ahmed et al., 2017). Nasreen et al. (2023) examined different treatment times for dyeability of cotton with haar singhar flower extract by using eco-friendly mordants. It was concluded that irradiation of cotton fabric for 30 min brought about excellent colour strength results when irradiated alkaline extract was used. Both the extract and the fabric which were irradiated for 20 min had excellent results in the aqueous medium, while both the extract and the fabric which were irradiated for 30 min had excellent results under acidic conditions. In the study, the combination of eco-friendly mordants and irradiated haar singhar flower also brought about enhanced washing and light fastness as well as improved dry and wet rubbing fastness (Nasreen et al., 2023).

6 Conclusion

With growing environmental awareness, interest in integrating sustainable technologies into textile production processes has been rising. In this regard, the use of natural dyes which are environmentally friendly alternatives to synthetic dyes in textile dyeing is becoming more widespread. However, some concerns about natural dyes for cotton dyeing, such as low dye exhaustion, limited colour range, poor fastness properties, and colour variability, limit their extensive use. Adaptation of radiation technologies, including plasma treatment, microwave, UV irradiation, and ultrasound technology, to the process is a highly promising approach to both overcome the aforementioned challenges and reduce the negative impact of the textile dye industry on the environment. In the literature, irradiation technologies can be used for fabric surfaces and even the powders and extracts of natural dyes. The efficiency and outcomes of these technologies in the natural dyeing of cotton systematically depend on several process-related parameters. Specifically, when plasma treatment is considered, the type of gas is one of the effective parameters. If the efficiency of different gas types is evaluated, ammonia plasma treatment has the potential to improve the adsorption of natural dye molecules onto cotton surfaces. In terms of plasma treatment time, short times may cause limited dye uptake, while prolonged treatment may cause an excessive etching effect on the fabric surface. When microwave irradiation is considered, irradiation time parameter stands out. If microwave irradiation is applied to natural dye sources for too short time, the colourant molecules will not be thoroughly extracted from the cell wall and dissolved in the solution, which may cause poor colouration during the dyeing process. If prolonged irradiation is applied, impurities except for colour substance may evolve into the solvent or denaturation of colourant molecules may occur, which can lead to low colourant yields and undesired colour shade. Besides irradiation time, microwave power is another attractive parameter in natural dyeing. As microwave power is increased, more dyes can be extracted from natural sources. However, higher power may cause fabric deterioration or uneven colour distribution. When ultraviolet irradiation is regarded, irradiation time is the most emphasized parameter. Short irradiation times are ineffective in activating the carboxyl and OH groups of cotton, while long irradiation times may deteriorate or dislocate the fibres. Another innovative technology is ultrasonic dyeing. Ultrasonic power level is the most discussed parameter for the natural dyeing of cotton. High power levels provide more effective dye–fibre interactions, while excessively high levels may degrade natural dye or deteriorate the cotton fibres. Treatment time is another critical parameter in the extraction of natural dye as it affects the efficiency of transferring colourants from the cell wall. While short irradiation time does not foster the rupture of the cell wall, prolonged irradiation treatment may contribute to the separation of other components affecting dye shade stability. Natural dyeing of cotton is a complex process, shaped by the type of natural dye, dye extraction conditions, fabric pre-treatment as well as dyeing conditions. Therefore, it is critical to conduct optimization studies on both process-related parameters and parameters of the dyeing process to determine the ideal conditions to obtain the best colourimetric results.

References

Adeel, S., Fazal-ur-Rehman, Hanif, R., Zuber, M., Ehsan-ul-Haq, & Muneer, M. (2014). Ecofriendly dyeing of UV-irradiated cotton using extracts of Acacia nilotica bark (Kikar) as source of quercetin. *Asian Journal of Chemistry, 26*(3), 830–834. https://doi.org/10.14233/ajchem.2014.15829

Adeel, S., Zuber, M., Fazal-ur-Rehman, & Zia, K. M. (2018). Microwave-assisted extraction and dyeing of chemical and bio-mordanted cotton fabric using harmal seeds as a source of natural dye. *Environmental Science and Pollution Research, 25*(11), 11100–11110. https://doi.org/10.1007/s11356-018-1301-2

Adeel, S., Ahmad, S., Habib, N., Saif, M. J., Batool, F., & Aftab, M. (2022a). Environmental friendly pollution free bio-dyeing of wool with haar singhar (coral jasmine) flower extract. *Environmental Science and Pollution Research, 30*(5), 12453–12465. https://doi.org/10.1007/s11356-022-22992-z

Adeel, S., Anjum, F., Zuber, M., Hussaan, M., Amin, N., & Ozomay, M. (2022b). Sustainable extraction of colourant from harmal seeds (Peganum harmala) for dyeing of bio-mordanted wool fabric. *Sustainability, 14*(19), 12226. https://doi.org/10.3390/su141912226

Adeel, S., Rehman, F.-U., Amin, A., Amin, N., Batool, F., Hassan, A., & Ozomay, M. (2023). Sustainable exploration of coffee extracts (*Coffea arabica* L.) for dyeing of microwave-treated bio-mordanted cotton fabric. *Pigment & Resin Technology, 52*(3), 331–340. https://doi.org/10.1108/PRT-02-2022-0024

Ahmed, H. M., Ahmed, K. A., Mashaly, H. M., & El-Halwagy, A. A. (2017). Treatment of cotton fabric with dielectric barrier discharge (DBD) plasma and printing with cochineal natural dye. *Indian Journal of Science and Technology, 10*(10), 1–10. https://doi.org/10.17485/ijst/2017/v10i10/100973

Ajmal, M., Adeel, S., Azeem, M., Zuber, M., Akhtar, N., & Iqbal, N. (2014). Modulation of pomegranate peel colourant characteristics for textile dyeing using high energy radiations. *Industrial Crops and Products, 58*, 188–193. https://doi.org/10.1016/j.indcrop.2014.04.026

Alegbe, E. O., & Uthman, T. O. (2024). A review of history, properties, classification, applications and challenges of natural and synthetic dyes. *Heliyon, 10*(13), e33646. https://doi.org/10.1016/j.heliyon.2024.e33646

Ali, S., & Noor, S. (2016). Central composite design approach for optimization of extraction and dyeing conditions of marigold colorant. *Natural Products Chemistry & Research, 04*(04). https://doi.org/10.4172/2329-6836.1000224

Arık, B., Canitez, E., & Kirtak, A. (2020). Investigation of dyeing properties of red cabbage to cotton fabrics in different pH and mordanting conditions. *Süleyman Demirel Üniversitesi Fen Bilimleri Enstitüsü Dergisi, 24*(2), 244–255. https://doi.org/10.19113/sdufenbed.639829

Arik, B., Dikkal, I., & Karacan, S. (2020). Evaluation of the dyeing and fastness properties of wool fabrics treated by chitosan in different molecular weights and dyed with mint. *Pamukkale University Journal of Engineering Sciences, 26*(7), 1186–1192. https://doi.org/10.5505/pajes.2020.47427

Ashraf, M. A., Amjad, A., Farooq, A., & Khan, J. U. (2019). Pretreatment of cotton with different radiations to improve colour strength and fastness properties in reactive dyeing. *Pakistan Journal of Scientific & Industrial Research Series A: Physical Sciences, 62*(3), 174–180. https://doi.org/10.52763/PJSIR.PHYS.SCI.62.3.2019.174.180

Atakan, R., Martínez-González, I., Díaz-García, P., & Bonet-Aracil, M. (2023). Sustainable dyeing and functional finishing of cotton fabric by Rosa canina extracts. *Sustainability, 16*(1), 227. https://doi.org/10.3390/su16010227

Ayesh, M., Horrocks, A. R., & Kandola, B. K. (2022). The impact of atmospheric plasma/UV laser treatment on the chemical and physical properties of cotton and polyester fabrics. *Fibers, 10*(8), 66. https://doi.org/10.3390/fib10080066

Baig, U., Khatri, A., Ali, S., Sanbhal, N., Ishaque, F., & Junejo, N. (2021). Ultrasound-assisted dyeing of cotton fabric with natural dye extracted from Marigold flower. *The Journal of The Textile Institute, 112*(5), 801–808. https://doi.org/10.1080/00405000.2020.1779907

Banna, B. U., Mia, R., Hasan, M. M., Ahmed, B., & Hasan Shibly, M. A. (2023). Ultrasonic-assisted sustainable extraction and dyeing of organic cotton fabric using natural dyes from Dillenia indica leaf. *Heliyon, 9*(8), e18702. https://doi.org/10.1016/j.heliyon.2023.e18702

Ben Ticha, M., Haddar, W., Meksi, N., Guesmi, A., & Mhenni, M. F. (2016). Improving dyeability of modified cotton fabrics by the natural aqueous extract from red cabbage using ultrasonic energy. *Carbohydrate Polymers, 154*, 287–295. https://doi.org/10.1016/j.carbpol.2016.06.056

Bokhari, T. H., Bano, S., Adeel, S., Fazal-ur-Rehman, Ahmed, B., Al Mahmud, M. A., Qayyum, M. A., & Khattak, S. P. (2023). Eco-friendly sustainable dyeing of cotton fabric using reactive violet 05 and direct violet 09 dyes. *Coatings, 13*(4), 677. https://doi.org/10.3390/coatings13040677

Bydoon, E. A., & Saad, H. A. (2023). Comparing various dyeing techniques using plant extracts on cotton fabrics. *Egyptian Journal of Agricultural Research, 101*(4), 1068–1098. https://doi.org/10.21608/ejar.2023.228357.1426

Correia, J., Mathur, K., Bourham, M., Oliveira, F. R., Siqueira Curto Valle, R. D. C., Valle, J. A. B., & Seyam, A.-F. M. (2021). Surface functionalization of greige cotton knitted fabric through plasma and cationization for dyeing with reactive and acid dyes. *Cellulose, 28*(15), 9971–9990. https://doi.org/10.1007/s10570-021-04143-8

Dhouibi, N., Baaka, N., Charradi, R., Bouine, I., Dhaouadi, H., & Dridi-Dhaouadi, S. (2021). Multi-fiber dyeing improvement using natural supercritical CO_2 extracts. *Fibers and Polymers, 22*(7), 1874–1882. https://doi.org/10.1007/s12221-021-0422-2

Elshemy, N. S. (2011). Unconventional natural dyeing using microwave heating with cochineal as natural dyes. *Research Journal of Textile and Apparel, 15*(4), 26–36. https://doi.org/10.1108/RJTA-15-04-2011-B004

Ennaceur, S., Bouaziz, A., Gargoubi, S., Mnif, W., & Dridi, D. (2022). Enhanced natural dyeing and antibacterial properties of cotton by physical and chemical pretreatments. *PRO, 10*(11), 2263. https://doi.org/10.3390/pr10112263

Fajraoui, A., Ben Nasr, J., Lacoste, C., Amar, M. B., Dony, P., Odof, S., & El Halouani, F. (2019). Coloration of the polylactic acid with the natural dye extracted from acacia cyanophylla flowers. *Polymer Testing, 78*, 105988. https://doi.org/10.1016/j.polymertesting.2019.105988

Gorjanc, M., Mozetič, M., Vesel, A., & Zaplotnik, R. (2018). Natural dyeing and UV protection of plasma treated cotton. *The European Physical Journal D, 72*(3), 41. https://doi.org/10.1140/epjd/e2017-80680-9

Guesmi, A., Ladhari, N., & Sakli, F. (2013). Ultrasonic preparation of cationic cotton and its application in ultrasonic natural dyeing. *Ultrasonics Sonochemistry, 20*(1), 571–579. https://doi.org/10.1016/j.ultsonch.2012.04.012

Haji, A. (2017). Improved natural dyeing of cotton by plasma treatment and chitosan coating; optimization by response surface methodology. *Cellulose Chemistry and Technology, 51*(9–10), 975–982.

Haji, A. (2019). Dyeing of cotton fabric with natural dyes improved by mordants and plasma treatment. *Colorants and Coatings, 12*(3), 191–201.

Haji, A., & Naebe, M. (2020). Cleaner dyeing of textiles using plasma treatment and natural dyes: A review. *Journal of Cleaner Production, 265*, 121866. https://doi.org/10.1016/j.jclepro.2020.121866

Hamida, S. B., Štěpánová, V., Zahedi, L., Kováčová, M., Nasadil, P., Valášková, K., Humpolíček, P., & Kováčik, D. (2024). Enhancement of the dyeability and antibacterial properties of cotton fabric by plasma assisted cationization using chitosan and quaternized poly[bis(2-chloroethyl) ether-alt-1,3-bis[3-(dimethylamino)propyl]urea]. *Cellulose, 31*(11), 7119–7136. https://doi.org/10.1007/s10570-024-06027-z

Han, S., & Yang, Y. (2005). Antimicrobial activity of wool fabric treated with curcumin. *Dyes and Pigments, 64*(2), 157–161. https://doi.org/10.1016/j.dyepig.2004.05.008

Hassabo, A., & El-Sayed, E. (2021). Recent advances in the application of plasma in textile finishing (A review). *Journal of Textiles, Coloration and Polymer Science, 18*(1), 33–43. https://doi.org/10.21608/jtcps.2021.67798.1050

Hassan, M. M., & Saifullah, K. (2023). Ultrasound-assisted sustainable and energy efficient pre-treatments, dyeing, and finishing of textiles—A comprehensive review. *Sustainable Chemistry and Pharmacy, 33*, 101109. https://doi.org/10.1016/j.scp.2023.101109

Hong, K. H. (2024). Effect of UV irradiation on color and functionality of wool and cotton fabrics finished with spent coffee extract. *Cellulose, 31*(9), 5931–5946. https://doi.org/10.1007/s10570-024-05948-z

Hoque, M. M. U., Rimu, S. H., Mallik, A. K., & Rahman, M. M. (2022). Ecofriendly pretreatment of cotton fabrics by ultrasonication and reusing bath chemicals. *Environmental Science and Pollution Research, 30*(13), 36904–36914. https://doi.org/10.1007/s11356-022-24947-w

Hossain, S., Jalil, M. A., Kamal, S. A. B., & Kader, A. (2021). A natural dye extracted from the leaves of *Mimusops elengi* Linn and its dyeing properties on cotton and silk fabrics. *The Journal of the Textile Institute, 112*(3), 455–461. https://doi.org/10.1080/00405000.2020.1763057

Hussaan, M., Raza, A., Habib, N., Adeel, S., Iqbal, N., & Javed, M. T. (2023). Microwave-assisted sustainable exploration of cocklebur leaves (Xanthium strumarium L.) as a novel source of distinct yellow natural colorant for dyeing cotton fabric. *Environmental Science and Pollution Research, 30*(14), 42246–42254. https://doi.org/10.1007/s11356-023-25296-y

Ibrahim, N. A., El-Gamal, A. R., Gouda, M., & Mahrous, F. (2010). A new approach for natural dyeing and functional finishing of cotton cellulose. *Carbohydrate Polymers, 82*(4), 1205–1211. https://doi.org/10.1016/j.carbpol.2010.06.054

Irfan, M., Gao, A., Irfan, H., & Hou, A. (2024). Low liquor microwave dyeing of cotton fabric with a minimal amount of salt and the effect of different reactive dye substituents. *Fibers and Polymers, 25*(9), 3471–3480. https://doi.org/10.1007/s12221-024-00653-x

Islam, S. R., Alam, M. K., Alassod, A., Ahmed, T., Yousif, A. H. D., Rashid, M. M., Kamal, S. A. B., & Mia, R. (2024). Natural dyes and pigments as a source of medicine. In *Renewable dyes and pigments* (pp. 177–232). Elsevier. https://doi.org/10.1016/B978-0-443-15213-9.00009-0

Ji, X., Zhao, Z., Ren, Y., Xu, F., & Liu, J. (2023). Dyeing properties, color gamut, and color evaluation of cotton fabrics dyed with Phellodendron amurense Rupr. (Amur Cork tree bark). *Molecules, 28*(5), 2220. https://doi.org/10.3390/molecules28052220

Ke, G., Zhu, K., & Chowdhury, M. H. (2021). Dyeing of cochineal natural dye on cotton fabrics treated with oxidant and chitosan. *Journal of Natural Fibers, 18*(3), 317–329. https://doi.org/10.1080/15440478.2019.1623742

Khattab, T. A., Abdelrahman, M. S., & Rehan, M. (2020). Textile dyeing industry: Environmental impacts and remediation. *Environmental Science and Pollution Research, 27*(4), 3803–3818. https://doi.org/10.1007/s11356-019-07137-z

Lara, L., Cabral, I., & Cunha, J. (2022). Ecological approaches to textile dyeing: A review. *Sustainability, 14*(14), 8353. https://doi.org/10.3390/su14148353

Li, Z., Zhang, Y., Xia, W., Tang, Y., & Li, Q. (2023). Producing natural-colored super-powerful antibacterial cotton with plasma-assisted fiber surface modification: A green and effective cotton process for medical and healthcare applications. *Materials Advances, 4*(3), 932–939. https://doi.org/10.1039/D2MA00701K

Lykidou, S., Pashou, M., Vouvoudi, E., & Nikolaidis, N. (2021). Study on the dyeing properties of curcumin on natural and synthetic fibers and antioxidant and antibacterial activities. *Fibers and Polymers, 22*(12), 3336–3342. https://doi.org/10.1007/s12221-021-0412-4

Ma, X., Wei, Y., Wang, S., Zuo, X., & Shen, B. (2020). Sustainable ultrasound-assisted ultralow liquor ratio dyeing of cotton fabric with natural turmeric dye. *Textile Research Journal, 90*(5–6), 685–694. https://doi.org/10.1177/0040517519878793

Meng, X., Wang, X., Wang, P., Miao, D., & Ning, X. (2021). Enhanced dyeability and wash fastness through a salt-free plasma-induced grafting of cationic monomers on cotton fabrics. *Fibers and Polymers, 22*(12), 3378–3384. https://doi.org/10.1007/s12221-021-1105-8

Mostafa, K., Ameen, H., & Medhat, A. (2024). Harnessing of low-temperature nitrogen plasma technique as an eco-friendly approach for dye-ability of cotton fabric with acid dye. *Pigment & Resin Technology, 53*(4), 425–433. https://doi.org/10.1108/PRT-08-2022-0098

Naebe, M., Haque, A. N. M. A., & Haji, A. (2021). The effect of plasma treatment on dyeing of natural fibers. In *Innovative and emerging technologies for textile dyeing and finishing* (pp. 191–212). Wiley. https://doi.org/10.1002/9781119710288.ch7

Nasreen, H., Adeel, S., Yameen, M., Amin, N., Ozomay, M., & Qayyum, M. A. (2023). Green application of ultrasonic waves for extraction of yellow colorant from Haar Singhar and its colouring behaviour in cotton dyeing. *Textile & Leather Review, 6*, 18–36. https://doi.org/10.31881/TLR.2022.67

Naveed, R., Bhatti, I. A., Adeel, S., Ashar, A., Sohail, I., Khan, M. U. H., Masood, N., Iqbal, M., & Nazir, A. (2022). Microwave assisted extraction and dyeing of cotton fabric with mixed natural dye from pomegranate rind (Punica granatum L.) and turmeric rhizome (Curcuma longa L.). *Journal of Natural Fibers, 19*(1), 248–255. https://doi.org/10.1080/15440478.2020.1738309

Otaviano, B. T. H., Sannomiya, M., de Queiroz, R. S., Sánchez, A. A. C., Freeman, H. S., Mendoza, L. E. R., Veliz, J. L. S., Leon, M. M. G., Leo, P., da Costa, S. A., & da Costa, S. M. (2023). Natural dye extracted from pomegranate peel: Physicochemical characterization, dyeing of cotton fabric, color fastness, and photoprotective properties. *Fibers and Polymers, 24*(4), 1321–1332. https://doi.org/10.1007/s12221-023-00138-3

Panda, S. K. B. C., Sen, K., & Mukhopadhyay, S. (2021). Sustainable pretreatments in textile wet processing. *Journal of Cleaner Production, 329*, 129725. https://doi.org/10.1016/J.JCLEPRO.2021.129725

Panda, S. K. B. C., Sen, K., & Mukhopadhyay, S. (2023). A sustainable low temperature combined pretreatment process for cotton assisted by ultraviolet-C radiation. *Sustainable Materials and Technologies, 37*, e00664. https://doi.org/10.1016/j.susmat.2023.e00664

Pei, L., Gu, X., & Wang, J. (2021). Sustainable dyeing of cotton fabric with reactive dye in silicone oil emulsion for improving dye uptake and reducing wastewater. *Cellulose, 28*(4), 2537–2550. https://doi.org/10.1007/s10570-020-03673-x

Pranta, A. D., & Rahaman, M. T. (2024). Extraction of eco-friendly natural dyes and biomordants for textile coloration: A critical review. *Nano-Structures & Nano-Objects, 39*, 101243. https://doi.org/10.1016/j.nanoso.2024.101243

Pruś, S., Kulpiński, P., Matyjas-Zgondek, E., & Wojciechowski, K. (2022). Eco-friendly dyeing of cationised cotton with reactive dyes: Mechanism of bonding reactive dyes with CHPTAC cationised cellulose. *Cellulose, 29*(7), 4167–4182. https://doi.org/10.1007/s10570-022-04521-w

Raslan, W., El-Halwagy, A., & Elsayad, H. (2020). Recent advances in plasma/nanoparticles treatments of textile fibers. *Journal of Textiles, Coloration and Polymer Science*. https://doi.org/10.21608/jtcps.2020.33748.1042

Rehman, F., Adeel, S., Hanif, R., Muneer, M., Zia, K. M., Zuber, M., Jamal, M. A., & Khosa, M. K. (2017). Modulation of marigold based lutein dye and its dyeing behaviour using UV radiation. *Journal of Natural Fibers, 14*(1), 63–70. https://doi.org/10.1080/15440478.2016.1146642

Sadeghi-Kiakhani, M., Tehrani-Bagha, A. R., Safapour, S., Eshaghloo-Galugahi, S., & Etezad, S. M. (2021). Ultrasound-assisted extraction of natural dyes from hawthorn fruits for dyeing polyamide fabric and study its fastness, antimicrobial, and antioxidant properties. *Environment, Development and Sustainability, 23*(6), 9163–9180. https://doi.org/10.1007/s10668-020-01017-0

Samant, L., Jose, S., Rose, N. M., & Shakyawar, D. B. (2022). Antimicrobial and UV protection properties of cotton fabric using enzymatic pretreatment and dyeing with *Acacia catechu*. *Journal of Natural Fibers, 19*(6), 2243–2253. https://doi.org/10.1080/15440478.2020.1807443

Sandanuwan, T., Hendeniya, N., Amarasinghe, D. A. S., Attygalle, D., & Weragoda, S. (2021). The effect of atmospheric pressure plasma treatment on wetting and absorbance properties of cotton fabric. *Materials Today: Proceedings, 45*, 5065–5068. https://doi.org/10.1016/j.matpr.2021.01.573

Saxena, S., & Raja, A. S. M. (2014). *Natural dyes: Sources, chemistry, application and sustainability issues* (pp. 37–80). https://doi.org/10.1007/978-981-287-065-0_2

Sinnur, H. D., Samanta, A. K., Verma, D. K., & Kaware, R. (2018). Studies on coloration and UV protective action of anar peel (pomegranate rind) as an effective natural colorant for cotton khadi fabric. *Journal of the Institution of Engineers (India): Series E, 99*(1), 9–26. https://doi.org/10.1007/s40034-017-0106-z

Sukemi, Pratumyot, K., Srisuwannaket, C., Niamnont, N., & Mingvanish, W. (2019). Dyeing of cotton with the natural dye extracted from waste leaves of green tea (*Camellia sinensis* var. *assamica*). *Coloration Technology, 135*(2), 121–126. https://doi.org/10.1111/cote.12381

Tanrıkulu, M. (2024). Türk Kırmızısı ve Turkuazın Türk Renk Kültüründeki Yeri ve Önemi. *Kültür Araştırmaları Dergisi, 21*, 332–353. https://doi.org/10.46250/kulturder.1469310

Trinh Thi Kim, H., & Le Song Thanh, Q. (2023). Investigating the effect of microwave-assisted method in dyeing 100% cotton fabric using colorants extracted from fresh turmeric. *BOHR Journal of Material Sciences and Engineering (BIJMSE), 1*(1), 20–24. https://doi.org/10.54646/bjmse.2023.04

Tudoran, C., Roşu, M., & Coroş, M. (2020). A concise overview on plasma treatment for application on textile and leather materials. *Plasma Processes and Polymers, 17*(8). https://doi.org/10.1002/ppap.202000046

Var, C., & Palamutcu, S. (2024a). Man-made bio-based and biodegradable fibers for textile applications. In S. S. Muthu (Ed.), *Sustainable manufacturing practices in the textiles and fashion sector. Sustainable textiles: Production, processing, manufacturing & chemistry* (pp. 229–280). Springer. https://doi.org/10.1007/978-3-031-51362-6_10

Var, C., & Palamutcu, S. (2024b). Sustainable approaches in textile-sizing process. In S. S. Muthu (Ed.), *Sustainable manufacturing practices in the textiles and fashion sector. Sustainable textiles: Production, processing, manufacturing & chemistry* (pp. 55–74). Springer. https://doi.org/10.1007/978-3-031-51362-6_3

Yameen, M., Adeel, S., Nasreen, H., Fazal-ur-Rehman, Ghaffar, A., Ahmad, T., & Inayat, A. (2022). Sustainable eco-friendly extraction of yellow natural dye from haar singhar (Nyctanthes arbortritis) for bio coloration of cotton fabric. *Environmental Science and Pollution Research, 29*(55), 83810–83823. https://doi.org/10.1007/s11356-022-21450-0

Yilma, B. B., Luebben, J. F., & Nalankilli, G. (2020). Cold plasma treatment in wet chemical textile processing. *Fibres and Textiles in Eastern Europe, 28*(6(144)), 118–126. https://doi.org/10.5604/01.3001.0014.3807

Yusuf, M., Mohammad, F., Shabbir, M., & Khan, M. A. (2017a). Eco-dyeing of wool with Rubia cordifolia root extract: Assessment of the effect of Acacia catechu as biomordant on color and fastness properties. *Textiles and Clothing Sustainability, 2*(1), 10. https://doi.org/10.1186/s40689-016-0021-6

Yusuf, M., Shabbir, M., & Mohammad, F. (2017b). Natural colorants: Historical, processing and sustainable prospects. *Natural Products and Bioprospecting, 7*(1), 123–145. https://doi.org/10.1007/s13659-017-0119-9

Zhang, Y., Shahid-ul-Islam, Rather, L. J., & Li, Q. (2022). Recent advances in the surface modification strategies to improve functional finishing of cotton with natural colourants—A review. *Journal of Cleaner Production, 335*, 130313. https://doi.org/10.1016/j.jclepro.2021.130313

Ancient (Pre-historic and Historic) Indian and European Dye Pigments: A Case Study

Charmini Balasubramanian and Parthiban Manickam

Abstract Textile and fashion industry is considered to be one of the most environmentally polluting industries. There are many processes involved in making a fashion garment, out of which coloration has adverse effects on environment and human health. Fashion industry is dynamic and hence coloration is inevitable. Most of the recent developments concentrate for sustainable process and techniques in the area of dyeing and printing. India being a land of cultural heritage is well known for its natural dyeing and techniques and hence sustainability is not a new phenomenon. Pre-historic period of Bimbedka caves in India and Altamira caves in Spain are considered to be one of the oldest cave in the world as per archeological findings. It is evident that the pigments used in Cave painting are derived naturally without causing any harm to the environment and has been eco-friendly and sustainable. The commonly found colors in the cave painting are red, black, white, and brown. During historic period, paintings such as Kalighat are done on fabric also called "patas." One of the major concerns in modern day natural dyeing is the fixation whereas the cave paintings withstand thousands of years beside all climatic changes.

This paper will provide a detailed analysis on the different color pigments used on caves and fabric from different natural sources like roots, flowers, leaves, flowers, fruits, tree barks, animal-tusks, horns, insects, and minerals. It addresses the composition of dye pigments used in India and Europe during historic and prehistoric period to understand the differences, insights on the challenges and drawbacks of using it in the present day dyeing process. This paper also emphasis on the evolution of ancient dye pigments used by our ancestors and to apply the knowledge on current sustainable coloration process in fashion industry.

Keywords Pigments · Dyeing · Paintings · Historic period · Natural sources

---

C. Balasubramanian (✉) · P. Manickam
Department of Fashion Technology, PSG College of Technology, Coimbatore, India
e-mail: bcm.fashion@psgtech.ac.in

S. S. Muthu (ed.), *Sustainable Coloration of Textiles*, Sustainable Textiles: Production, Processing, Manufacturing & Chemistry,
https://doi.org/10.1007/978-3-031-91217-7_10

1 Dyes and Pigments in Textiles

Dyes and pigments are essential components in textile production, providing color that is not only decorative but also culturally and functionally significant. They have been used since ancient times, evolving from natural sources to synthetic innovations, reflecting technological advances and cultural exchanges. Dyes are soluble substances that impart color to textiles by chemically bonding with fibers. They require the use of water or another solvent to facilitate absorption into the material, often with the aid of a mordant—"a chemical that ensures the dye adheres to the fabric and enhances its durability." Common natural dyes include indigo, derived from plants like *Indigofera*, and madder, which produces red hues. Synthetic dyes, introduced in the nineteenth century, revolutionized the industry by providing consistent and vibrant color options (Tamburini, 2021; Zhang et al., 2020). In contrast, pigments are insoluble particles that provide color by coating the textile's surface. Unlike dyes, they do not chemically bond with fibers but require a binding agent to adhere. Pigments are often derived from minerals, such as ochre and ultramarine, or are synthesized for modern applications (Sharma & Puri, 2017). The use of dyes and pigments in textiles spans functional, esthetic, and cultural domains. Ancient civilizations used plant-based dyes for ceremonial garments, while mineral pigments were applied to fabrics and other materials for art and decoration. In modern times, dyes and pigments cater to industrial needs, with synthetic variants offering improved stability, variety, and cost-effectiveness (Fontani, 2022).

2 Natural and Synthetic Dye Classification

Fiber is classified into two types: natural and synthetic dye as shown in Fig. 1.

2.1 Natural Dyes

Derived from plants, animals, and minerals, natural dyes are among the oldest methods of coloring textiles. **Plant-based dyes** are obtained from different parts of the plants such as roots, bark, stem, leaves, flowers, and fruits. Some examples of plant-based dyes are Indigo (*Indigofera tinctoria*) which is a deep blue dye, historically significant in Indian, African, and European textile traditions. Madder (*Rubia tinctorum*) produces red hues, widely used in Europe and the Middle East. Turmeric(Curcuma longa) yields yellow shades which is commonly found in South and Southeast Asia. **Animal-based dyes** are extracted from different parts of different animals. Insects, snails, algae and bees can also serve as a source of natural dyes. Also the fluids from different animals are used as binders and fixations for dyeing. Cochineal and Murex sea snails are best examples of producing red dye and tyrian

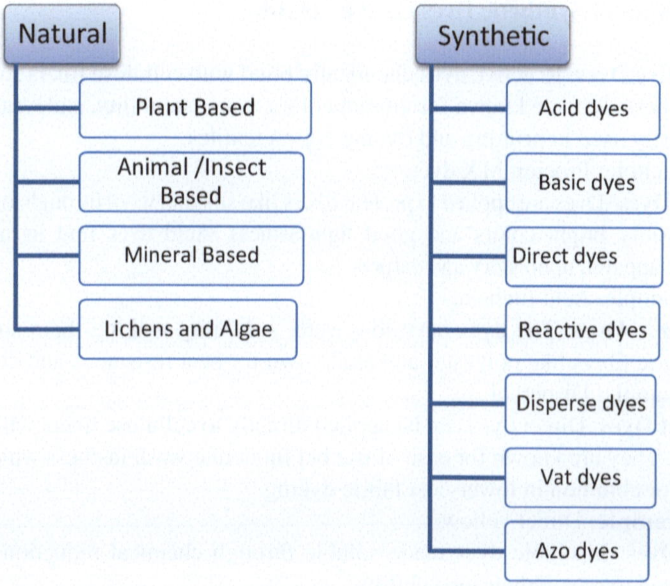

Fig. 1 Classification of dyes. (Source: Natural Dyes: Sources, Chemistry, Application and Sustainability Issues by Sujata Saxena and A. S. M. Raja)

purple, respectively. **Mineral-based dyes** are obtained from different mineral stones available on the earth. Limestone, manganese, cinnabar, azurite, and malachite are some examples of mineral stones. Lichens and algae also have dye producing components. Hematite and ochre were used in early textile traditions, especially in prehistoric cave paintings (Cardon, 2007; Tamburini, 2021).

2.2 *Synthetic Dyes*

Synthetic dyes have transformed the textile industry since their invention in the mid-nineteenth century. Their introduction provided cost-effective, vibrant, and durable coloring solutions that surpassed the limitations of natural dyes. The widespread adoption of synthetic dyes was catalyzed by their diversity, ease of application, and compatibility with various fiber types. The industrial production of synthetic dyes such as aniline black, indigo, and azo dyes revolutionized textile coloration. The industrial revolution facilitated the large-scale production and global distribution of these dyes, integrating them into mainstream textile manufacturing (Perkin, 1860; Zhang et al., 2020). By the mid-nineteenth century, scientific discoveries shifted the focus to synthetic dyes. The accidental synthesis of mauveine in 1856 marked the beginning of synthetic color production, which expanded rapidly during the industrial revolution. Despite this, traditional natural dyes remained in use for their esthetic and cultural significance (Fontani, 2022; Tamburini, 2021).

2.2.1 Types of Synthetic Dyes Used in Textiles

1. **Reactive Dyes**: Reactive dyes chemically bond with cellulose fibers such as cotton. These dyes are known for vibrant colors, water solubility, and wash fastness primarily used in printing and dyeing cotton textiles.
   Example: Procion MX dyes.
2. **Acid Dyes**: Dyes are applied to protein fibers like silk and wool through ionic bonds. They offer bright colors and good lightfastness. Acid dyes find its application in fine apparel, upholstery and carpets.
   Example: Acid fuchsin.
3. **Disperse Dyes**: The dyes have low water solubility making them suitable for synthetic fibers like polyester and are known for heat resistance and durability.
   Example: Dispersol.
4. **Direct Dyes**: Direct dyes can be applied directly to cellulose fibers without mordants. They are known for ease of use but moderate wash fastness which is suitable for common in lower-cost fabric dyeing.
   Example: Direct yellow.
5. **Vat Dyes**: Insoluble dyes made soluble through chemical reduction and then oxidized to regain their insolubility.
   Example: Synthetic indigo, widely used for denim.
6. **Azo Dyes:** It comprise a large group of dyes characterized by azo bonds (-N=N-) known for their versatility and wide color range.
   Example: Congo red, methyl orange (Fontani, 2022; Sharma & Puri, 2017).

2.2.2 Advantages of Synthetic Dyes

- **Lower Cost of Production**: Synthetic dyes are produced on an industrial scale using readily available raw materials, reducing production costs compared to the labor-intensive extraction of natural dyes (Fontani, 2022).
- **Scalability**: Synthetic dyes can be mass-produced, meeting the high demand of the growing textile industry during and after the Industrial Revolution.
- **Uniform Coloration**: Synthetic dyes offer consistent shades and quality, which was difficult to achieve with natural dyes due to variations in plant or animal sources (Tamburini, 2021).
- **Predictable Properties**: Natural dyes could vary based on environmental conditions, harvest timing, or processing techniques, while synthetic dyes ensure predictable performance.
- Synthetic dyes provide a far wider range of colors than natural dyes, including vibrant and saturated shades that were previously unattainable (Zhang et al., 2020).
- Synthetic dyes exhibit superior fastness properties, including resistance to washing, light, and abrasion. For example, reactive dyes form covalent bonds with fibers, enhancing their durability (Sharma & Puri, 2017).

- Synthetic dyes are suitable for a broad range of natural and synthetic fibers, including polyester, nylon, and acrylic, which are not readily dyeable with natural dyes (Kant, 2012).
- The application of synthetic dyes is more straightforward, with many not requiring mordants, thus simplifying textile production and reducing labor intensity (Sharma & Puri, 2017).
- The availability of natural dyes was subject to agricultural cycles, climate, and regional limitations. Synthetic dyes ensured year-round production and global distribution (Fontani, 2022).
- The advent of the chemical industry, supported by advancements in organic chemistry, provided the technical knowledge and infrastructure needed for synthetic dye production. The discovery of mauveine in 1856 marked a turning point, showcasing the potential of synthetic colorants (Perkin, 1860).
- The decline of traditional knowledge and techniques for producing and applying natural dyes contributed to their replacement by synthetic dyes.
- Additionally, the slow production process of natural dyes could not meet the demands of industrial-scale textile manufacturing.

2.2.3 Challenges Associated with Synthetic Dyes

- The production and disposal of synthetic dyes generate harmful waste, including toxic effluents and heavy metals.
- Synthetic indigo production for denim contributes significantly to water pollution.
- Certain azo dyes release carcinogenic aromatic amines, posing health risks to workers and consumers.
- Stricter environmental regulations,such as REACH in Europe pose significant challenges for the synthetic dye industry.

2.2.4 Various Applications of Dyes

Dyes have a wide range of applications in every industry as consumer prefers esthetics along with functionality. Some of the areas in which dyes are used widely are Textiles, Cosmetics, Food products, Decoration, Artifacts art, Solar cells, UV protection, Medicines and drugs, Antimicrobial agent, pH Indicator.

3 Overview of Ancient Dye Pigments

Dye pigments have played a significant role in human history, originating from natural sources such as plants, minerals, and insects. Their use dates back to prehistoric times, with evidence found in cave paintings and artifacts that reveal a deep understanding of color and its applications. Mineral-based pigments are inorganic pigments such as ochre and lapis lazuli were used widely in ancient art. Ochre, derived from iron-rich clay, was used for its reddish and yellowish hues, while lapis lazuli provided the brilliant blue color, highly valued in European and Indian artworks. Organic Pigments were extracted from plants, insects, and mollusks. For example, indigo from the *Indigofera* plant and Tyrian purple, derived from the *Murex* sea snail, were prized for their vibrant and lasting colors. The use of cochineal, derived from insects, also became prominent in later centuries, particularly in textile dyeing. Ancient civilizations developed advanced techniques for processing these pigments. Mordanting is an organic dyes often required mordants—substances that bind dye to fabrics. This technique enabled vibrant and long-lasting colors. Organic dyes were precipitated onto an inorganic substrate to create pigments for painting, a technique particularly common in Europe and India. The Silk Road and other trade networks facilitated the exchange of pigments like cinnabar, ultramarine, and madder, enriching the palettes of artists across continents. Mordants enhance dye-fiber affinity, especially for natural dyes. Types of mordants include metallic salts (alum, iron, and tin), tannins, and organic acids. Mordanting is usually used as pre-mordanting, post-mordanting, and simultaneous mordanting.

4 Methods of Dyeing

4.1 Conventional Dyeing Systems

1. **Direct Dyeing**: Fabrics like silk and wool are dyed by immersing them in dye solutions without additional agents.
2. **Mordant Dyeing**: For fabrics like cotton, mordants (chemical agents) are used to improve dye binding and fastness. Mordants can be applied before dyeing (pre-mordanting), after dyeing (post-mordanting), or simultaneously with dyeing.
3. **High-Temperature Dyeing**: Uses heated dye baths to promote dye penetration into textiles, commonly applied to polyester and synthetic fibers.

4.2 Advanced Dyeing Systems

1. **Gel Dyeing**: Utilizes chemically formed gels for creating precise color designs and patterns, compatible with traditional dyeing equipment.
2. **Ultrasonic Dyeing**: Employs ultrasonic waves to enhance dyeing efficiency through cavitation, which accelerates dye dispersion, diffusion, and liquid agitation.
3. **Plasma Treatment**: A surface-modification technique improving dye uptake and fastness while being environmentally friendly.

4. **Radiation-Assisted Dyeing**: Methods such as UV or gamma radiation improve color fastness and efficiency.

5 Indian Dye Pigments

India is known for its rich heritage from ancient times. The famous caves like Ajanta ellora have rich paintings. Paintings are not only seen in caves but in paper forms like Mughal miniature paintings, Kalighat paintings, and Madhubani paintings. The dyes used for these paintings are sourced naturally for each color. India was known for its mastery of organic dyes, with techniques such as block printing and resist dyeing creating intricate patterns on textiles. In Europe, mineral pigments like lead white and verdigris were prevalent in art and manuscript illumination (Tamburini, 2021).

5.1 Red and Orange Dye

5.1.1 Plant-Based

Madder is a perennial and herb-aceous climber plant having very long, cylindrical, and flexuous root with thin red barks. Contains alizarin and purpurin, offering bright red tones used widely in India for textile dyeing since ancient times. It was a staple in Harappan civilization, providing lasting color through mordanting with alum. The botanical name of madder plant is Rubia cordifolia. Parpata or Jatuka or Parpat where the botanical name is Oldenlandia umbellate belongs to a family of Rubiaaceace is also called as Indian madder or chay root. The dye pigments are present in roots at the age of 2 years. This plant is seen widely in Tropical India. Different shades like brown, purple, orange are produced when the dye is combined with milk, water, cowdung, rice gruel, etc. **Manjistha** is commonly found in Punjab Himalayan region and a native of Syria, Palestine, and Egypt or Siberia. This is considered as creeping plant as per Atharveda. This plant is full of dye pigments especially in the stem. The red coloring matter of the dye is chemically known as Alzirin. This dye was not only used for dyeing textile fabrics but also leather and hair. Kampillaka or Kampilla is called as KAmela powder in Indian context. Mallotus philippensis is its botanical name of Euphorbiaceae family. The dye color is orange red and is seen in the fruits of the plant. The fruit ripes in the month of Feb-March every year and can get a yield of 1.4–3.7% weight of the fresh fruit. Chalcones are the major coloring component in Kamala powder. **Safflower** (***Carthamus tinctorius***): Produces both red (carthamin) and yellow (carthamidin). The red is used for textiles like silk. Seeds of the plant are used for oil and flowers especially petals for dyeing. The water-soluble yellow dye, carthamidin, and a water-insoluble red dye, carthamin, which is readily soluble in alkali, can be obtained from safflower florets. It produces shades of yellow pink which is easily soluble in water, whereas red has higher durability when compared. Saffron: Kashmira and Vahlika (Balkh) were two places of cultivation of this plant. **Henna leaves,** beetroot, annote are also sources of shades of red.

5.1.2 Animal-Based

Lac (*Laccifer lacca*) and cochineal (*Dactylopius coccus*) produce carminic acid, which yields shades from red to scarlet. They were widely used in India and exported globally in the medieval period. The dyes are produced by female insects on the tree twigs. Based on different literature reviews, Lac is found to be seen in Maharashtra. The dye was believed to produce permanent tints to fabric if it was boiled and made cool before being applied to the desired article. Silk industry primarily used the lac dyes in ancient times. Now, Lac dye is officially used as food preservative. **Kermes Vermillo**: Dried bodies of the female of the genus Kermes produce Red dye. These insects are the native of Mediterranean region. Many countries such as Egyptians, Mesopotamians, Indians, Greeks, Romans, and Iranians have used these dyes. It falls under the crimson shade of red. Silk and wool fabrics were used for Kermes dye. In some places, it is believed that dye is from the blood of female insect who pours this during the period of hatching. The color, which the kermes dye produced, was scarlet, for which in the Middle Ages it was traded as "scarlet-grains" or "venetian scarlet."

5.1.3 Mineral-Based

Cinnabar (mercury sulfide) gives bright scarlet, while red ochre provides earthy reds. Cinnabar, also known as vermillion, refers to common bright scarlet to brick-red form of mercury sulfide, a common source ore for refining elemental mercury and serves directly as dyeing pigment. Cinnabar is found in several hues, which ranges from yellow to deep orange or brown. The red ochre is found to have it traces in many of the ancient Indian paintings. Gairikha is derived from clay pigmented hematite which has a high durability. It stays through ages without decaying or washing away. Since it is mineral it can be commonly found in valley edges, eroding cliffs, caves, and soil. This color was abundantly found during Paleolithic period. They had a wide range of application in the ancient times such as rock, leather, clothing, tools, and indeed the human body. The high usage of this color is because of its easy preparation. When the mineral is picked up it stains in its hand. This is easy to grate into a powder form and can be mixed with water, saliva or egg whites and can be turned into pigmented paint. As per Indian context, this dye gained its popularity during Post Vedic Period. Sindura is also called a Red lead. This had its traces in Medieval period and found its application in artworks, including wall paintings, polychrome sculptures, manuscript Illuminations, and canvas paintings. Mordants like alum enhance the binding of plant-based red dyes.

Iron mordants create deeper reds and browns. For lac and cochineal, boiling the insect extract with mordants produces vivid and durable shades (Nasreem Begum, 2022).

5.2 Yellow Dyes

5.2.1 Plant-Based

India is one of the largest producers of turmeric in the world and also termed as Indian solid goal. Turmeric produces a warm yellow colour which is widely used in India since the vedic period and is suitable for cotton, silk and wool. It is favored for its medicinal and esthetic properties. The dried stigmas of Crocus sativus (saffron) yield a yellow pigment primarily due to the presence of crocin (Crocus sativus) produces yellow through its dried stigmas. It was cultivated under royal supervision in Kashmir and Balkh. The plant is described in the Ain-i-Akbari, a historical text from the Mughal period attributed to Emperor Akbar's court, as reaching an approximate height of one-quarter yard (about 23 cm). The flower stands on the top of the stalk and consists of six petals and six stamens. Three of the six petals have a fresh lilac color and stand round about the remaining three petals. The stamens are similarly placed, three of a yellow color standing round about the other three, which are red. The latter yield saffron. Mordants and materials like cream of tartar, oxalic acid, formic acid, acetic acid helps in better fixation. Marigold and Parijata are even used now in many places and adopted by many designers. Turmeric and saffron do not require mordants but can be enhanced with alum for brighter hues. They are suitable for cotton, silk, and wool (Nasreem Begum, 2022).

5.2.2 Animal-Based

Gorocana Extracted from cow urine, used for religious and dyeing purposes in early India.

Green dyes are obtained by overlaying blue (indigo) and yellow (turmeric) dyes. Textiles are sequentially dyed, first in yellow and then in blue, or vice versa, to create green hues. Other direct sources of green are Tulsi, lily, and Nettles.

5.3 *Blue Dyes*

5.3.1 Plant-Based

Indigo (Indigofera tinctoria)
A hallmark of ancient Indian dyeing, indigo was processed into cakes and exported globally. The dye is primarily extracted from the *Indigofera tinctoria* and *Suffruticosa* plants. The earliest evidence of indigo dyeing is found in the Harappan civilization. Archeologists have recovered remains of blue-dyed cloth dating back to 1750 BCE from Mohenjo-Daro, now located in present-day Sindh, Pakistan. The Greeks named the dye "Indikon," meaning "product from India," emphasizing its Indian origins. Traditionally, the process of indigo dye preparation involves soaking leaves in water to extract the coloring material through fermentation. The solution is then beaten and exposed to air, which induces oxidation and forms the indigo dye. The resulting blue substance is dried and pressed into pellets or balls. Yarn dyed with indigo is sundried to deepen the color. Unlike other dyes, indigo does not require mordants and is entirely natural, relying on fermentation. To achieve deeper shades, textiles are dipped multiple times. Fermentation vats are essential for this process, during which the fabric turns greenish upon immersion and transforms into a vibrant blue upon oxidation in the air. Other plants, such as woad and water lilies, can also be used to extract blue pigments for dyeing various materials (Nasreem Begum, 2022).

5.4 *Black Dyes*

5.4.1 Plant-Based

Extracts from gallnuts and other tannin-rich plants are used in the preparation of black dyes. Twigs and wood from the tamarind tree, when burned in a closed vessel, produce a fine, homogeneous black powder. Similarly, almonds and coconuts are charred to obtain a similar black pigment. This is not only used for dyeing textile material but also used as Kajal, a beauty statement of Indian women which is prepared by burning oil in a lamp and depositing the soot on an earthen bowl. Haritaki known as Kadukkai in Tamil is also a source of Black and was popular in the Medieval period. In some areas, custard apple, alder, harda are used for black dye. Iron salts such as ferrous sulfate are the sources of mineral-based black dyes.

5.4.2 Animal Based

Bone black is prepared by charring animal bones. The black formed is not as intense as plant based. Animal-based charred black finds its application drawing outlines rather than painting Tannins are applied to textiles, followed by treatment with iron-rich solutions to achieve a black hue. This method is common for cotton and leather (Nasreem Begum, 2022).

5.5 *Violet/Purple*

Hibiscus flowers yield shades of red, pink, and purple, depending on the dyeing method, mordants used, and fiber type. The dye is primarily extracted from the petals of the hibiscus flower. These petals are rich in anthocyanins, which are the key coloring compounds. Fresh hibiscus flowers are collected, washed, and boiled in water to release the color pigments. The boiling process may include the addition of natural acids (like lemon juice or vinegar) to enhance the brightness and stability of the dye. The flowers are simmered in water for 30–60 min. This releases the anthocyanins into the water, creating a concentrated dye solution. The boiled mixture is filtered to remove the petal residues, leaving behind the liquid dye extract. The extracted solution may be further boiled down to concentrate the dye, making it suitable for deeper coloration. The extracted dye is used directly on fabrics, often with alum as a mordant to fix the color and improve its durability. Different pH conditions can produce varied shades, such as red or purple. The hibiscus dye is mainly used for coloring cotton, silk, and wool fabrics. Apart from textiles, it is also employed in cosmetics and food coloring (Iheoma, 2023). The coloring compounds include anthocyanins, which are responsible for the red and purple hues. These colors may vary slightly due to pH levels during dyeing. Cabbage and bougainvillea can be used for dyeing shades of purple (Aggarwal, 2021).

5.6 *Other Sources*

Dye-bearing lichen produce a wide range of greens, oranges, yellows, reds, browns, and bright pinks and purples. The lichen Rocella tinctoria was found along the Mediterranean Sea and was used by the ancient Phoenicians. In recent times, lichen dyes have been an important part of the dye traditions of Wales, Ireland, Scotland, and among native peoples of the southwest and Intermontane Plateaus of the United States. Scottish lichen dyes include cudbear (also called archil in England and litmus in Holland), and crottle.

Miriam C. Rice, (1918—2010) of Mendocino, California, pioneered research into using various mushrooms for natural dyes. She discovered mushroom dyes for a complete rainbow palette. Swedish and American mycologists, building upon Rice's research, have discovered sources for true blues (Sarcodon squamosus) and mossy greens (Hydnellum geogenium). Hypholoma fasciculare provides a yellow dye, and fungi such as Phaeolus schweinitzii and Pisolithus tinctorius are used in dying textiles and paper.

6 European Dye Pigments

European dye pigments have the similarities with Indian natural dyeing. Madder roots, Kermes, Cochineal are the sources of red. Weld leaves and stem produced bright yellow, popular for creating green when combined with woad. Stigmas of saffron crocus flowers produced rich yellow to orange hues, highly valued for luxury garments Woad (*Isatis tinctoria*) leaves common in Britain and Europe, used for pale to vibrant blues Secretion from sea snails Tyrian Purple (*Murex* species) exclusively used by the elite in Roman society for vibrant purple hues. Orchil (*Roccella tinctoria*) which are lichens processed with ammonia fermentation creates soft purples and pinks, popular in Mediterranean and Celtic dyeing traditions. Galls on oak trees processed with iron mordants produce black and dark brown dye for ink and textiles. Scottish Lichens provide reddish broen hues which are widely seen in Scotland. Lapiz lazuli is a mineral stone that produces violet lavender color when ground and is exported to different countries. Traces are found in Mughal painting in India reflecting the high economy and trading during that period.

6.1 Key Differences Between Indian and European Dyes

Table 1 clearly explains the difference between Indian and European dyes

Table 1 Difference between Indian and European dyes

Factors	Indian	European
Plant sources	Indigo, turmeric, sandalwood	Woad, Madder, Weld dominant
Flower source	Marigold, hibiscus, and safflower	Yarrow, Bloodroot, Black-eyed susan
Animal-based dyes	Animal-based dye usage is less compared to European	Snails and Kermes
Minerals	Gairika (ochre), Hingula (cinnabar), Sphatika (alum), Shweta churna (chalk), Raktashila (red sandstone), Kasis (iron-rich stones), Tamra Shila (copper rich), Vermillion stones	Malachite, Lapiz lazuli, Azurite are some stones found uniquely in European countries in addition to cinnabar, chalk, Vermillion, iron oxide which are also found in India
Cultural context	Religious	Textiles as a status symbol.
Techniques	Resist dyeing techniques like bandhani, ikat	Fermentation process and guild protected methods
Geographic influence	India's biodiversity provide broader spectrum of color palatte	Temperate plants leading to limited color palatte
Color palatte	Bright colors due to tropical plants	Subtle palatte due to temperate plants
Usage	Everyday rituals and artisan trades	Symbol of luxury and exclusivity
Export	Major export of indigo madder dye to Europe	Lapiz lazuli were exported to India

7 Application in Different Mediums

Three mediums that are commonly used in the ancient period for the usage of dye are paper in the form of manuscripts and miniature paintings in periods like mughal era, textiles or clothing and wall paintings or murals. A same natural dye can be used in all three mediums but with additives and fixations.

Paper: Paper typically requires natural dyes to be mixed with binders (like gum arabic or starch) to adhere to the smooth surface. Absorbent surfaces are pre-treated to prevent excessive spreading of dyes and thin layers of color are applied, and drying is quick. Brushes or tools like sponges are used to create precise artwork. Watercolor-style washes or intricate designs are common in artworks like illuminated manuscripts or modern natural dye art. Susceptible to fading if exposed to light or moisture. Used mainly for decorative art or temporary applications.

Textile: In textiles, dyes need mordants like alum for strong adhesion and color fastness. Techniques like tie-dye, block printing, or hand painting involve soaking or resist treatments. Fabrics like cotton or silk are pre-washed and treated to absorb dye better. Freehand painting or block printing adds unique patterns. Longevity depends on the mordant and dye used. Natural dyes may fade with repeated washing or exposure to sunlight without proper fixation.

Wall: Dyes are often mixed with lime or other bases for durability on plaster in wall painting. The preparation includes a primed wall (e.g., mud, lime, or cement) to help the dye adhere. Pigments need additional binders for weather resistance. Dyes are applied with wide brushes, rollers, or by hand for murals. Fresco painting found in ajanta caves, Sittanavasal temple use natural pigments mixed with wet plaster are live examples of durability which is withstanding for centuries. Patterns often require stencils or freehand techniques more durable due to the addition of lime or casein as a binder. Fading and weathering occur over time, but restoration techniques can revive colors.

8 Future Perspective

Paleo-inspiration, the process of mimicking properties of specific interest (mechanical, optical, and structural) observed in ancient and historical systems, is proposed for innovative chemical conception. The inspiration is gained from an advanced study of ancient materials that were often synthesized in soft chemical ways, using low energy resources, and sometimes rudimentary manufacturing equipment" (Bertrand). India is a country with vibrant textile colors from ancient times and its high time to preserve the techniques and start applying the process in the current scenario. Many technological developments are growing rapidly. Sustainability has become the mantra of every Textile apparel brands and reviving ancient techniques will serve the purpose. Collaborating the ancient techniques with new advancements without affecting the environment will bring out new innovative colors and process.

9 Conclusion

The exploration of ancient dye pigments from Indian and European traditions reveals intersection of artistry, science, and cultural heritage. The rich knowledge of natural dyeing practices, spanning prehistoric cave paintings to historic textile art, illustrates how human has utilized the best out of nature mainly derived from plants, minerals, and animals. The dyes offered not only esthetic appeal but also functional benefits such as durability and environmental harmony.

This chapter highlights the ingenuity of ancient civilizations in extracting and applying dyes. Indian traditions, with their use of diverse natural sources such as indigo, turmeric, and madder, reflect the country's biodiversity and deep cultural ties to sustainability. European techniques, characterized by innovations such as fermentation and the exclusive use of dyes like Tyrian purple, demonstrate the integration of natural resources with artistry and social stratification.

The comparative analysis of Indian and European dyeing traditions reveals several key differences. India's tropical climate allowed for brighter palettes and a broader spectrum of natural colors, while Europe's temperate environment resulted in subtler hues. The cultural significance also differed where Indian dyes were deeply intertwined with religious rituals and daily life, whereas European dyes often symbolized luxury and exclusivity.

The transition to synthetic dyes in the nineteenth century marked a turning point for the textile industry. Synthetic dyes brought consistency, scalability, and economic efficiency but at the cost of environmental degradation and health risks. This chapter underscores the urgency of addressing these challenges by revisiting ancient techniques that prioritized ecological balance.

Modern innovations in dyeing processes, such as plasma treatment and ultrasonic dyeing, provide opportunities to integrate traditional natural dyes into contemporary applications. Techniques like mordanting and fermentation, foundational in ancient dyeing practices, can be enhanced with advanced technologies to create vibrant, durable, and sustainable coloration solutions.

This chapter emphasizes the importance of preserving traditional knowledge and applying it to contemporary challenges. Reviving ancient dyeing techniques, combined with modern sustainability initiatives, offers a path toward eco-friendly innovation in the textile industry. This approach aligns with the global movement toward sustainability and provides a culturally enriched framework for future practices. Researchers, designers, and industry leaders can embrace the wisdom of the past while innovating for the future by honoring the traditions of natural dyeing and adapting them to modern needs where we can forge a more sustainable, ethical, and vibrant future for the textile and fashion industries.

References

Aggarwal, S. (2021). Indian dye yielding plants: Efforts and opportunities. *Natural Resources Forum, 45*(1), 63–86.

Begum, Nasreem. Traditional Knowledge and Practices of Natural Dyes in South Asia. *Journal of Ethnobotany and Traditional Medicine*, vol. 5, no. 2, 2022, pp. 45–59.

Cardon, D. (2007). *Natural dyes: Sources, tradition, technology and science*. Archetype Publications.

en.wikipedia.org/wiki/Natural_dye

Fontani, M. (2022). The modernity of ancient pigments: A historical approach. *Colorants, 1*(3), 307–346. https://doi.org/10.3390/colorants1030019

https://www.homesciencejournal.com/archives/2023/vol9issue1/PartB/9-1-25-268.pdf

Iheoma, C. (2023). Hibiscus sabdariffa natural dye extraction process with central composite. *Sustainable Chemistry for the Environment, 2*(21), 100008. https://doi.org/10.1016/j.scenv.2023.100008

Kant, R. (2012). Textile dyeing industry: Environmental impacts and remediation. *Natural Science, 4*(1), 22–26. https://doi.org/10.4236/ns.2012.41004

Perkin, William Henry. On the Purple Dye Obtained from Coal-Tar. *Journal of the Chemical Society*, vol. 13, 1860, pp. 53–54.

Sharma, P., & Puri, S. (2017). Natural and synthetic dyes: A comparative review of textile dyeing techniques. *Textile Science & Engineering, 5*(3), 1–8. https://doi.org/10.4172/2165-8064.1000228

Tamburini, D. (2021). Colour analysis: An introduction to the power of studying pigments and dyes in archaeological and historical objects. *Heritage, 4*(4), 4366–4371. https://doi.org/10.3390/heritage4040241

Zhang, Y., Li, H., & Wang, X. (2020). Sustainable dyeing and finishing processes for textiles: An overview. *Journal of Cleaner Production, 123*, 1245–1260. https://doi.org/10.1016/j.jclepro.2020.1245

Optimisation of Knit Dyehouse: Pathways Towards 'Zero' Water Requirements

Md. Akhtarujjaman, Kazi Farhan Hossain Purba, Md. Sakib Al Montasir, K. M. Abdun Noor, M. R. Ullah Rashed, and Mohammad Abbas Uddin

Abstract The Textile and Apparel industry has been important in developing Bangladesh's economy. Textile wet processing industry is considered the backward linkage of the apparel industry, and the most significant regarding water, energy, chemical consumption, reprocessing, and productivity. Thus, the optimisation of any dyehouse for textile processing has been a 'key challenge' for many dyehouse managers. Depending on the production process, textile dyeing can be divided into continuous, semicontinuous, or exhaust methods. It can also be classified as knit dyeing, woven dyeing, yarn dyeing, denim dyeing, and garment dyeing/washing based on the colouration of products. Amongst these, woven and denim dyeing is more technology-driven, needing a huge machine setup with major investments. On the other hand, knit, yarn, and garment washing are much more skill-oriented. In this chapter, based on extensive experiences in the Bangladesh knit dyeing industry, various practical optimisation techniques for dyeing are discussed, a path that will lead to the least amount, *aka* 'Zero' water requirements.

Keywords Textile wet processing · Dyehouse · Water consumption · Process optimisation · Sustainable dyeing · Decarbonization

---

M. Akhtarujjaman · K. F. H. Purba · M. S. Al Montasir · M. R. U. Rashed
Reed Consultancy, Dhaka, Bangladesh

K. M. A. Noor
Faculty of Agricultural, Life and Environmental Science, Department of Human Ecology, University of Alberta, Edmonton, AB, Canada

M. A. Uddin (✉)
Department of Dyes and Chemical Engineering, Bangladesh University of Textiles, Dhaka, Bangladesh
e-mail: abbas.shiyak@dce.butex.edu.bd

S. S. Muthu (ed.), *Sustainable Coloration of Textiles*, Sustainable Textiles: Production, Processing, Manufacturing & Chemistry,
https://doi.org/10.1007/978-3-031-91217-7_11

1 Introduction

Textiles and clothing have been inexorably intertwined with human society throughout history, serving as basic needs and expressions of culture and identity (Sanders et al., 2021). Over the centuries, the textile industry has turned into a global economic powerhouse for many countries. Transformation of raw fibres to finished garments involves a sequence of steps, including fibre processing, spinning, weaving, knitting, dyeing, and finishing. Among these steps, textile dyeing is a crucial process that contributes colour to fibres, yarns, fabrics, and significantly enhances their quality, aesthetic appeal, and market value (Ammayappan et al., 2016). Dyeing is not only a mere application of colour but it is integral to the fashion industry, where the colours often dictate trends and consumer preferences. The ability to produce vibrant, consistent, and durable colours on textiles is a must for manufacturers striving to meet the demands of designers and consumers alike. This makes dyeing a creative, technical operation and economic driver within the global textile sector.

The significance of dyeing extends far beyond adding colour; it plays a vital role in distinguishing textiles, contributing to brand identity, and enabling product diversification (Cassidy, 2019) in the fashion industry, where the right colour palette is key to success, whether in high fashion or everyday apparel. However, despite its importance, textile wet processing has remained one of the most environmentally challenging production steps (Drumond Chequer et al., 2013; Kane et al., 2020), as it is highly water and energy-intensive, contributing significantly to water pollution and resource depletion. Unless properly treated, textile effluents contain residual dyes, heavy metals, salts, and acids and their reaction products (Al-Tohamy et al., 2022).

Given the profound environmental and social impacts, there is an urgent need to modify traditional dyeing methods, especially in fibre, yarn, knit, woven, and denim dyehouses, to minimize their ecological footprint. Moving towards more sustainable practices, such as achieving near-zero water usage in dyeing processes, is essential for both environmental protection and the long-term viability of the industry. This shift addresses the immediate need for reducing harmful emissions and presents an opportunity for the textile sector to innovate and lead in sustainable production, ensuring that the creation of beautiful textiles does not come at the cost of the planet's health.

This chapter will address strategies for optimizing knit dyehouses, focusing on pathways towards achieving near-zero water usage in dyeing. By exploring innovative approaches and sustainable practices, this chapter aims to contribute to the ongoing efforts to improve resource efficiency in textile wet processing, paving the way for a more sustainable future in the industry.

This chapter started with the basic requirements of textile colouration, and the role of water as solvent dyeing. Then, it discusses water-based and waterless dyeing techniques followed by water consumption in exhaust-based dyeing techniques, which is the main focus of the study. It discusses various factors of water consumption, which were highlighted hardly before, such as type of machine, colour range, and colour depths in a dyehouse. Then, this chapter extensively discusses optimisation possibilities to reduce water consumption towards zero.

2 Colouration of Textile Fibres

In general, dyeing is a process in which a textile substrate absorbs the dye molecules from dye liquor and resists the release of the dye back into the solution (Grishanov, 2011). The dye–fibre interaction is a complex process, which is influenced by the nature of the dyes and fibres and dyeing conditions, such as time, temperature, and pH. The primary requirements of any textile colouration technique is:

- High colour depth with lowest amount of dye/pigment
- High fixation with textile fibre
- Uniform colouration and levelness
- Required fastness, such as wash, rubbing.

Colouration in textile sectors requires various textile fibres to be coloured which can be classified as natural or man-made. Cotton, wool, and silk are natural fibres, while man-made fibres include polyester, nylon, acrylic, and regenerated cellulose fibres, such as viscose, modal, and lyocell. Each type of fibre has distinct characteristics that influence its affinity for different types of dyes. Other common synthetic dyes and their applicable fibre types are shown in Fig. 1.

Understanding dye–fibre interactions is crucial as these interactions vary depending on the fibre type and dye class. Reactive dyes form covalent bonds with cellulose fibres such as cotton, providing good wash and light fastness because the dye is chemically bonded rather than merely absorbed on the surface. Disperse dyes, used for synthetic fibres such as polyester, are physically trapped within the fibre structure, requiring high penetration temperatures and good fastness properties. Acid dyes, used for protein fibres, such as wool, silk, and nylon, form ionic bonds with the fibres, resulting in vibrant colours and good fastness properties (Lewis, 1996). Basic dyes are used for dyeing jute, wool, silk, acrylic, and mod-acrylic fibres

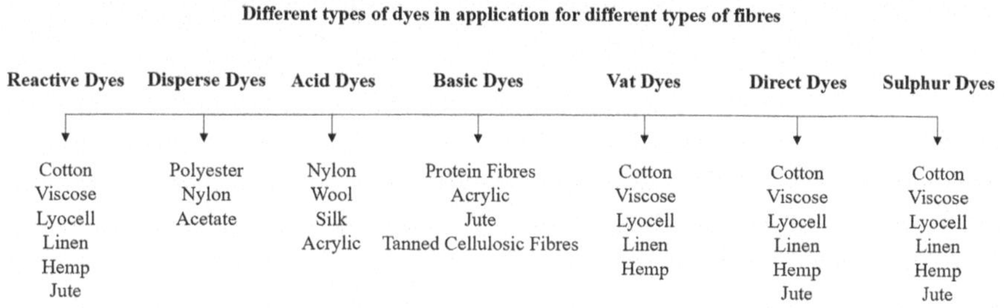

Fig. 1 Dyes and their applicable fibre types

(Chakraborty et al., 2005). Vat dyes are water-insoluble dyes used for dyeing cellulosic or cotton fibres, which, before application, are reduced under alkaline conditions to make them water soluble so that the dye can be absorbed by the cellulose (Richards, 2015). Direct dyes, inherently substantive to cellulosic fibres such as cotton, are water-soluble compounds, and requires electrolytes such as sodium chloride or sodium sulphate for absorption (Sekar, 2011). Sulphur dyes are also water-insoluble dyes, which are reduced to make water-soluble and substantive to cellulosic fibre. Both vat and sulphur dyes are oxidized to return to their initial insoluble form within the fibre (Chakraborty, 2011).

According to the primary medium used to dissolve and transport dyes, dyeing can be divided into water-based dyeing and waterless dyeing. Digital printing on textiles, while not entirely waterless, also reduces water usage by applying dyes directly to the fabric using inkjet technology, minimizing excess water and chemical waste.

Although these technologies are discussed, they cannot replace the conventional water-based dyeing technology due to various limitations of application. Therefore, although these new technologies have been of particular interest to many researchers over decades, we focussed mainly on water-based optimisation of knit dyehouse pathways towards ‘zero’ water requirements, and our dyeing technologies are predominantly water-based; we will discuss water-based dyeing in detail in the following sections.

3 Water-Based Dyeing

Water-based dyeing technology refers to the process of applying dyes and pigments to textile substrates using water as the primary medium to dissolve and transport dyes. This method involves submerging textiles in a water-based dye bath, where the dye molecules interact with the fibres under controlled conditions, including temperature, time, and pH. The water acts as a solvent that facilitates the dye’s dissolution, exhaustion, adsorption, diffusion, and fixation onto the textile substrate. Water-based dyeing is the most common method used in the textile industry due to its effectiveness in producing various colours and finishes on different types of fibres. Water-based textile dyeing methods can be categorized into three main types: continuous, semicontinuous, and exhaust. The total dyeing time required in actual production depends on several factors, such as, nature of textile material to be dyed, type of colour, depth of shade, type of dyestuff used, and dyeing techniques and machine employed. Colouration involves multiple stages, including pretreatment, dyeing, and finishing. Multiple washing and rinsing cycles are required to remove processing chemicals and unfixed and/or hydrolysed dyes from the substrate; the process is water-intensive.

Pretreatment The textile materials need to be ready for the subsequent dyeing colouration section. Thus, materials undergo several processes collectively called pretreatment, which includes singeing/desizing for woven fabric, scouring/bleaching for fibre, yarn, knit and woven fabric, garment and mercerising optionally for both woven and knit fabric.

Dyeing After pretreatment, the ready to dye fabrics go through dyeing, either in batch or continuous methods, using exhaust or padding techniques.

Finishing Finishing involves treatments applied after dyeing to enhance textiles' appearance, feel, and performance. Generally, knit fabrics go through slitting, stentering, and drying after the completion of dyeing. Common finishing processes include softening, resin treatments, or water-repellency applications, depending on the end-use requirements. The finishing stage may also include mechanical processes, such as calendaring or heatsetting, to stabilize the fabric dimensions and surface characteristics.

3.1 Exhaust Dyeing Method

Exhaust dyeing, also known as batch dyeing, is where a known amount of textile substrate has been dipped or in touch with a known amount of liquor, which contains dyes, chemicals, and auxiliaries for a certain time under temperature and pressure with or without circulation of materials and/or liquor or both. Dye gradually moves onto the surface of the substrate, known as 'adsorption,' and to the inside substrate, known as 'absorption'. This process of gradual movement of dyes from liquor to substrate is called 'Exhaust' dyeing. Over time, a steady state is achieved and dye–fibre–liquor system is saturated, and it is called 'Exhaustion' is complete in a given dyeing scenario. Eventually, dye gets fixed with the fibre in various mechanisms, such as salt linkage, covalent bond, hydrogen bond, physical forces, or mechanical entrapping inside the fibre, known as 'Fixation'. The rate of dyeing—adsorption, absorption, exhaustion, and fixation is controlled by temperature and pressure, liquor and/or substrate movement, pH of the bath, use of chemicals such as electrolytes, and levelling agents. The typical flow chart of exhaust dyeing of cotton fabric is shown in Fig. 2.

The exhaust dyeing method is carried out in various machines, as shown in Fig. 3. The winch dyeing machine can be used for medium to heavyweight fabric in rope form to withstand tension created during the dyeing cycle. The jet dyeing machine imposes much less lengthway tension than the winch machine and is therefore used for delicate or synthetic fabrics, which can be damaged under high tension. Jigger dyeing machine is used for woven fabric, mostly in local settings, or for washing only after scouring, bleaching, or dyeing.

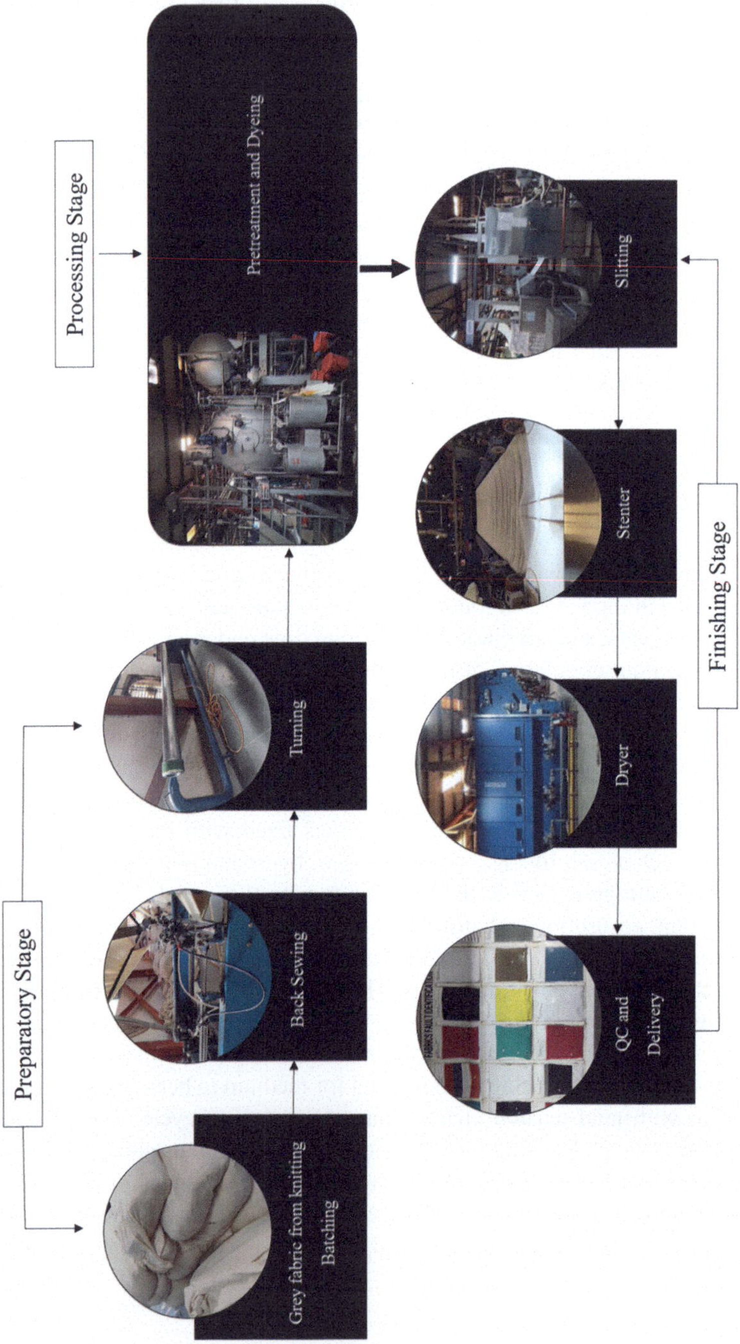

Fig. 2 Commonly used sequence of water-based cotton knit fabric dyeing

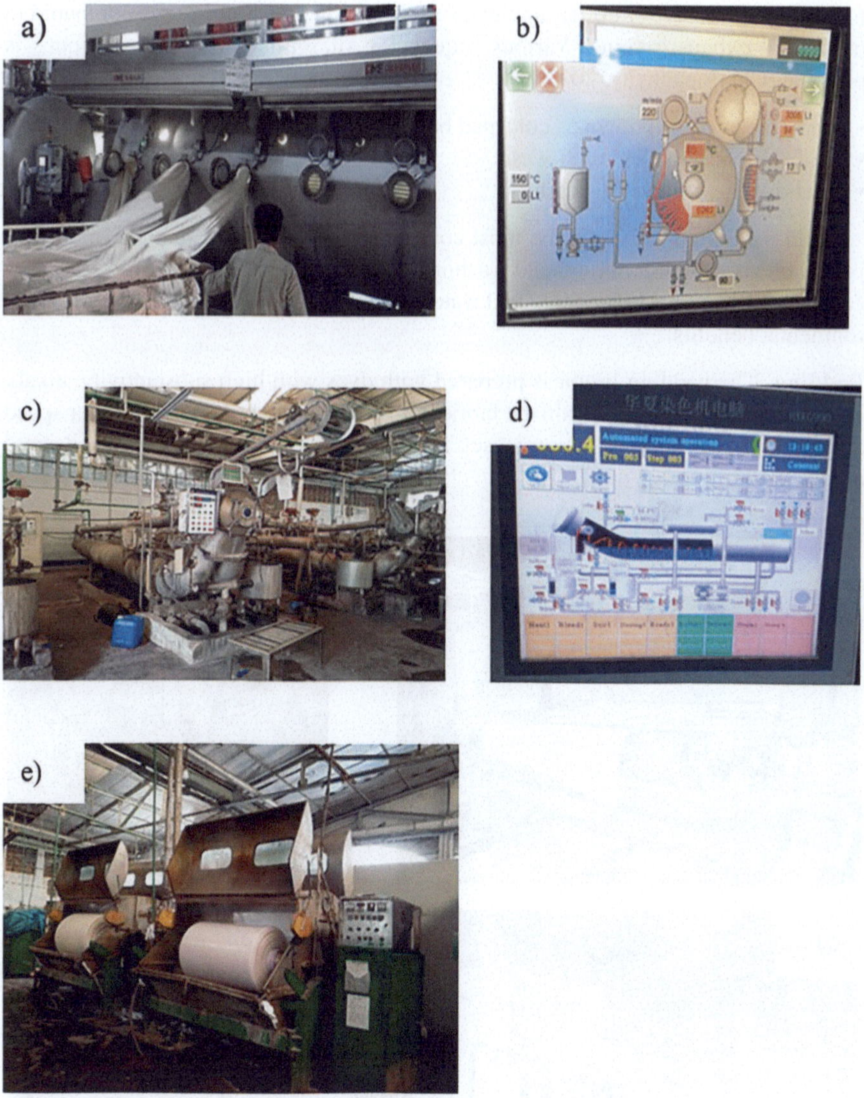

Fig. 3 Various batch dyeing machines for knit—(**a**) fabric loading in Winch dyeing machine, (**b**) control panel of winch dyeing machine, (**c**) Jet dyeing machine, (**d**) control panel of jet dyeing machine, (**e**) Jigger dyeing machine

3.2 Semicontinuous Dyeing Method

Semicontinuous dyeing combines exhaust and pad dyeing features, making it efficient for specific fabric types such as woven and knit materials. In this method, the fabric is first impregnated with dye using a padding machine and then stored or aged to allow the dye to gradually fix to the fibres before washing. This approach allows for precise control over colour application while reducing water usage compared to traditional exhaust dyeing. Various sequencing of semicontinuous techniques is possible (Chattopadhyay, 2011).

- Pad-batch-wash (known as cold-pad batch).
- Pad-dry-pad (alkali)-batch-wash.
- Pad-dry-jigger develop-wash.

Cold-pad batch dyeing is the most common semicontinuous method, including pretreatment, padding, fixing, and washing steps as shown in Fig. 4. This process is economical, requires less energy and water, and is known for its flexibility and environmental benefits.

Padding The padding liquor is prepared with dyes with high substantivity, alkali, and other auxiliaries applied on the fabric with the padding mangle. The roller speed and padding mangle pressure is set based on the fabric quality and desired colour shade.

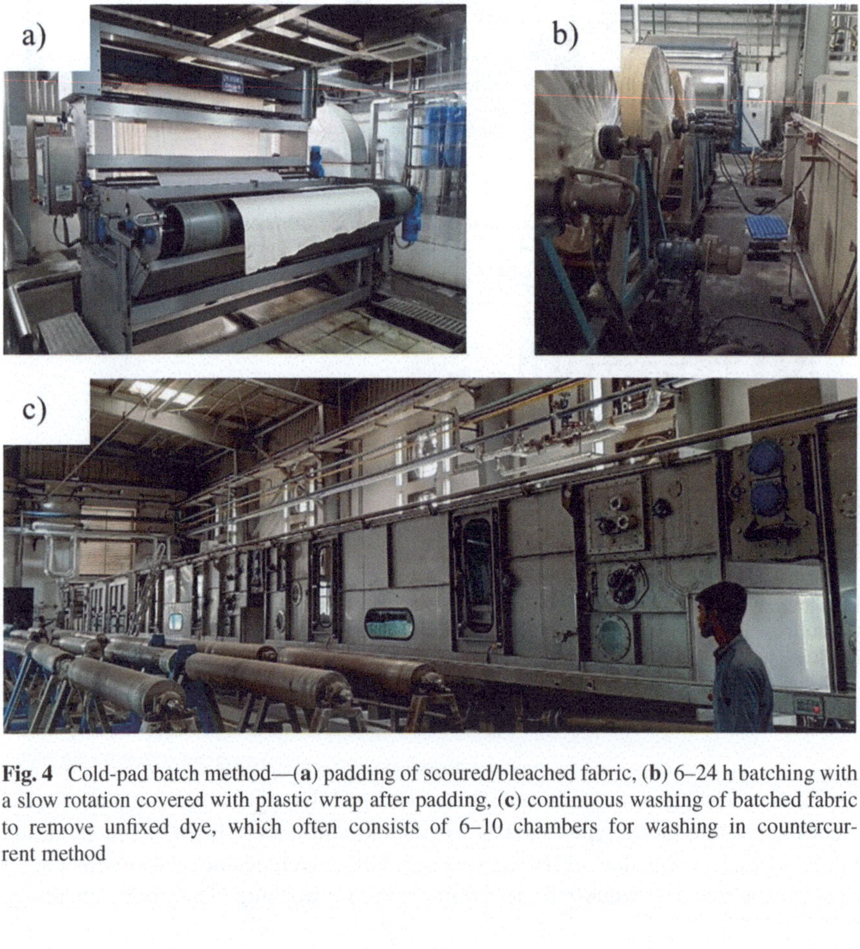

Fig. 4 Cold-pad batch method—(**a**) padding of scoured/bleached fabric, (**b**) 6–24 h batching with a slow rotation covered with plastic wrap after padding, (**c**) continuous washing of batched fabric to remove unfixed dye, which often consists of 6–10 chambers for washing in countercurrent method

Batching Fixing: The fabric is rolled up while coming out of the padding mangle and placed in the rotating batching machines for the fixation of the applied dyes. Usually, the batching time is 6–24 h, depending on the dyes, fabric, and shade types.

Washing After the dye fixation is completed in the batching process, the fabric is washed off in a continuous dyeing machine to remove the unfixed dyes. The washing machine typically contains 4–8 wash chambers, and the water consumption is 10–12 L/kg depending on the shade percentage of colour.

3.3 Continuous Dyeing Method

In continuous dyeing range includes a setup with a pretreatment range such as singeing-desizing machine, scouring-bleaching machine with washing, optional mercerising, dyeing range such as pad-dry-pad-thermosol or pad-dry-steam dyeing range, Fig. 5, and finishing range such as stenter, sanforising, and other dry finishing application. Continuous dyeing can process fabrics at speeds ranging from 50 to 250 m/min, making it highly efficient for mass production of woven fabrics. In many ways, it is similar to a semicontinuous process where the process continues one after another in different machines, whereas in the case of cold-pad batch, dyeing is carried out in padding followed by an extended batching technique. Continuous method is known for its efficiency, consistency, and cost-effectiveness, especially when dealing with large volumes of fabric.

In the continuous method, an open-width fabric roll passes continuously through a padding trough containing dye liquor and then through a padding mangle, which controls the amount of dye percentages to be absorbed for the required shade of colour. The padding mangle is critical in shade uniformity from selvedge to selvedge or centre to selvage shade variation. The desired wet pick-up can be controlled by dipping depth in the padding trough and squeezing pressure in padding mangles, the amount of dyes in dye liquor, added auxiliaries in liquor and speed of the fabric. The dye fixation is achieved by drying and curing or steaming by passing through a variety of sequence, such as pad-dry-fix dyeing, pad-dry steam dyeing method), or pad-steam dyeing methods.

Various sequences are possible in this technique, such as (Chattopadhyay, 2011).

- Pad-dry-pad (alkali)-steam-wash.
- Pad-dry-cure-wash.
- Pad-dry-hot alkali pad-wash.
- Pad-dry thermosol.

Pad-dry-thermosol machine follows the pad-dry-cure technique where the fabric is padded first, pre-dried in infrared (IR) dryer, dried and cured by thermofixation, and cooled in cooling drums.

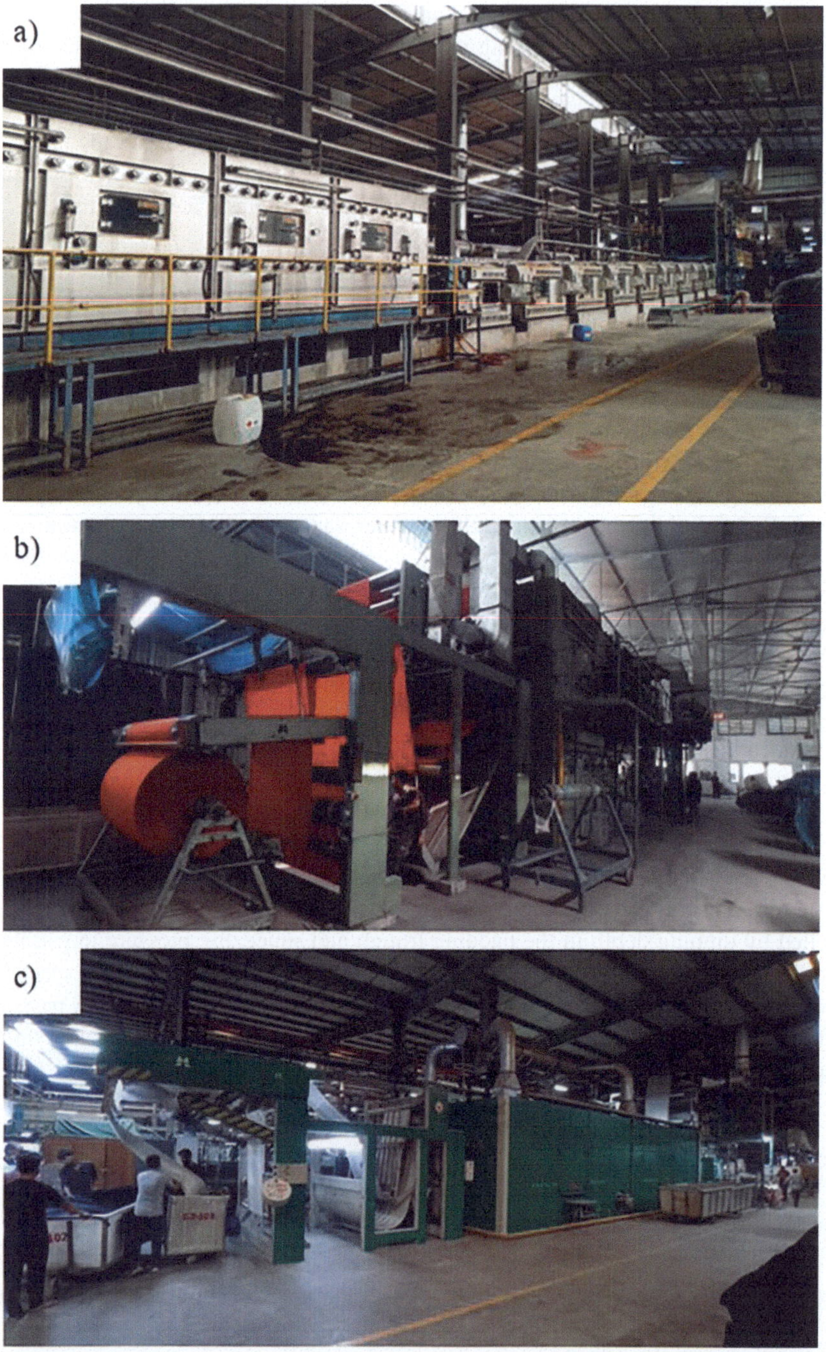

Fig. 5 Different types of continuous dyeing machine—(**a**) pad-steam machine, (**b**) pad-dry-steam machine, (**c**) pad-dry-pad-thermosol machine

3.4 Comparison Between Exhaust and Continuous Dyeing

Pad dyeing is generally more water- and energy-efficient than batch processes, requiring less time and involving fewer heating and cooling cycles. A comparison between exhaust and continuous dyeing techniques is given in Table 1. Water consumption is comparatively higher than continuous dyeing. Water consumption is relatively less than exhaust dyeing.

Table 1 Comparison between exhaust and continuous dyeing

	Exhaust dyeing	Continuous dyeing
Meaning	Exhaust dyeing, immerse textiles in a dye bath, where absorption is controlled by adjusting time, temperature, and pH	Continuous dyeing involves passing fabric through a dye solution followed by fixation
Machines used	Winch dyeing, jet dyeing, jigger, beam dyeing, package dyeing, tops dyeing, to name a few	Pad-dry-thermosol, pad-steam, pad-dry-cure, pad-dry-pad-steam
Productivity	Each batch can be 50–2000 kg, depending on the machine size. Pilot dyeing from 1 kg is possible	Suitable for medium to large-scale production from 1000 m and above. Pilot dyeing of 20 m is possible in sample machine
Fabrics dyed	Most suitable for delicate, specialty, or small-batch fabrics such as knits, wool, silk, and certain synthetic blends	Most suitable for woven fabrics.
Form of dyeing	Fabric is usually in rope form but can be open width (e.g., jigger), also in fibre, yarn, and garment form	Open-width fabric, open-width yarn (in denim dyeing)
Affinity of dyes towards substrate	Good affinity dyes are preferred for exhaust dyeing.	Less affinity dyes are preferred for padding.
Pretreatment and after-treatment of the fabric	Pretreatment, dyeing, and after-treatment can be done using the same machine	Pretreatment happens in separate machines: Singeing and desizing, scouring and bleaching, and mercerising. Aftertreatment is carried out using a continuous washing machine.
Advantages	Exhaust dyeing, typically used for smaller batches and complex shades, offers deep dye penetration and uniformity	Continuous process is favoured for its speed and efficiency in large-scale production. High colour yield is possible
Challenges	Formation of crease marks, uneven dyeing, hairiness of surface, the tension in the fabric causing high shrinkage value, high consumption of water, steam, chemicals, dyestuff, use of salt and other chemicals causing problems in effluent treatment	The uniform application of dye from liquor to the fabric is the most critical part. Listing and tailing for long-length fabric has been the most challenging, along with colour development in steaming/curing process

Table 2 Stepwise water requirements in continuous, semicontinuous, and exhaust processes

	Dyeing method		
Stages	Continuous	Semicontinuous	Exhaust
Pretreatment, L/kg	14–22	15–22	20–36
Dyeing, L/kg	~1	~1	4–20
Aftertreatment, L/kg	15–40	12–24	20–50
Total water consumption (L/kg)	**30–63**	**28–47**	**44–98**

Table 2 presents a comparison of water consumption across three dyeing methods—continuous, semicontinuous, and exhaust—measured in litres per kilogram (L/kg) of fabric processed. In the pretreatment stage, continuous and semicontinuous methods show a lower water usage range (14–22 and 15–22 L/kg, respectively) compared to the exhaust method, which consumes significantly more water (24–36 L/kg). During the dyeing stage, all three methods utilize minimal water, with continuous and semicontinuous methods consuming 1 L/kg and the exhaust method ranging from 4 to 12 L/kg. However, in the aftertreatment stage, water consumption rises significantly, especially for the exhaust method (32–50 L/kg), whereas continuous and semicontinuous methods are more efficient, using 25–40 and 12–24 L/kg, respectively. Overall, continuous dyeing is the most water-efficient, consuming 40–63 L/kg, while the exhaust method, at 60–98 L/kg, is the least efficient regarding total water usage.

4 Waterless Dyeing

Innovative waterless dyeing technologies have emerged as alternatives to traditional water-based methods, claimed to reduce water and energy consumption, and environmental pollution significantly in the textile industry. Water as 'solvent' is substituted by different other solvent media such as carbon dioxide (CO_2), air, powder, foam, and many more.

4.1 Supercritical Carbon Dioxide (CO_2) Dyeing

One of the most advanced waterless dyeing methods is supercritical carbon dioxide (CO_2) dyeing, where CO_2 is heated and pressurized beyond its critical point of 31.1 °C and 74 bar, allowing it to exhibit both liquid and gas properties (Agrawal, 2015). CO_2 is chosen as the dyeing medium due to its non-flammability, non-toxicity, and abundance. In its supercritical state, CO_2 has liquid-like densities that allow it to dissolve hydrophobic dyes effectively and gas-like viscosities that enable rapid diffusion in synthetic fibres, especially polyester, resulting in shorter dyeing times than water-based methods. This process is highly efficient and sustainable, as

CO_2 can be recovered and reused, and the fabric does not require drying due to the gaseous nature of CO_2 at room temperature (Mahmud & Kaiser, 2020). However, the high initial equipment cost and the limited compatibility with natural fibres challenge its broader adoption. The application of supercritical CO_2 dyeing is primarily limited to synthetic fabrics due to the low polarity of supercritical CO_2, which poses challenges for dyeing polar hydrophilic fibres like cotton.

4.2 *Air-Dyeing*

The air-dye process is another innovative approach to reducing water usage in textile dyeing, Fig. 6. In this method, dyeing liquor is atomized and combined with high-pressure air, then sprayed directly onto the fabric (Mahmud & Kaiser, 2020). This technique reduces water usage since the dyeing solution directly contacts the fabric with minimal reliance on water as a solvent. Beyond water conservation, the air-dye process also eliminates the risk of water pollution and the generation of hazardous waste, making it an environmentally superior alternative to traditional dyeing methods. However, the air-dye process, similar to supercritical carbon dioxide dyeing, is primarily effective on synthetic materials such as polyester and unsuitable for natural fibres.

4.3 *Electromagnetic and Plasma-Based Dyeing*

Another promising technology is electromagnetic and plasma-based dyeing, which utilizes energy from electromagnetic fields or plasma to modify the surface properties of textile fibres, enhancing their dye absorption capabilities. In

Fig. 6 Air-dyeing machine—(**a**) Imogo digital spray dyeing machine, (**b**) Control panel of Imogo digital spray dyeing machine. (Source: Reed Consultancy, Bangladesh)

electromagnetic dyeing, fields such as microwaves excite dye molecules and fibre surfaces, facilitating efficient dye penetration with minimal or no water usage. Plasma-based dyeing involves exposing fibres to ionized gases, which alter their surface at a molecular level, improving dye uptake (Zille, 2020). These methods offer significant reductions in water usage and energy efficiency, but their complexity and the technical knowledge required for operation can be barriers to widespread implementation.

4.4 *Foam Dyeing*

Foam dyeing is an innovative technique that utilizes foam to apply dyes and chemicals to textile substrates, significantly reducing water usage and improving process efficiency. Foam dyeing is suitable for synthetic and natural fibres and can accommodate various fabric structures. It also enhances colour strength and dye fixation, as demonstrated in studies using reactive dyes on cotton and acid dyes on wool. In foam dyeing, air is mixed with a dye solution containing a foaming agent to create a stable foam. The primary factors influencing the foam dyeing process were identified as the level of wet pick-up, the choice of fixation agent, and the stability of the foam stabilizer (Yu et al., 2014). This foam, consisting of tiny gas bubbles dispersed in a liquid matrix, is then uniformly applied to the fabric. The foam's characteristics—such as bubble size, stability, and viscosity, are critical to achieving consistent dye application (Hoque et al., 2023). The process begins with preparing a dye liquor, which is then aerated using a foam generator to create a foam with the desired properties. This foam is applied to the fabric using direct or indirect applicators, ensuring thorough penetration and even distribution of dye. After application, mechanical processes such as drying and fixation are performed to set the dye.

Foam dyeing uses a foam matrix to carry dye and chemicals onto the fabric, spreading them evenly with minimal water use and reducing the energy required for drying. One such technology is Oasis foam dyeing (Heller, 2015). Oasis is a special pretreatment on cellulosic fabrics that enhances the dye attraction to the fabric substrate. It allows the fabric to dye without salt or alkali and requires less energy.

5 Water Consumption in Exhaust Dyeing for Cotton Knit Fabric

The discussion of water consumption in exhaust dyeing is paramount because of our focus on optimisation and the predominance of the cotton knit dyeing sector in Bangladesh. According to a recent study, the water required to process 1 kg of textile materials was as high as 285 L/kg for a knit dyeing factory (Uddin et al., 2023). Water consumption in exhaust dyeing for knit fabric is influenced by different factors. The key factors are dyeing machine technology, colour range, and colour depths. The type of dyeing machine, such as those with innovations in liquor ratio and fabric tension, significantly impacts water usage. Additionally, the dominance of dark shades in production, driven by market demand and colour fastness, and the varying shade percentages of dyes used, from light to extra dark, play a crucial role. Finally, the number of dyeing steps based on colour depth directly correlates with water consumption, highlighting the need for optimized processes to reduce water use.

5.1 Dyeing Machine Factor

A variety of machine brands is used, which employ exhaust dyeing in rope form with a variety of innovations in liquor ratio, minimum tension on textiles, movement of liquor, amount of liquor to be used for processing, special washing or rinsing technique, energy efficiency, dosage of dye and chemicals, and operation control. Some of the most popular machine brands are Athena of Sclavos (Greece), Signature and iMaster series of Thies (Germany), Chronoflow HT of MCS (Italy), Eco, Then and Allfit of Fongs (Hong Kong), TECHSmart of Canlar (Turkey), HT of Dilmenler (Turkey), as shown in Fig. 7.

Another dyeing technology was invented based on the jet system, which has an integrated mechanical transport system instead of water and can run with a low to extremely low liquor ratio. A conveyor drive system is used to move the fabric forward, so water is only needed for pipe circulation of fabric and its dye pick up, shown in Fig. 8. The conveyor drive ensures constant speed of the fabric flow, improving dyeing evenness with less water and low tension. It is claimed that some of these types of machines can operate with a liquor ratio as low as 1:2.5, which, in practice, is found to be in the range of 1:3–1:4. A shower rinsing system ensures lower water use in rinsing processes. The machines have a modular approach; each tube can handle 250–300 kg, and the capacity can be increased by putting several tubes together as much as 2000 kg, which can then be run independently or together depending on the number of control panels. The machine loading capacity is based on length rather than weight; therefore, the machine can be loaded more than 100% and operate smoothly even when nozzle pressure is under 0.5–1 kg/cm^2. This system is very suitable for spandex and low-tension fabric.

Fig. 7 Different types of exhaust dyeing machines and their dyebath liquor ratio. (**a**) Dilmenler, Minimum dyebath liquor ratio used by the factories = 1:6–1:8. (**b**) Canlar, Minimum dyebath liquor ratio used by the factories = 1:4.5–1:6. (**c**) Sclavos, Athena 2, Minimum dyebath liquor ratio used by the factories = 1:5–1:6. (**d**) Sclavos, Athena 3A, Minimum dyebath liquor ratio used by the factories = 1:3–1:6. (**e**) Sclavos, Athena 4, Minimum dyebath liquor ratio, claimed ~1:3. (**f**) Thies iMaster H_2O, Minimum dyebath liquor claimed ~ 1:3. (**g**) MCS, Minimum dyebath liquor ratio used by the factories = 1:4–1:6. (**h**) Fong's ECO-38-3T, Minimum dyebath liquor ratio used by the factories = 1:5–1:8. (Source: Reed Consultancy, Bangladesh)

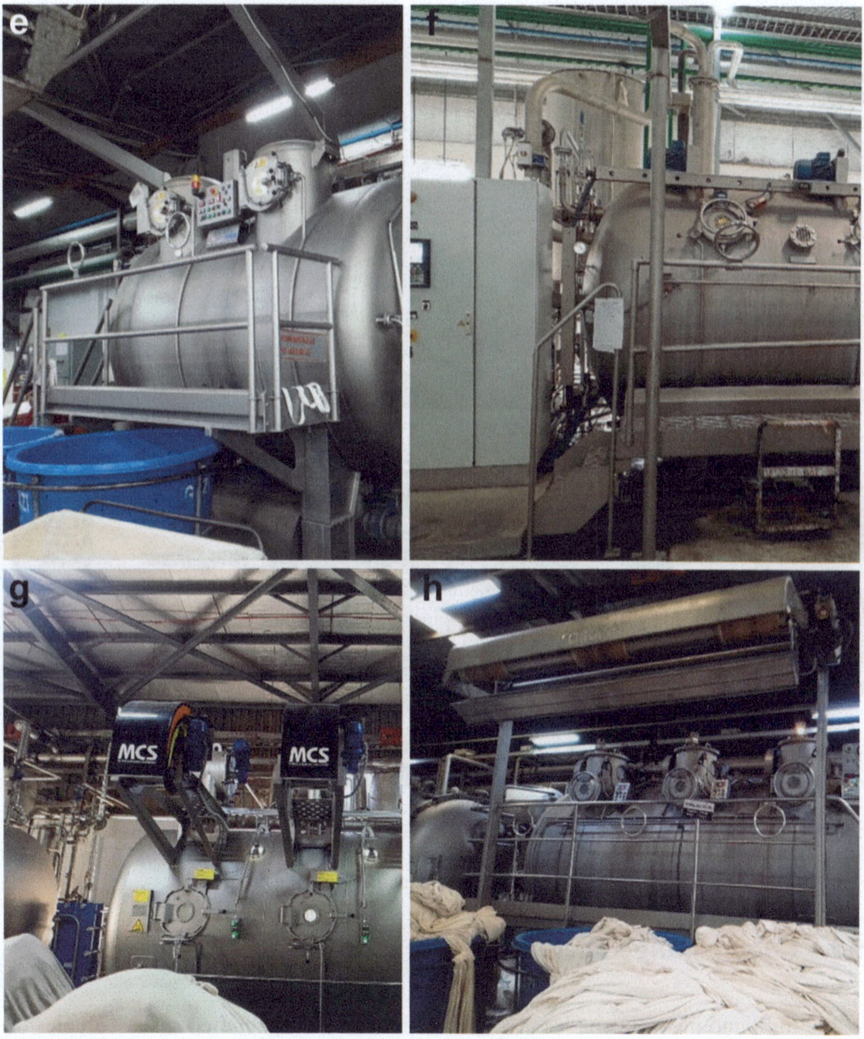

Fig. 7 (continued)

Fig. 8 Low pressure dyeing machine for very light or delicate fabric—(**a**) AK Dyeing machine from Taiwan with liquor ratio from 1:4, (**b**) ACME conveyor belt dyeing machine from Taiwan with liquor ratio from 1:2.5–1:4. (Source: Reed Consultancy, Bangladesh)

5.2 Colour Range Factor

While light and medium shades are also produced, the dominance of dark shades in the knit dyeing industry reflects the interplay of market demand and production efficiency.

The knit dyeing industry often prioritizes the production of dark shades, such as black, navy, and bleach/grey, as the dark colours, are frequently favoured in fashion trends due to their versatility and perceived elegance, especially for professional or formal attire. Darker shades generally have better colour fastness and can be used as a base for a variety of lighter shades through bleaching or dyeing techniques, allowing manufacturers to produce a wide range of colours from a smaller inventory of dyes.

The water requirement for knit dyeing of various shades for a particular factory and a particular machine with a liquor ratio of 1:6 is shown in Fig. 9.

(a)

Process Steps	L/kg
White process	6
Wash at room temperature	4
Wash at room temperature	4
Neutralization + Peroxide killing + Biopolishing	4
Wash at room temperature	4
Total	**22**

(b)

Process Steps	L/kg
Bioscouring	6
Wash at 60-90°C	4
Wash at room temperature	4
Dyeing	4
Wash at room temperature	4
Wash at room temperature	4
Wash at room temperature	4
Neutralization + Soaping	4
Wash at 60-90°C	4
Wash at room temperature	4
Wash at room temperature	4
Wash at room temperature	4
Total	**50**

(c)

Process Steps	L/kg
Polyester part dyeing + Acid based reduction	6
Wash at room temperature	4
First reduction clearing	4
Second reduction clearing	4
Cotton dyeing with enzyme	4
Rinse wash at room temperature	4
Rinse wash at room temperature	4
Wash at room temperature	4
Neutralization + Soaping	4
Wash at 60-90°C	4
Wash at room temperature	4
Wash at room temperature	4
Wash at room temperature	4
Fixing	4
Total	**58**

Fig. 9 Processing steps in knit dyeing factory of dominant colours and corresponding water consumption in a liquor ratio of 1:6—(**a**) White colour for cotton fabric, (**b**) Black or Navy colour for cotton fabric, (**c**) Black or navy colour for cotton-polyester blended fabric

Table 3 Typical standard of colour depth with associated shade% and average production in those colour depths in knit factory

Colour depth	Initial dye concentration or Shade% (owf)	Average dyeing production percentage in knit factory
White with or without OBA	<0.5	~27
Light	>0.5	~16
Medium	>1.5	~6
Dark	>3.0	~33
Extra dark	>5.0	~18

5.3 *Colour Depths or Shade Percentage Factor*

Depending on the colour depths or shade % requirements, textile materials to be dyed are divided into four categories—white, light, medium, and dark. Shade percentage (%) is fundamental to any textile dyeing, which defines as the amount of dye required in kg (or gram) to produce a certain colour depth for 100 kg (or 1 kg) of textile substrate. For example, 2% shade means 2 kg of dye will be required to produce a certain colour on 100 kg of fabric, expressed as 2% on the weight of the fabric or simply 2% owf. Shade % is always the initial concentration of dye to be used for dyeing, and there is no relation with fixation or how much dye is bonded with the textile fibre. Relation with shade % to light, medium, and dark varies depending on the types of dyes used from the factory to factory, as not all dyes in the same concentration provide the same colour. Based on experience, in reactive dyeing of cotton fabric, typical colour depth with associated shade % along with dyeing production percentage based on colour depth in knit factory is given in Table 3.

Table 4 Water requirements in various colour depths

Process	Material liquor ratio at different colour depths		
	Light	Medium	Dark
Scouring and bleaching	1:6	1:6	1:6
Wash at 60–90 °C	1:4	1:4	1:4
Rinse wash at room temperature	1:6	1:6	1:6
Peroxide cleaning, neutralizaion and enzyme	1:4	1:4	1:4
Rinse wash at room temperature	1:4	1:4	1:4
Dyeing	1:4	1:4	1:4
Rinse wash at room temperature	–	1:8	1:8
Wash at 60–90 °C	1:4	1:4	1:4
Rinse wash at room temperature	1:8	1:8	1:8
Rinse wash at room temperature	–	–	1:6
Acid wash	1:4	1:4	1:4
Soaping	1:4	1:4	1:4
Rinse wash at room temperature	1:4	1:4	1:6
Rinse wash at room temperature	–	–	1:4
Softening/fixing	1:4	1:4	1:4
Total	56	64	76

The colour depth in a dyehouse is important as it determines the number of steps to produce that particular colour and is also responsible for number of reprocessing. Both of these factors hugely influence water consumption in textile processing. For example, as given in Table 4, for dyeing of cotton fabric, typical processing for light depth requires 12 steps with a theoretical water consumption of 56 L/kg, medium shade requires 13 steps with a theoretical water consumption of 64 L/kg, and dark shade consists requires 15 steps for a theoretical water consumption of 76 L/kg.

A recent study on more than 20 water-based knit dyeing factories has collected the water consumption data of the dyeing process, analysed the consumption stages and causes of inefficiencies of those stages, and recommended an improvised process flow with less water consumption (RCB, 2024). The significance of the study is that rather than focusing on drastic infrastructural changes, it emphasizes the existing setup with the usage of multifunctional chemicals and auxiliaries, optimizing process steps and reusing the processed water.

6 Optimisation Possibilities in Knit Dyeing Factories

The dyeing of a knit fabric with the exhaust dyeing method includes up to 15 processing steps depending on the colour and type of the fabric. It may consume up to 90 L/kg of water depending on the number of required processing steps, the type of machine, and the fabric loading percentage. Water consumption in knit fabric dyeing factories can be reduced through process optimization measures such as shortening the processing step, reducing the liquor ratio, reusing process step baths in multiple batches, and implementing semicontinuous or continuous dyeing technologies. The process optimization measures are discussed below.

6.1 *Shortening of Processing Step*

Reducing processing steps is a viable option in exhaust knit dyeing process optimization to reduce water and other resource consumption. The processing steps can be reduced by eliminating some optional steps in pretreatment, combining multiple processing steps in a single bath with multifunctional or special chemicals and enzymes with special features, and reducing the rinsing or washing steps. The following industry best practices for exhaust knit dyeing can be implemented in the knit dyehouses to reduce processing steps:

1. No scouring/bleaching or semi-scouring in pretreatment;
2. Implementation of bioscouring in pretreatment with or without bleaching;
3. Use of dyebath/acid cellulase enzyme for biopolishing;
4. Neutralization, peroxide killing, and dyeing after pretreatment in the same bath;
5. Reduction clearing of polyester fabric in the dye bath after dyeing with disperse dyes (Reduction clearing is a process of removing the unfixed disperse dyes from the polyester fabric with reducing agents);
6. Cotton-polyester dyeing in the same bath;
7. Reduction of wash/rinse steps after pretreatment, dyeing, and aftertreatment.

6.1.1 Implementation of Bioscouring in Pretreatment with or Without Bleaching

Typically, the scouring of cotton fabric is carried out with sodium hydroxide (NaOH), which increases the pH of the fabrics, so they need to be neutralized. This scouring can also be carried out with some enzymes that will not require neutralization and can be washed off easily. The application of enzymes in the scouring of cotton is known as bioscouring. Pectinase, protease, and α-amylase enzymes are used in the bioscouring process to remove the impurities of the cotton fibre and to improve water absorbency. According to industry practice, this process is suitable for dyeing medium to dark shade colours, which does not require bleaching.

The conventional scouring process requires four processing steps, each consuming a single bath, whereas the bioscouring process requires only two processing steps, Fig. 10, each consuming a single bath in the pretreatment before dyeing.

(a)

Process Steps
Scouring
Wash at 60-90°C
Wash at room temperature
Neutralization

(b)

Process Steps
Bioscouring
Wash at room temperature

Fig. 10 Scouring of cotton fabric—(**a**) conventional process, (**b**) with bioscouring process

(a)

Process Steps
Scouring-bleaching
Wash at 60-90°C
Wash at room temperature
Neutralization
Peroxide killing
Neutralization
Wash at room temperature
Dyeing

(b)

Process Steps
Scouring
Wash at 60-90°C
Wash at room temperature
Neutralization

(c)

Process Steps
Demineralization/Detergent wash
Wash at 60-90°C
Wash at room temperature

Fig. 11 Pretreatment of cotton fabric before dyeing—(**a**) conventional, can be used for all purposes, (**b**) without bleaching applicable for medium to dark colour, (**c**) without scouring-bleaching, with only detergent wash, applicable to extra dark colour such as black

6.1.2 No Scouring/Bleaching or Semi-Scouring in Pretreatment

Raw cotton fibres contains impurities such as fats, waxes, colours, and other organic compounds, which is removed through scouring and bleaching process to increase the absorbency of the cotton fabrics and to increase dye uptake. Scouring and bleaching are carried out with alkali, such as, sodium hydroxide (NaOH), and bleaching agent, such as hydrogen peroxide (H_2O_2), which consume many chemicals and water. In Fig. 11, it can be seen that a typical scouring and bleaching process involves five steps, each requiring one bath of water. The first bath treats the fabric with sodium hydroxide and hydrogen peroxide to remove the impurities from the fabric, the second bath washes the fabric with hot water, the third bath washes the fabric with room temperature water, the fourth bath treats the fabric with a reducing agent to deactivate the residual hydrogen peroxide in the fabric, the fifth bath treats the fabric with acid to neutralize its pH, and the sixth bath conducts a wash to clean it before dyeing.

Figure 11 shows that the conventional pretreatment process can be optimized for dark shades to reduce the process steps by replacing the treatment bath with sodium hydroxide and hydrogen peroxide with a detergent or demineralization wash. Eliminating the sodium hydroxide and hydrogen peroxide treatment bath reduces the peroxide killing and neutralization steps in the pretreatment. In some cases, the use of bleaching chemical, hydrogen peroxide can be avoided in the pretreatment bath for dark colours, which will also eliminate the peroxide killing step. It is being

(a)

Process Steps
Scouring-bleaching
Wash at 60-90°C
Wash at room temperature
Peroxide killing
Neutralization
Wash at room temperature
Biopolishing
Wash at room temperature
Dyeing

(b)

Process Steps
Scouring-bleaching
Wash at 60-90°C
Wash at room temperature
Peroxide killing
Neutralization
Wash at room temperature
Biopolishing-dyeing

Fig. 12 Biopolishing process of cotton fabric—(**a**) conventional process with acid or neutral cellulase enzyme, (**b**) combined biopolishing and dyeing process with neutral enzyme

implemented in many dyehouses to save water, energy, chemical, and time consumption in cotton knit fabric dyeing.

The pretreatment without scouring and bleaching requires only three processing steps, consuming three water baths, which is around 14–20 L/kg. The pretreatment without bleaching requires four processing steps, consuming four water baths and around 18–26 L/kg of water. Both process step reduction measures in pretreatment for extra dark shade dyeing of knit fabric significantly reduce water consumption.

6.1.3 Use of Dyebath/Acid Cellulase Enzyme for Biopolishing

Biopolishing is an enzymatic treatment process that improves the handfeel by removing protruding fibres from the surface of cellulose-based fabric. Two types of cellulase enzymes are used for biopolishing—acid cellulase enzyme operates within the pH range of 4.5–6, and neutral cellulase enzyme operates within the pH range of 6–8 (Sheikh & Bramhecha, 2018). In the conventional process, a separate bath is used for biopolishing, followed by another washing bath during pretreatment. Combining this biopolishing step with dyeing can save these two baths in the pretreatment. The neutral cellulase enzyme needs to be used in the combined dyeing and biopolishing process. Overall, it reduces the process of baths from eight to six in pretreatment with biopolishing, as shown in Fig. 12.

In the combined biopolishing and dyeing process, the fabric is treated with the neutral cellulase enzyme at the beginning, and the dyeing process starts in the same bath after the enzyme treatment is completed. The water consumption in pretreatment with conventional biopolishing is around 34–50 L/kg, depending on the liquor ratio of the machine and the loading percentage of fabric in the machine loading. Meanwhile, the combined biopolishing and dyeing process consumes around 26–38 L/kg of water.

(a)

Process Steps
Scouring-bleaching
Wash at 60-90°C
Wash at room temperature
Peroxide killing
Neutralization
Wash at room temperature
Dyeing

(b)

Process Steps
Scouring-bleaching
Wash at 60-90°C
Wash at room temperature
Peroxide killing + Neutralization + Dyeing

Fig. 13 Pretreatment and dyeing process—(**a**) conventional steps, (**b**) combined process including peroxide killing, neutralization, and dyeing steps in a single bath

6.1.4 Neutralization, Peroxide Killing, and Dyeing After Pretreatment in Same Bath

Pretreatment of cotton fabric with sodium hydroxide and hydrogen peroxide requires the deactivation of the residual hydrogen peroxide with reducing agents and neutralization with the acid of the fabric before dyeing. Conventionally, the peroxide killing, neutralization, and dyeing are carried out in separate process baths, which can be combined to shorten the process steps, Fig. 13.

This process can be modified by combining the peroxide killing, neutralization, and dyeing steps in a single bath. In this combined process, the dyeing is started in the same bath after the peroxide killing and neutralization. The pH and residual peroxide of the bath are checked before starting the dyeing process. This combined process shows acceptable results, and water consumption reduces to 18–24 L/kg from 30 to 42 L/kg in conventional process.

6.1.5 Dyebath Reduction Clearing in Polyester Part Dyeing

The polyester fibre is dyed with disperse dyes that require reduction clearing to remove the unfixed dyes after the dyeing step. This reduction clearing is carried out for polyester or its blend. Conventional reduction clearing typically uses sodium dithionate under alkaline conditions, which is sensitive to air oxidation, leading to excess usage and environmental concerns (Kiruthika et al., 2019). The conventional process requires an alkaline condition, which needs to be neutralized after clearing. The process steps for polyester dyeing and reduction clearing can be reduced if the reduction clearing is carried out in the polyester dyebath. Acidic reduction clearing agents, alternatives to sodium dithionate, can be used in the same acidic dyebath of polyester. Applying these acid-reduction clearing agents also reduces additional steps for pH change after-treatment. The combined steps of reduction clearing reduce processing steps, Fig. 14, which can be reduced up to 8 L/kg water.

(a)

Process Steps
Polyester dyeing
Reduction clearing
Wash at 60-90°C
Wash at 60-90°C
Neutralization

(b)

Process Steps
Polyester dyeing-Reduction clearing
Wash at 60-90°C
Wash at room temperature
Neutralization

Fig. 14 Polyester dyeing—(**a**) with conventional reduction clearing step, (**b**) with reduction clearing at dyebath without bathdrop

6.1.6 Cotton-Polyester Dyeing in the Same Bath

Cotton-polyester blended fabrics contain two fibres whose dyeing cycle and dye type are entirely different. Therefore, the conventional approach requires dyeing these two fibres separately. The polyester part is dyed first, followed by reduction clearing, then dyeing of the cotton part takes place, followed by neutralization, soaping, and fixing, Fig. 15. Single bath dyeing of the cotton and polyester parts of the cotton-polyester blended fabrics shortens the process and saves water. Using compatible reactive and disperse dye combinations and auxiliaries, processing steps can be reduced from 17 to 13.

6.1.7 Reduce Wash Steps After Pretreatment, Dyeing, and Aftertreatment

In the exhaust dyeing of knit fabrics, reducing the number of wash steps to an optimum number is a best practice to minimize water consumption. There could be various options, such as:

1. Use of high fixation dyes, e.g. 80–90% fixation rate, which leaves fewer unfixed dyes after dyeing and requires less water to wash off;
2. Use of a strong alkali-neutralizing agent that neutralizes the fabric from its core in the pretreatment, which requires less washing steps after soaping;
3. Use of block wash instead of rinse wash as a 5 min rinse wash consumes around 9–10 L/kg of water in a winch dyeing machine, which, with a block wash, can be carried out with 4–5 L/kg of water;
4. Changing the sequence of polyester and cotton dyeing, which also reduces the number of steps of washing in blended dyeing, as shown in Fig. 16;
5. Strong monitoring of the dyeing process and recipe card, as it is often seen in the dyeing machine operators carry out some extra wash steps to improve the quality of dyeing without understanding the impact.

(a)

Process Steps
Scouring-bleaching
Wash at 60-90°C
Wash at room temperature
Neutralization
Polyester dyeing
Reduction clearing
Wash at 60-90°C
Wash at room temperature
Neutralization
Cotton dyeing
Bath drop wash
Bath drop wash
Neutralization
Soaping
Wash at 60-90°C
Wash at 60-90°C
Fixing and softening

(b)

Process Steps
Scouring-bleaching
Wash at 60-90°C
Wash at room temperature
Neutralization
Polyester and cotton dyeing
Bath drop wash
Bath drop wash
Bath drop wash
Soaping and reduction clearing
Wash at 60-90°C
Wash at 60-90°C
Wash at room temperature
Fixing and softening

c)

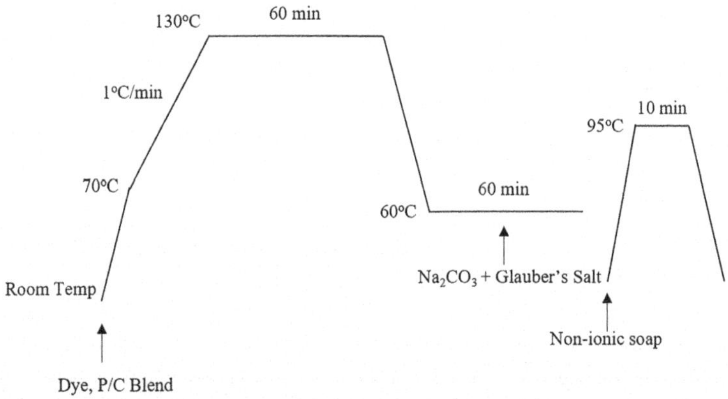

Fig. 15 Cotton-polyester blended dyeing—(**a**) conventional process where dyeing of two fibres carried out separately, (**b**) single bath process where polyester and cotton blended fibre is dyed together, (**c**) dyeing curve of single bath combined process with both reactive and disperse dyes

(a)

Process Steps
Scouring-bleaching
Wash at 60-90°C
Wash at room temperature
Peroxide killing + Neutralization
Polyester dyeing
Reduction clearing
Wash at 60-90°C
Neutralization
Cotton dyeing

(b)

Process Steps
Polyester dyeing
Scouring-bleaching
Wash at 60-90°C
Wash at room temperature
Peroxide killing + Neutralization
Cotton dyeing

Fig. 16 Polyester-cotton blend dyeing—(**a**) conventional process with pretreatment first, followed by polyester-part dyeing, reduction clearing, and cotton-part dyeing in separate baths, (**b**) modified sequence, where polyester-part is dyed first followed by scouring-bleaching and dyeing of cotton part

Process Steps
Cotton dyeing
First bath drop wash
Second bath drop wash
Third bath drop wash
Neutralization
Soaping
Wash at 60-90°C
Wash at 60-90°C
Fixing and softening

Fig. 17 Reuse of wash bath from last to first bath after storing in a reservoir

6.2 Reduction of Liquor Ratio

The liquor ratio is another fundamental point for water consumption, which varies depending on the machine, types of fibres, washing type, loading percentage of substrates in the machine. The liquor ratio can be reduced by:

1. Use of low liquor ratio dyeing machines;
2. Increase fabric loading percentage in the machine to 80% or 90% of its loading capacity, as there are usually minimum water requirements in any machine.

6.3 Reuse of Process Baths

There are many options to reuse the process baths in the exhaust dyeing process of the knit fabrics, which can reduce significant water consumption in the dyehouses.

1. Reuse of the soaping bath in multiple batches of the same colour if stored in a reservoir;
2. Reuse of the wash bath, particularly the last wash bath in a sequence, Fig. 17, could be the first washing step of subsequent batches of the same colour.

6.4 Use of Multifunctional Chemicals

Multifunctional chemicals can perform multiple functions and, thus, reduce process steps, leading to water saving. Many examples of using multifunctional chemicals to reduce water in the dyehouses exist, for example:

1. Chemicals that combine several wash steps;
2. Soaping agent that can work in a saline dyebath after dyeing, or acidic environment that can also work for neutralization;
3. Biopolishing enzyme that can perform both biopolishing and peroxide killing in a single bath.

6.5 Reuse of Cooling Water of Dyeing Machines

In the knit fabric dyehouses, dyeing polyester or polyester-blend is carried out at a higher temperature of about 130 °C in the dyeing machines. After the end of the dyeing process, to drain the liquor, the tempearature has to be reduced to about 80 °C through circulating freshwater via heat exchanger attached to the dyeing machine. In the process, freshwater gets warmer, with a temperature range around 50–60 °C, which is generally wasted in the drain. This cooling water is high-quality soft water that can be used in the processes or for any other use.

6.6 Ensuring RFT

Ensuring Right First Time (RFT) dyeing is crucial for maintaining high-quality and consistent dyeing outcomes. Key factors in achieving RFT are optimized processes, standardized work culture, appropriate dyestuff selection, suitable machinery, and thorough documentation. Monitoring substrate quality, water pH, hardness, and precise process control is essential to maintain reproducibility. Quality infrastructure and practices, such as regular equipment calibration, proper storage, handling of dyes and chemicals, and adherence to standardized procedures, are important in preventing dyeing defects. Optimized parameters such as liquor ratio, machine parameters, and dosing systems significantly impact dye uptake and shade consistency. Ensuring substrate readiness, including absorbency, pH balance, and base whiteness, is essential for uniform dyeing.

The checkpoints in machine/process that need to be considered:

- Optimum liquor ratio.
- Rope length.
- Correct weight of fabric.
- Optimum speed of machine.

- Even pressure on heads.
- Reel speed.
- Even magnet sensor reading.
- Dosing system.
- Calibration of the dosing system, tank/side tank, temperature sensor, steam and compressed air pressure.
- Machine cleaning.

Consistency between lab and bulk dyeing ensures that laboratory parameters translate effectively into bulk production, maintaining shade accuracy and quality. By focusing on RFT, the industry can improve production quality, reduce reprocessing rates, optimize resource utilization, and ensure timely delivery, ultimately lowering costs and boosting efficiency.

Several major quality issues may arise during exhaust dyeing, including uneven dyeing, dye spots, layering and haziness, rope-to-rope variations, and white patches. Dyeing faults cause significant fabric waste of around 10% of the fabrics. For example, uneven dyeing requires reprocessing, which is complicated because it involves removing the colour through stripping. In the case of colour spots, if the number of spots is small, the fabric can be used by separating the affected panels before sewing. However, it is often better to re-dye the fabric to reduce waste.

7 Conclusion

The textile dyeing industry holds immense economic growth and sustainable development potential for many developing countries. However, the environmental consequences of traditional dyeing processes, characterized by excessive water and energy consumption, cannot be ignored. The path towards optimizing dyehouses, especially knit dyeing facilities, requires a comprehensive approach integrating technological advancements, resource efficiency, and sustainable practices. When decarbonising textile and apparel sector is a priority agenda globally, achieving near-zero water usage in dyeing processes is desirable and essential for staying competitive and relevant in the industry.

By focusing on innovative techniques and sustainable alternatives, dyehouses can significantly reduce their environmental impact while improving productivity and cost-efficiency. As highlighted in this chapter, optimizing dyeing processes can lead to a more responsible industry that aligns with global sustainability goals and addresses the urgent need for environmental protection. The future of textile dyeing depends on the industry's ability to innovate and adapt to these challenges, ensuring that the vibrant colours and quality textiles demanded by consumers do not come at the expense of the planet.

Ultimately, the shift towards near-zero water usage is a technical achievement and a necessary step in securing a sustainable and responsible future for textile dyeing, benefiting both businesses and the environment.

References

Agrawal, B. J. (2015). Supercritical carbon-dioxide assisted dyeing of textiles: An environmental benign waterless dyeing process. *International Journal of Innovative Research and Creative Technology, 1*(2), 201–206.

Al-Tohamy, R., Ali, S. S., Li, F., Okasha, K. M., Mahmoud, Y. A.-G., Elsamahy, T., Jiao, H., Fu, Y., & Sun, J. (2022). A critical review on the treatment of dye-containing wastewater: Ecotoxicological and health concerns of textile dyes and possible remediation approaches for environmental safety. *Ecotoxicology and Environmental Safety, 231*, 113160. https://doi.org/10.1016/j.ecoenv.2021.113160

Ammayappan, L., Jose, S., & Arputha Raj, A. (2016). *Sustainable production processes in textile dyeing* (pp. 185–216). Springer. https://doi.org/10.1007/978-981-10-0111-6_8

Cassidy, T. D. (2019). Colour forecasting. *Textile Progress, 51*(1), 1–137. https://doi.org/10.1080/00405167.2019.1659564

Chakraborty, J. N. (2011). Sulphur dyes. In *Handbook of textile and industrial dyeing* (pp. 466–485). Elsevier. https://doi.org/10.1533/9780857093974.2.466

Chakraborty, S., De, S., DasGupta, S., & Basu, J. K. (2005). Adsorption study for the removal of a basic dye: Experimental and modeling. *Chemosphere, 58*(8), 1079–1086. https://doi.org/10.1016/J.CHEMOSPHERE.2004.09.066

Chattopadhyay, D. P. (2011). Chemistry of dyeing. In *Handbook of textile and industrial dyeing* (pp. 150–183). Elsevier. https://doi.org/10.1533/9780857093974.1.150

Drumond Chequer, F. M., de Oliveira, G. A. R., Anastacio Ferraz, E. R., Carvalho, J., Boldrin Zanoni, M. V., & de Oliveir, D. P. (2013). Textile dyes: Dyeing process and environmental impact. In *Eco-friendly textile dyeing and finishing*. InTech. https://doi.org/10.5772/53659

Grishanov, S. (2011). Handbook of textile and industrial dyeing principles, processes and types of dyes. In M. Clark (Ed.), *Handbook of textile and industrial dyeing principles, processes and types of dyes* (Vol. 1). Woodhead Publishing. https://doi.org/10.1533/9780857093974.1.28

Heller, T. S. (2015). *Method of dyeing cellulosic substrates* (Patent US 9,091,021 B2). https://patentimages.storage.googleapis.com/cd/e0/6b/2393e771bbbad8/US9091021.pdf

Hoque, E., Acharya, S., Shamshina, J., & Abidi, N. (2023). Review of foam applications on cotton textiles. *Textile Research Journal, 93*(1–2), 486–501. https://doi.org/10.1177/00405175221107400

Kane, F., Shen, J., Morgan, L., Prajapati, C., Tyrer, J., & Smith, E. (2020). *Innovative technologies for sustainable textile coloration, patterning, and surface effects* (pp. 99–127). Springer. https://doi.org/10.1007/978-3-030-38545-3_4

Kiruthika, S. D., Ugamoorthi, R., Venkatachalapathi, C., & Ramarethinam, S. (2019). Sustainable production by modifying reduction clearing in polyester dyeing. In *Functional textiles and clothing* (pp. 383–403). Springer. https://doi.org/10.1007/978-981-13-7721-1_30

Lewis, D. M. (1996). Dyestuff—Fibre interactions. *International Journal of Cosmetic Science, 18*(3), 123–135. https://doi.org/10.1111/j.1467-2494.1996.tb00142.x

Mahmud, I., & Kaiser, S. (2020). Recent progress in waterless textile dyeing. *Journal of Textile Science & Engineering*. https://doi.org/10.37421/jtese.2020.10.421

RCB. (2024). *Programme on water saving in the textile and garment industries (WaSaTex)*.

Richards, P. R. (2015). Fabric finishing. In *Textiles and fashion* (pp. 475–505). Elsevier. https://doi.org/10.1016/B978-1-84569-931-4.00019-2

Sanders, D., Grunden, A., & Dunn, R. R. (2021). A review of clothing microbiology: The history of clothing and the role of microbes in textiles. *Biology Letters, 17*(1), 20200700. https://doi.org/10.1098/rsbl.2020.0700

Sekar, N. (2011). Direct dyes. In *Handbook of textile and industrial dyeing* (pp. 425–445). Elsevier. https://doi.org/10.1533/9780857093974.2.425

Sheikh, J., & Bramhecha, I. (2018). Enzymes for green chemical processing of cotton. In *The impact and prospects of green chemistry for textile technology* (pp. 135–160). Elsevier. https://doi.org/10.1016/B978-0-08-102491-1.00006-X

Uddin, M. A., Begum, M. S., Ashraf, M., Azad, A. K., Adhikary, A. C., & Hossain, M. S. (2023). Water and chemical consumption in the textile processing industry of Bangladesh. *PLOS Sustainability and Transformation, 2*(7), e0000072. https://doi.org/10.1371/journal.pstr.0000072

Yu, H., Wang, Y., Zhong, Y., Mao, Z., & Tan, S. (2014). Foam properties and application in dyeing cotton fabrics with reactive dyes. *Coloration Technology, 130*(4), 266–272. https://doi.org/10.1111/cote.12088

Zille, A. (2020). Plasma technology in fashion and textiles. In *Sustainable technologies for fashion and textiles* (pp. 117–142). Elsevier. https://doi.org/10.1016/B978-0-08-102867-4.00006-2

Sustainable Innovations in Textile Coloration: Addressing Environmental Challenges

Jesica Roshima A, Selvakumar A, and Vasanth Kumar D

Abstract Natural textile colourants have historically been sourced from insects, plants, and animals, with their use dating back to 3500 BC. Until W.H. Perkins discovered synthetic dyes in 1856, and these natural dyes were used to colour cellulosic and protein-based fibres. Today, the textile industry poses a significant environmental threat by producing vast amounts of fabric and consuming enormous quantities of water annually. During the dyeing process, approximately 10–15% of the unfixed dye is released into water bodies as wastewater, significantly contributing to pollution. Currently, 90% of our clothes are dyed using synthetic dyes. Addressing the environmental challenges related to textile dyeing and treatment and exploring new sustainable technologies are crucial. Textile dyeing depletes freshwater resources and continuously pollutes the surface water. For decades, researchers have developed waterless and pollution-free dyeing technologies. For more than a decade, research has been striving for developing no-water and pollution-free dyeing technologies. The successful development of air and waterless dyeing techniques and the process of dyeing by using microbes have all sprung from these efforts, from pre-autoclaving the fabrics to avoid contamination. When compared to traditional methods, these methods have several advantages, such as lower energy consumption, shorter processing times, and reduced waste and greenhouse gas emissions. They are easy to access, reusable, non-flammable, odourless, and non-toxic, thus rendering them both ecologically sound and economical. To address these urgent concerns and interests, this study gives an overview of the status of sustainable research and development on the colouration of textile materials.

Keywords Air dyeing · Bacterial dyeing · Greenhouse gas emission reduction · Sustainable textile coloration · Waterless · Pollution-free dyeing

J. R. A · S. A · V. K. D (✉)
VIT Fashion Institute of Technology (VFIT), Vellore Institute of Technology (VIT), Chennai, Tamil Nadu, India
e-mail: vasanthkumar.d@vit.ac.in

S. S. Muthu (ed.), *Sustainable Coloration of Textiles*, Sustainable Textiles: Production, Processing, Manufacturing & Chemistry,
https://doi.org/10.1007/978-3-031-91217-7_12

1 Introduction

Due to increasing demands from the consumers and the environmental crisis, sustainable practices in the textile industry are gaining traction. As consumers put an increased emphasis on sustainability and ethics in production, it becomes apparent that this is a greater trend towards more sustainable and eco-friendly methods of production. Sensing the wider concern of sustainable textile production, companies have adopted practices that encompass eco-friendly dyeing methods and innovations of less chemical and less water (Niinimäki & Hassi, 2011). Sustainability refers to stability which would help in the preservation of continuity among generations without the haunts of deprivation met upon them. It is a complicated quest, calling for the use of many sides. There is growth of awareness about the ecological preservation of natural resources, which is contained with the central theme of sustainability (Önlü, 2017).

It is commonly recognized that dyeing textiles can negatively affect water bodies, culminating in widespread harm attributable to the production and use of synthetic dyes. Serious public health issues arise when hazardous dyeing effluents travel through the food chain and into communities near textile operations (Islam et al., 2021). The ability of sustainably made alternatives to economically threaten industrial-dyed textiles depends on consumer awareness and the perception of the advantages of natural dyes over their synthetic cousins. More informed consumers will generate greater market demand for environmentally and human-biological-friendly natural dyes, thereby driving the changes into more sustainable practices within the textile industry (Mabuza et al., 2023). In light of heightened environmental awareness and regulations, the textile industry is fitting itself into a more sustainable outlook. This now means that it increasingly focuses on developing substitutes for synthetic dyes and mordents, which pose high threats to health and ecology. It calls for more rigorous promotion of safe and ecologically sound dyeing methods in the textile sector (Pranta & Rahaman, 2024).

Given the current emphasis on sustainable practices in food, pharmaceutical, and textile products, modern technologies as well as alternative solvents for deriving bio-based food colouring have gained significant attention (de Souza Mesquita et al., 2021). Natural dyes obtained from plants, animals, and minerals have been widely accepted as an alternative to synthetic dyes for colouring textiles because of the regenerative nature and ecological conscience associated with them (Bauer et al., 2001). For production of textile dyestuffs, colourants must be derived from ecologically viable sources, including plants, insects, and microorganisms. Among the plant-related materials, many components such as seeds, fruits, bark, leaves, stems, flowers, and certain microorganisms can produce vibrant natural colours. In most modern applications, various extraction methods are used to obtain these natural colours. This selection also reflects rising awareness of using renewable and sustainable resources to quench the rising demand for ecological pixels in the textile industry (Vickers, 2017).

Water use is a major challenge in the textile industry. Environmental and sustainability issues are made worse by the high-water requirements of the dyeing and finishing processes. Initiatives to reduce water use and implement more efficient practices are essential to mitigate these impacts and enhance the sustainability of the textile supply chain as a whole. Air dyeing is a method of adding colour to textiles without the use of water by distributing colours using air or other non-aqueous substances. A number of tactics have been created to lessen the impact on the environment (Pizzicato et al., 2023). Years of intense discussion on sustainability have resulted from its complexity. Significant changes must be made, particularly by major organizations, to achieve sustainable growth. A key basis for this is the Triple Bottom Line, which was first put forth by economist John Elkington in 1997. This strategy emphasizes the interdependence of the three main pillars of sustainability economic, social, and environmental performance in order to facilitate significant change (Akanyange et al., 2021).

The research on natural dyes has witnessed enormous strides in the other endwing. These include the optimally tempering of dye colours, ameliorating dye extraction methods and processes of dyeing, and formulation of dyes with minimum environmental effect. All these advances will cut down pollution and increase the volume and production capacity of natural dyes. Moreover, through these new extraction processes, the textile industry can adopt greener practices, thereby meeting the escalating demand for sustainable options (de Oliveira et al., 2021). As scientists explore these methods further, the possibility of making colourful, eco-friendly dyes from natural sources are becoming more real (Hossain et al., 2024). Recently, the research study on natural based dyes has seen many developments. These include adjusting colours, improving extraction methods, refining dyeing processes, and creating environmentally friendly mixtures. This helps reduce environmental harm and in turn increases the effectiveness of natural dyes. The adoption of greener practices is also now possible through these new extraction techniques. The growing demand for sustainable solutions is met through this approach. This study looks at the latest developments and future possibilities in natural dye research, especially for use in textiles. There's a clear trend towards making textiles in a more sustainable way, thanks to improvements in natural dyes. This trend shows a growing awareness of the environmental problems caused by traditional dyeing methods. Natural dyes offer a way to significantly cut down on water and chemical use while keeping colours bright and vibrant.

2 Methods of Sustainable Innovations in Textile Coloration

1. Air Dyeing
2. Bacterial Dyeing
3. Waterless Dyeing

4. Pollution-Free Dyeing
5. Greenhouse Gas Emission Reduction

2.1 Air Dyeing

Earth is unique in our solar system for its life-sustaining conditions, supporting a wide range of organisms, from microbes to humans. The biosphere sustains life while withstanding human impact, with the "environment" referring to the social, economic, biological, physical, and chemical factors that shape human existence and guide environmental management. The impact of population growth on environmental degradation, such as resource depletion and habitat loss, analysing the consequences of deforestation, soil erosion, species extinction, and climate change; evaluating the role of industrialization, particularly in the textile industry, in contributing to environmental challenges; and exploring sustainable solutions to preserve biodiversity and mitigate the adverse effects of human activities on the environment (Tiwari, 2022). Water is a crucial resource in traditional dyeing methods, but air dyeing uses a novel and environmentally friendly way to colour textiles. There are some of the air dyeing techniques and its standout features showed in Fig. 1.

2.1.1 Air Dyeing Process Applications and Its Uses

This process also reduces the reliance on harmful chemicals by utilizing a more efficient dye application system, which minimizes chemical waste. Additionally, it is more energy-efficient, as it eliminates the need for water heating and lowers the overall energy consumption of the dyeing process. By cutting down on energy use, air dyeing contributes to the reduction of greenhouse gas emissions and air pollution. Furthermore, it generates less waste by minimizing both water and chemical use, reducing the volume and toxicity of effluents. Table 1 outlines the techniques used in various methods to reduce water usage by applying dye directly to fabric without the need for water-based solutions.

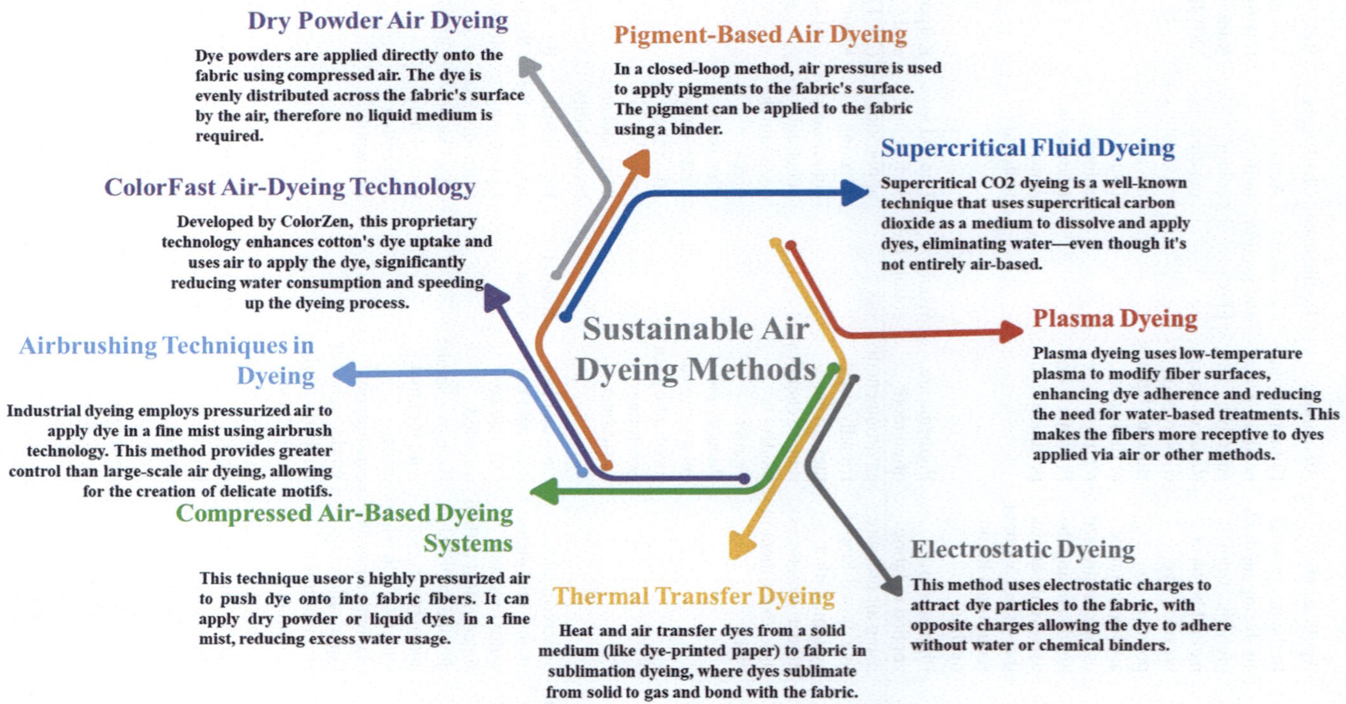

Fig. 1 Sustainable air dyeing methods. (Authors own creation)

Table 1 Sustainable air dyeing applications, uses and its challenges

Techniques of air dying	Application	Advantages	Textile uses	Challenges	Ref.
Dry powder air dyeing	In the dry powder air dyeing method, pigments from natural herbs or recycled dye powders are applied to fabric using compressed air, dispersing the dye evenly across the surface without requiring a liquid medium	Eliminates the need for water entirely Reduces energy consumption since no heating or drying is required Prevents water pollution	Cellulosic and protein fibres	Achieving uniform colour penetration, especially on thicker fabrics Issues with colour fastness in some materials	Mortensen (2006), Kant (2012)
Pigment-based air dyeing	Pigments are applied to fabric using air pressure in a closed-loop system, with a binder sometimes used to fix them in place	Pigments are often more environmentally friendly since they do not require chemical binders or mordents Reduces the environmental footprint by not generating wastewater	Wool, cotton, and silk	Limited to surface application, meaning colour doesn't penetrate deep into the fibres The use of binders may affect the fabric's hand-feel and breathability	Karpova et al. (2021), Khan (2016)
Supercritical fluid dyeing	Supercritical CO_2 dyeing eliminates water by using supercritical carbon dioxide to dissolve and apply dyes, with CO_2 behaving like both a gas and liquid, allowing efficient fibre penetration	No water is used, and the CO_2 can be recycled High dye uptake with minimal energy consumption Works well with synthetic fabrics like polyester	Natural fibres	High initial setup cost for equipment Currently less effective on natural fibres like cotton and wool	Banchero (2013), Montero et al. (2000)
Plasma dyeing	Plasma dyeing uses low-temperature plasma to modify fibre surfaces, making them more dye-receptive and reducing the need for water-based treatments	Uses minimal water Enhances dye penetration without the need for traditional chemical treatments	Wool, cotton, silk, nylon, polyester, and acrylic	High cost of plasma equipment. May not be suitable for all types of fabrics	Haji and Naebe (2020), Zaidy et al. (2019)

(continued)

Table 1 (continued)

Techniques of air dying	Application	Advantages	Textile uses	Challenges	Ref.
Electrostatic dyeing	This method uses electrostatic charges to attract dye particles to the fabric, with opposite charges allowing the dye to adhere without water or binders	Reduces the use of water and energy Precise application of dye, resulting in less waste	Natural fibres	Limited to certain types of dyes and fabrics Requires careful calibration to avoid uneven dye distribution	Hu et al. (2013), Uğur and Sariişik (2011), Zhang et al. (2012)
Thermal transfer dyeing	Heat and air transfer dyes from a solid medium, such as dye-printed paper, to fabric in a process called sublimation dyeing, where dyes sublimate directly from solid to gas and bond with the fabric	Works well for synthetic fabrics, particularly polyester No water is involved, and the process is fast	Polyester, nylon, and synthetic blends	Limited to polyester and synthetic blends Does not work well on natural fibres like cotton or wool	Ito et al. (1996), Kim et al. (2021), Xiu et al. (2014))
Compressed air-based dyeing systems	This technique uses highly pressurized air to apply dye to the fabric, similar to dry powder air dyeing, but it can also incorporate liquid dyes in a fine mist, minimizing excess water use	Significantly reduces water use compared to traditional methods Lower energy consumption due to the absence of drying stages	Natural (cellulosic and protein yarns), synthetic and natural and synthetic blends	Requires specialized machinery Effective for some fabrics, but less effective for those needing deeper dye penetration	Damrath et al. (2021), Acar et al. (2022)
Airbrushing techniques in dyeing	Airbrush technology is adapted for industrial dyeing, allowing dye to be sprayed onto fabrics in a fine mist using pressurized air. This method offers greater control than large-scale air dyeing and enables more intricate designs	Allows for detailed dyeing, gradients, and custom patterns Minimal water and chemical use	Natural and synthetic	May not be practical for large-scale fabric production Requires skill and precision to achieve even dye coverage	Bell et al. (2021), Zahid et al. (2018)

(continued)

Table 1 (continued)

Techniques of air dying	Application	Advantages	Textile uses	Challenges	Ref.
Colour fast air dyeing technology	Developed by ColorZen, this proprietary technology enhances cotton's dye uptake and uses air for application, drastically reducing water consumption and speeding up the dyeing process	Substantial reduction in water usage Faster dyeing process with less waste	Natural and synthetic	The technology is new and not yet widely available	Mahmud and Kaiser (2020)

2.2 *Bacterial Dyeing*

Every element of our life is influenced by colour, including the food we consume, the furnishings in our houses, and the clothes we wear. The term "microbial pigments" was employed, despite the fact that the term refers to colours and pigments that come from both fungi and bacteria (Manikprabhu & Lingappa, 2013).

2.2.1 Sources of Pigment by Bacteria

Filamentous fungi are known to produce an astonishing variety of pigments, including carotenoids, melanins, flavones, phenazines, quinones, monascins, violacein, and indigo (Dufossé et al., 2014). The use of Monascus for producing ang-kak (red mould rice) is one of the oldest recorded uses of fungal pigments. Monascus produces yellow (ankaflavin, monascine), orange (rubropunctatin, monascorubrin), and purple (rubropunctatin, monascorubrin) pigments, commonly found in Oriental foods, especially in Southern China, Japan, and Southeast Asia. Over 50 Monascus pigments have been identified, and more than 50 global patents have been issued regarding their use in food (Dufossé et al., 2005; Shirata et al., 2000). Genera such as Eurotium, Fusarium, Curvularia, and Drechslera have been reported to contain pigments like anthraquinones, naphthaquinones, dihydroxy naphthalene, melanin, flavin, anthraquinone, chrysophanol, cynodontin, helminthosporin, tritisporin, and erythroglaucin (Nigam & Pandey, 2009). Figure 2 showed some of the bacterial pigments and its sources.

2.2.2 The Colour Benefits and Its Sources

Natural colourants known as bacterial pigments are created by different bacteria as a by-product of their metabolic activities. In addition to their numerous uses in textiles, food, cosmetics, and pharmaceuticals, these pigments are prized for their

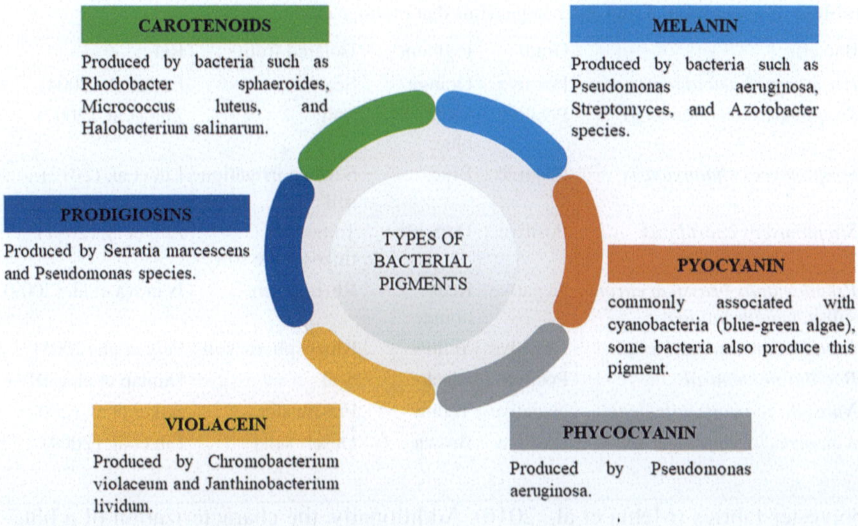

Fig. 2 Types of bacterial pigments and its sources. (Authors own creation)

potential environmental sustainability in comparison to synthetic colours. Table 2 presents a discussion of various microorganisms and their associated colour benefits.

Carotenoid-derived red and yellow pigments have been successfully applied to dye wool fabrics, showing excellent dye absorption and colour fastness. Researchers are encouraged to further investigate these pigments for potential industrial applications, as detailed chemical characterization could drive innovations in the dyeing industry and help reduce the environmental impact of synthetic dyes (Morales-Oyervides et al., 2017). The production of melanin in *Streptomyces* offers promising applications for the printing and dyeing of wool fabrics. The natural pigment offers an eco-friendly alternative for textile coloration, providing durability and UV protection, while minimizing the need for synthetic dyes and harmful chemicals (Amal et al., 2011; Liu et al., 2020; Melo, 2023). The colour fastness of textiles dyed with Streptomyces purpureus pigments is comparable to that of vegetable dyes, while retaining the natural softness typical of organic pigments. Studies on the distribution and stability of the purple pigments revealed that silk fabrics dyed with them exhibit (Darshan & Manonmani, 2015). Textile fabrics, encompassing cotton varieties such as linen, baft, gabardine, and jersey, as well as synthetic materials such as chiffon, satin, Dacron, and polyester, were effectively dyed with prodigiosin pigment extracted from the marine bacterium *Serratia rubidaea*. This pigment produced a diverse array of vibrant colour shades across the different fabric types (Metwally et al., 2021).

Violacein is utilized to dye both natural fibres, which are silk, cotton, and wool, and synthetic fibres like nylon and vinylon (Durán et al., 2012). Riboflavin is effective in dyeing both viscose and cotton fabrics rich in yellow colour (Iyer et al., 2019; Kumar et al., 2015). Various research studies have shown the dyeing abilities of green, yellow, and red pigments produced by *Leifsonia aquatica*, *Pseudomonas aeruginosa*, and *Serratia marcescens*, respectively, on cotton, viscose, and

Table 2 Enumeration of new microorganisms that produce colours

Bacteria	Gram	Pigment	Isolated from	Ref.
Paracoccus haeundaensis	Positive	Orange	Sea water	Lee et al. (2004)
Streptomyces vietnamensis	Positive	Violet–blue	Soil	Zhu et al. (2007)
Streptomyces shaanxiensis	Positive	Blue	Sewage irrigation soil	Lin et al. (2012)
Streptomyces caerulatus	Positive	Dark blue	Tomato rhizosphere soil	Zhu et al. (2011)
Pseudomonas brassicacearum subsp. *neoaurantiaca*	Negative	Red–orange	Rhizosphere	Ivanova et al. (2009)
Pseudomonas argentinensis	Negative	Yellow	Rhizospheric soil	Peix et al. (2005)
Bacillus nakamurai	Positive	Black	Soil	Dunlap et al. (2016)
Nubsella zeaxanthinifaciens	Negative	Yellow	Freshwater	Asker et al. (2008)
Kineococcus xinjiangensis	Positive	Brown	Desert sand	Liu et al. (2009)

polyester fabrics (Mehta et al., 2016). Additionally, the characterization of a blue-green pigment extracted from *Pseudomonas aeruginosa* has been performed, emphasizing its potential applications in dyeing textiles and paper (Sengupta & Bhowal, 2023).

2.2.3 Utilization of Pigments in the Textile Industry

The textile industry uses around 1.3 million tons of synthetic dyes and dye precursors each year, with an estimated 200,000 tons lost as effluents during dyeing and finishing operations (Venil et al., 2013). Regrettably, many of these dyes resist conventional wastewater treatment due to their high stability against factors such as light, temperature, water, detergents, bleach, and perspiration, resulting in environmental persistence. As the global trend shifts towards environmentally friendly and biodegradable products, the demand for natural colourants is increasing. Natural pigments are derived from various sources, including ores, insects, plants, and bacteria. Among germs, bacteria have enormous potential to produce varied bioproducts and one such bioproduct is pigments. The utilization of bacterial pigments as natural colourants and their manufacture have been the subject of investigation for multiple researchers (Venil, 2009).

The technique requires additional improvement to suit commercial expectations, the usage of pigments derived from bacteria has considerable promise for a range of industrial applications. Since bacterial pigments are less harmful, biodegradable and can be made from renewable resources, they present a sustainable substitute for synthetic dyes. To scale up bacterial pigments for wider industrial application, however, issues including colour stability, pigment yield optimisation, and production process optimisation need to be resolved. The potential of these pigments in food, medicine, textiles, and cosmetics can be realized by stepping up study on them (Abdulkadir, 2017). Figure 3 showed the utilization and applications of bacterial pigments within the textile industry.

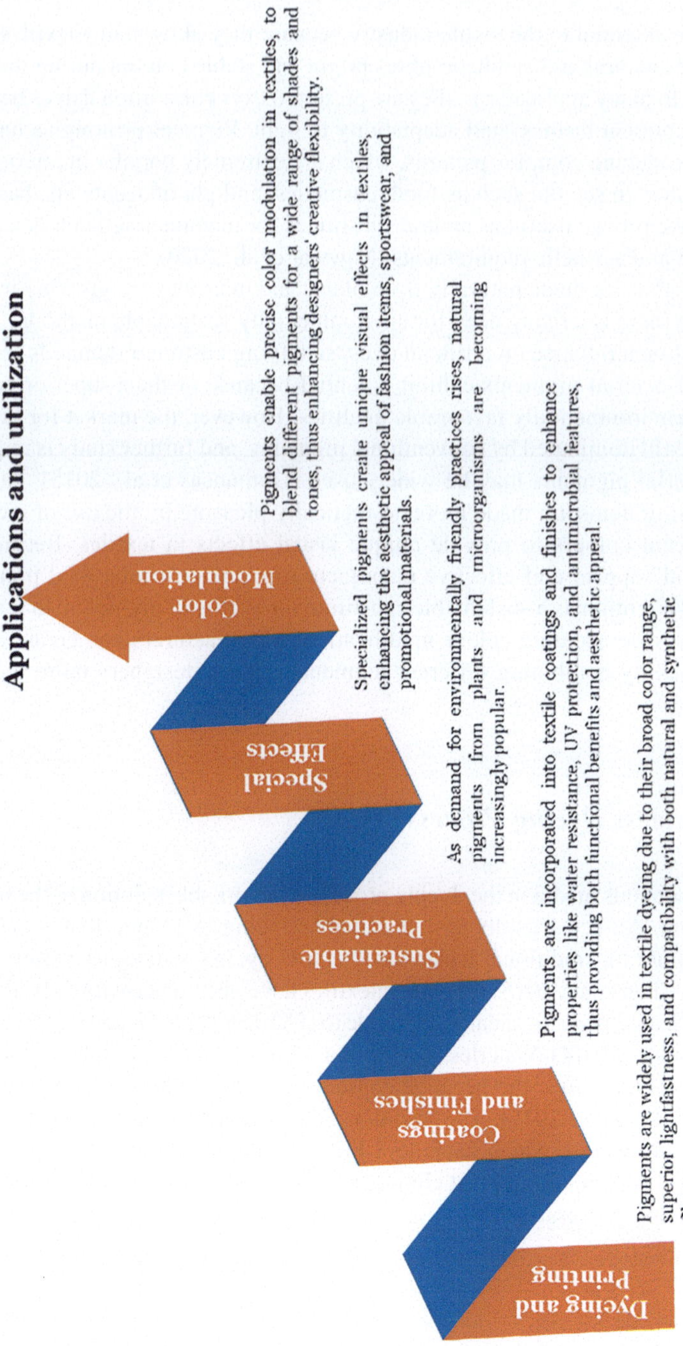

Fig. 3 Bacterial colourants in the textile industry. (Authors own creation)

2.2.4 Bacterial Colourants and Utilization in Recent Years

Pigments are essential to the textile industry because they allow many textiles to be dyed. It gives natural and synthetic fibres bright and stable colours during the dyeing process. In many applications, they are preferred over conventional dyes because of their exceptional fastness and adaptability to light. Pigment printing techniques also allow producing complex patterns, which are extremely popular in interior and fashion textiles. In sectors such as food, cosmetics, and pharmaceuticals, bacterial pigments have proven useful as natural substitutes for manufactured colours, meeting practical and aesthetic requirements (Nawwar et al., 2020).

Pigments that are made naturally from plants and microbes are growing in popularity since there is a rising need for environmentally sustainable methods. These substitutes lessen reliance on artificial dyes, satisfying customer demands for sustainability. Bacterial pigments exhibit potential because of their superior performance and environmentally favourable qualities. However, the market for organic pigments is still dominated by conventional pigments, and further study is required before bacterial pigments may be widely used (Dumancas et al., 2015). Fashion and advertising items are made more aesthetically pleasing by the use of metallic and fluorescent colours to provide unique visual effects in textiles. Because of their "natural" appeal and effective manufacturing methods, microbial pigments have gained attention as a sustainable substitute for synthetic pigments. In textiles, pigments provide accurate colour modification. Manufacturers can create a vast range of hues by combining different pigments, giving designers more creative freedom.

2.3 Waterless Dyeing Process

Using non-aqueous media in the dyeing process is a workable solution to the hydrolysis problem. An eco-friendly and sustainable dyeing technique that provides a green substitute for traditional water-based textile dyeing techniques is non-aqueous dyeing (Chen et al., 2015). Synthetic textiles have been successfully dyed using solvents such as supercritical carbon dioxide ($ScCO_2$) and chlorinated hydrocarbons (Hori & Kongdee, 2014). Waterless dyeing is a creative and eco-friendly way for the textile industry to reduce the negative consequences of traditional dying methods (Agrawal & Agrawal, 2015). The traditional dyeing method uses a lot of water, chemicals, and energy, which leads to a lot of waste and pollution. Table 3 lists waterless dyeing technologies that eliminate or drastically reduce the need for water during the dying process.

Table 3 Waterless dyeing advancements

Types of dyeing	Advantages	Challenges	Requirements	Ref.
Supercritical CO_2 dyeing	Water conservation and effluent reduction, energy efficiency compared to traditional methods Recyclability of CO_2	High upfront investment costs, the necessity for precise temperature and pressure control, and the requirement for cost-effective equipment are needed	Cost-effective equipment and improve scalability	Bai et al. (2019)
Foam dyeing	Water conservation, energy efficiency, reduced chemical usage and sustainability	Foam stability, even dye distribution, compatibility with dyes, process optimization and foam breakage	Optimized foam formulation and specialized equipment	Abate and Tadesse (2021)

2.3.1 Supercritical CO_2 Dyeing

Using the special qualities of supercritical carbon dioxide ($scCO_2$), supercritical impregnation technology incorporates one or more chemicals into polymeric materials. $scCO_2$ can function as a solvent and a diffusion enhancer because it can dissolve a broad variety of compounds and swell different polymers. In comparison to conventional liquid-based impregnation techniques, the absorption of $scCO_2$ into the polymer matrix not only improves the diffusivity of the penetrant molecules but also drastically reduces the required contact time. Because of its two functions, $scCO_2$ is a useful medium for improving the impregnation process in polymer applications (Goñi et al., 2018). Supercritical dyeing—a novel method for dyeing textile materials possesses umpteen of benefits such as, significantly reduced water consumption, energy saving hence that doesn't demand a separate effluent treatment to process the wastewater, also the dye remains could also be reestablished with the aid of supercritical carbon dioxide ($scCO_2$). This has resulted a keen implementation of the aforesaid method by textile processing industries where it is under soft launch and collaborations among industry and academia (Goñi et al., 2017). The dyeing of textile fibres and polymers using $scCO_2$ appears among the promising applications (Bach et al., n.d.). The technique sees increased colourant penetration and even distribution because it has resulted in a dying technique with such brilliance >bright and consistent tones. The use of $scCO_2$ technology advances an earth-friendly approach in the textile industry as it cuts down all the water utilizations and eliminates any necessity for hazardous chemical agents. Furthermore, $scCO_2$ minimizes environmental impact through easy retrieval of leftover dyes (Da-In Kim, 2024). Disperse dyes are typically used for polyester fabrics; however, Table 4 highlights the composition of disperse dyes applied in a waterless process for wool fabrics.

Table 4 Disperse dyes for supercritical carbon dioxide dyeing (sustainable dyes for wool) (Zheng et al., 2017)

Type of Dyes	Molecular formula	Molecular weight	Chemical structure
Disperse Red 153	$C_{18}H_{14}Cl_3N_5S$	438.76	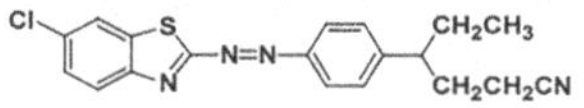
Disperse Blue 148	$C_{19}H_{19}N_5O_4S$	413.45	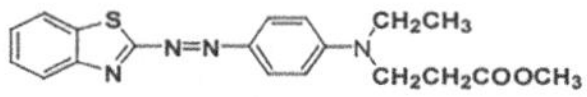
Disperse Red 54	$C_{19}H_{18}ClN_5O_4$	415.83	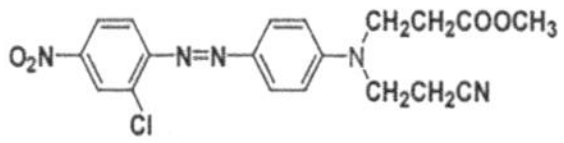
Disperse Blue 183	$C_{20}H_{21}BrN_6O_3$	473.32	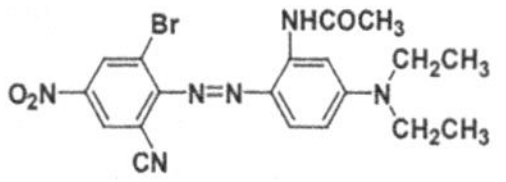
Disperse Red 92	$C_{24}H_{22}N_2O_7S$	482.51	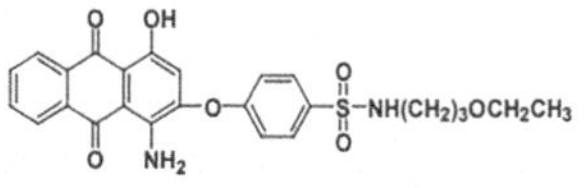
Disperse Yellow 82	$C_{20}H_{19}N_3O_2$	333.38	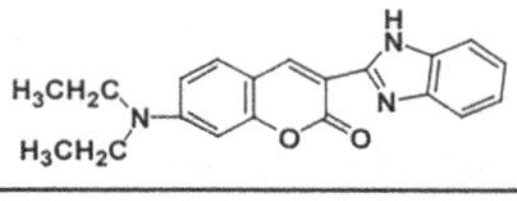

2.3.2 Foam Padding

Using foam as a carrier to apply chemicals like dyes and finishing compounds to textile substrates is known as foam technology, which is a minimally additive method. Foam serves as the medium in place of water, saving water and improving application process efficiency. By using less water and energy for drying, this method minimizes the impact on the environment while allowing chemicals to be distributed uniformly throughout the fabric (Goñi et al., 2018). In textile processing, foam technology is a cutting-edge, low-add-on technique that applies chemicals and colourants to textiles using foam rather than water. This approach significantly reduces water and energy consumption by replacing water with air. Foam technology typically operates at a wet pick-up of 20–40%, compared to the 60–100% range in traditional pad dyeing. This approach is promising in the push for sustainable textile manufacturing because it also enables single-side treatment of fabrics, improving accuracy and efficiency (Zhou et al., 2018). Energy conservation and emissions reduction have become essential needs in the textile dyeing business due to growing environmental concerns. Foam dyeing technology has drawn a lot of attention since it supports national sustainable development goals and uses little liquid, energy, or emissions (Zhu et al., 2022).

Application Process of Foam Padding

In foam dyeing, the generated foam is injecting air under pressure into a liquor using a dynamic foam generator. A pigment solution containing such as sodium dodecyl sulphate and hydroxyethyl cellulose is placed in a liquid container, then

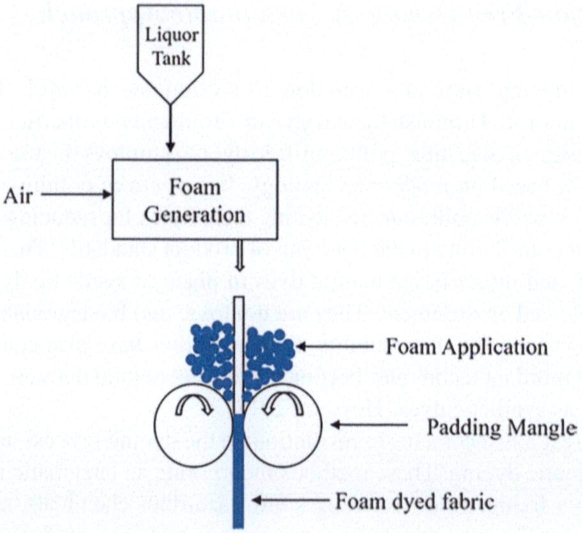

Fig. 4 Schematic of horizontal double-sided foam padding machine (Rahman et al., n.d.)

air and the solution are mixed under pressure to create foam. The foam is applied on both the sides of the fabric utilizing a horizontal padding machine. Drying and curing proceed in the same manner as traditional dyeing methods. A schematic diagram illustrating horizontal double-sided foam padding process is presented in Fig. 4 (Rahman et al., n.d.). Foams can be applied to textiles through both direct and indirect methods. In the direct pressurized system, foam is applied to the fabric under pressure via a slot applicator or rotary screen, with the fabric pressed against a backing roller. In the non-pressurized direct system, although the foam reservoir is not pressurized, some pressure is still applied during the process. In the indirect system, a carrier, such as a drum or blanket, facilitates the transfer of foam to the fabric upon contact. The application of foam padding to the textile surface may be precisely controlled with this technique (Elbadawi & Pearson, 2003).

Recent developments in foam dyeing technology have made is as a strong substitute for the textile industry's wet processing sector. This approach improves both efficiency and sustainability compared to conventional methods. The important advantages of this method though still under investigation includes less water, energy, and chemical use. With the continued research and development, Foam dyeing is expected to play a key role in textile production in the future. This method is more desirable compared to the traditional dyeing techniques as it is more affordable and environmentally friendly by aiding textile productions with reduced water usage and less wastewater creation.

2.4 *Pollution-Free Dyeing: A Sustainable Approach*

Pollution-free dyeing aims at lower down or eliminate harmful chemicals and superfluous water use. Diminish the adverse environmental properties like chemical discharges or water exhaustion, pollution-free dyeing employs the use of superlative dyeing methods based on modern technology. In this era of pollution, the world is highly inching towards pollution-free dyeing techniques for reducing its ecological impact without contributing to the lowering of product standards. This is done using plant, mineral, and insect-based natural dyes in place of synthetic dyes, which are safer for people and environment. They are nontoxic and biodegradable, answering customers' aspiration for sustainability. Recent studies have also concluded that if extraction and mordant techniques become advanced, natural dyes are used with the same efficacy as synthetic dyes (Hossain, 2018).

The next great advancements to revolutionize the dyeing process are biotechnology and enzymatic dyeing. These methods incorporate an enzymatic treatment that enhances colour fixing, saves water, uses few hazardous chemicals, and has all the practical benefits for the demand of newer minimal impact dyeing techniques (Chatha et al., 2017). Digital textile printing technology is one of the innovative solutions that have started gaining considerable attention. This method involves digitally printing the dye on the fabric, making colour application highly precise, and is far more environmental-friendly compared to conventional dyeing processes, and utilizes less water and creates a no-waste system (Lee et al., 2004). The overview states that pollution-free dyeing techniques are the area that should be seriously researched, as it is important that brands and manufacturers understand the imperativeness of these strategies to advance their profiles. This report includes numerous issues of pollution-free dyeing technologies and processes, mentions detail of principal advancements, along with the techniques being develop, and ramifications in the textile sectors.

2.4.1 Recent Developments in Pollution-Free Dyeing Technologies

Conventional dyeing practices involve the voicing of either hazardous chemicals, large volumes of water, or let loose toxic effluents into the environment. While the modern advancement of dyeing techniques is creating meaningful changes towards environment protection and sustainable future.

Natural Dyes: A Revival of Tradition

Natural colours made from plants, minerals, and insects are appealing as ecological alternatives to synthetic dyes. Recent studies have aimed at improvement of the colourfastness and application of natural dyes (Hossain et al., 2024) backing up the improved application technique and mordant pour to perform better for natural dyes in different textiles. Economically, it becomes a boon to communities engaged in natural dye manufacturing by promoting local resources and environmental sustainability. Furthermore, the advances in extraction methods, such as supercritical CO_2 extraction, allow more efficient synthesis of natural dyes. This technique is highly beneficial for reducing the degrees of waste and energy and water consumed during the dyeing process (Mahmud & Kaiser, 2020). Such advancements create competition for natural dyes by offering customers and brand something environmentally friendly.

Biotechnology: Innovations in Enzymatic Dyeing

Biotechnology, especially by virtue of enzymatic approaches to dyeing, is increasingly advancing the field of textile. Using enzymes to promote dye fixation of the fibres, thus reducing dependence on harsh chemicals and water. Enzymatic processes can replace or even outperform classical dye processes, as recent studies suggest (Chatha et al., 2017). One advance worth mentioning is the utilization of lactic acid enzymes in dyeing; these enzymes facilitate some further oxidation of dyes and improve their ability to adhere to textiles. This is a novel way catering to an enterprise that is inclined towards minimizing the impact of operations on the environment because of the concomitant reduction not just in pollution but also an increase in the longevity of the dye (Lee et al. 2021).

Digital Printing: Minimizing Waste and Chemical Use

Digital textile printing is the advancement of eco-friendly dyeing methods. Unlike traditional dyeing, which often consumes a lot of water and chemical treatments, digital printing employs dye only where it is needed. Because of its precision, less chemicals and water are utilized, which lowers the amount of effluent produced (Varadarajan & Venkatachalam, 2016). One recent advancement in digital printing technology that helps achieve environmental aims is the introduction of water-based inks free of hazardous substances. Additionally, digitally created programmes that optimize print patterns and enable more efficient production processes save waste. Because digital printing provides a modern solution to classic dyeing issues, fashion companies who are committed to sustainability are increasingly using it.

2.5 Greenhouse Gas Emission Reduction Sustainable Dyeing Practice

The goal of sustainable dyeing techniques is to lessen the environmental impact of traditional textile dyeing techniques in order to mitigate greenhouse gas emissions. These environmentally friendly tactics make use of natural dyes made from plants, minerals, and microbes as substitutes for synthetic dyes, which are frequently petroleum-based and pollute the air (Chanaphoo et al., 2021). Water and energy consumption are reduced by advanced waterless dyeing methods as digital textile printing, foam-based dye application, and supercritical CO_2 dyeing. Additionally, creating low-impact or biodegradable dye formulas and incorporating renewable energy into dyeing processes are essential steps in reducing emissions. These developments push the textile industry towards a more circular economy model by improving sustainability and reducing greenhouse gas emissions (Yousaf & Aqsa, 2023).

Table 5 referred several strategies that can reduce carbon emissions in textile dyeing. Waterless Dyeing Techniques like foam and supercritical CO_2 dyeing save water and energy, cutting emissions. Natural and Biodegradable Dyes: Replacing synthetic dyes with natural or biodegradable alternatives lessens environmental impact. Renewable Energy: Incorporating solar, wind, or other renewable energy sources in manufacturing reduces fossil fuel dependence. Closed-Loop Systems: Applying circular economy principles in dyeing and production reduces waste and emissions. Energy-Efficient Technologies: Innovations like low-impact dyes and energy-efficient equipment lower carbon footprints.

Table 5 Various strategies within the textile industry can aid in lowering greenhouse gas emissions and enhancing sustainability

Distinct aspects of sustainable dyeing practices	Strategies can be implemented to lower greenhouse gas emissions in the textile industry	Ref
Comparative study of natural and synthetic dyes in greenhouse gas emission reduction	Investigating the lifecycle carbon footprint of natural dyes versus synthetic dyes in textile production	Grisales et al. (2019)
The impact of waterless dyeing techniques on reducing carbon emissions	Evaluating the environmental effects and carbon reduction capabilities of waterless dyeing techniques, such as supercritical CO_2 dyeing and foam dyeing	Gobalakrishnan et al. (2020)
Integration of renewable energy in dyeing processes to lower greenhouse gas emissions	Exploring how renewable energy sources, such as solar or wind power, can be incorporated into textile dyeing facilities to reduce carbon emissions	Schnitzer et al. (2007)
Biodegradable and low-impact dyes: A sustainable option for reducing emissions	Exploring the efficacy and scalability of biodegradable dyes for emission reduction in textile manufacturing	Kumar and Choudhury (2018)
The carbon sequestration potential of dyeing with microbial and plant-based dyes	Investigating how to derive microbial and plant-derived dyes can contribute to carbon sequestration during the dyeing process	Haque et al. (2022), Fried et al. (2022)

(continued)

Table 5 (continued)

Distinct aspects of sustainable dyeing practices	Strategies can be implemented to lower greenhouse gas emissions in the textile industry	Ref
Optimizing dyeing processes to minimize energy consumption and greenhouse gas emissions	Developing energy-efficient dyeing techniques and assessing their impact on greenhouse gas reduction in textile manufacturing	Hasanbeigi and Lawrence (2013), Ozturk et al. (2020)
Lifecycle assessment of eco-friendly dyeing practices for greenhouse gas reduction	Conducting a cradle-to-grave lifecycle assessment of sustainable dyeing techniques to quantify their impact on greenhouse gas emissions	Grisales et al. (2019)
The impact of digital printing as an alternative dyeing method for reducing industrial carbon footprint	Exploring the adoption of digital textile printing and its potential to reduce greenhouse gas emissions compared to traditional dyeing methods	Yılmaz et al. (2024)
The role of policy and regulation in promoting low-emission dyeing technologies	Assessing the influence of government policies, industry standards, and incentives on the adoption of sustainable dyeing practices to reduce greenhouse gases	Dewi et al. (2019), Zhu et al. (2018)
Closed-loop dyeing systems and circular economy strategies for reducing greenhouse gas emissions	Examining the potential of closed-loop dyeing systems to minimize emissions and enhance resource efficiency in the textile industry	Ribul et al. (2021), Vignali et al. (n.d.)

3 Conclusion

One of the biggest issues facing the populace today is inadequate or erratic access to drinking water and sanitary services. The United Nations (UN) reports that over 80% of wastewater produced by human activities is dumped into rivers without the main contaminants being adequately removed. This condition poses a major threat to human health and environmental sustainability. By 2030, it is anticipated that over 47% of the world's population would reside in regions with limited access to drinking water (Connor, 2015). This illustrates the growing worry about water scarcity and its impact on sustainable development and public health.

The textile industry is associated with grave environmental concerns, especially during the colouring process, which traditionally involves enormous quantities of water, chemicals, and energy, and results in immense air pollution. However, recent developments in environment-friendly dyeing techniques have offered promising prospects that mitigate these problems. Air dyeing is, hence, a new process that is characterized by great chemical waste and directed towards reducing water consumption while also boosting energy efficiency. The result being the lowered pollution discharge along with preservation of essential resources that lead to shine cleanliness in our waterways and impact less upon the environment. The biotechnology approach encourages an eco-friendlier colouring mode and minimizes the need for toxic substances also endorsing the basic sustainable chemistry principles. Water-free dyeing methods crown sustainability farther by taking water out of the whole dyeing loop. On an added front, these technologies save energy associated with heating and acquiring water, thus chipping in the global struggle against climate change by establishing a measure to control greenhouse gas emissions. In treating pigments under the supercritical phase of carbon dioxide, the process becomes pollution-free; hence, such techniques are required to usher in completeness in the tortuous line of textile dyeing along sustainable ethics.

Of these various techniques, air, bacterial dyeing and printing, waterless dyeing, and pollution-free dyeing can be combined to address with great efficiency the environmental issues flowing from textile dyeing practices. These activities can seriously reduce greenhouse emissions, cut air pollution, reduce the use of water and other chemicals, and save energy. Moving forward into an age of sustainability, it is crucial that these innovations find their way into customary practice. It is the convergence of manufacturers, researchers, and policymakers that would spur this change and ensure that they all work towards making sure that the textile industry is a concerned player in the global efforts for environmental conservation and climate change mitigation. These acts could really assist in establishing a much more righteous and sustainable textile industry that is supportive of both people and the environment.

References

Abate, M. T., & Tadesse, M. G. (2021). Airflow, foam, and supercritical carbon dioxide dyeing technologies. In *Innovative and emerging technologies for textile dyeing and finishing* (pp. 137–164). Wiley. https://doi.org/10.1002/9781119710288.ch5

Abdulkadir, N. (2017). Bacterial pigments and its significance. *MOJ Bioequivalence & Bioavailability, 4*(3), 285–288. https://doi.org/10.15406/mojbb.2017.04.00073

Acar, C., Dincer, I., & Mujumdar, A. (2022). A comprehensive review of recent advances in renewable-based drying technologies for a sustainable future. *Drying Technology, 40*(6), 1029–1050. https://doi.org/10.1080/07373937.2020.1848858

Agrawal, B., & Agrawal, B. J. (2015). Supercritical carbon-dioxide assisted dyeing of textiles: An environmental benign waterless dyeing process. *IJIRCT1201043 International Journal of Innovative Research and Creative Technology, 201*. www.ijirct.org. https://www.researchgate.net/publication/331312923

Akanyange, S. N., Lyu, X., Zhao, X., Li, X., Zhang, Y., Crittenden, J. C., et al. (2021). Does microplastic really represent a threat? A review of the atmospheric contamination sources and potential impacts. *Science of the Total Environment, 777*, 146020. https://doi.org/10.1016/j.scitotenv.2021.146020

Amal, A. M., Abeer, K. A., Samia, H. M., Nadia, A. E.-N. H., Ahmed, K. A., & El-Hennawi, H. M. (2011). Selection of pigment (melanin) production in Streptomyces and their application in printing and dyeing of wool fabrics. *Research Journal of Chemical Sciences, 1*(5), 22–28.

Asker, D., Beppu, T., & Ueda, K. (2008). Nubsella zeaxanthinifaciens gen. nov., sp. nov., a zeaxanthin-producing bacterium of the family Sphingobacteriaceae isolated from freshwater. *International Journal of Systematic and Evolutionary Microbiology, 58*(3), 601–606. https://doi.org/10.1099/ijs.0.65493-0

Bach, E., Cleve, E., & Schollmeyer, E. (n.d.). *Past, present and future of supercritical fluid dyeing technology—an overview. Historical survey*. Wiley.
Bai, T., Kobayashi, K., Tamura, K., Jun, Y., & Zheng, L. (2019, October). Supercritical CO_2 dyeing for nylon, acrylic, polyester, and casein buttons and their optimum dyeing conditions by design of experiments. *Journal of CO2 Utilization, 33*, 253–261. https://doi.org/10.1016/j.jcou.2019.05.013
Banchero, M. (2013). Supercritical fluid dyeing of synthetic and natural textiles—A review. *Coloration Technology, 129*(1), 2–17. https://doi.org/10.1111/cote.12005
Bauer, C., Jacques, P., & Kalt, A. (2001). Photooxidation of an azo dye induced by visible light incident on the surface of TiO_2. *Journal of Photochemistry and Photobiology A: Chemistry, 140*(1), 87–92. https://doi.org/10.1016/S1010-6030(01)00391-4
Bell, F., Hong, A., Danielescu, A., Maheshwari, A., Greenspan, B., Ishii, H., et al. (2021). Self-destaining textiles: Designing interactive systems with fabric, stains and light. In *Conference on Human Factors in Computing Systems—Proceedings*. https://doi.org/10.1145/3411764.3445155
Chanaphoo, J., Yuttitham, M., Vanitchung, S., & Hanpattanakit, P. (2021). Greenhouse gas emission from energy consumption in dyeing factory at Samut Prakan Province, Thailand. *Chemical Engineering Transactions, 89*, 73–78. https://doi.org/10.3303/CET2189013
Chatha, S. A. S., Asgher, M., & Iqbal, H. M. N. (2017, July). Enzyme-based solutions for textile processing and dye contaminant biodegradation—A review. *Environmental Science and Pollution Research, 24*(16), 14005–14018. https://doi.org/10.1007/s11356-017-8998-1
Chen, L., Wang, B., Ruan, X., Chen, J., & Yang, Y. (2015). Hydrolysis-free and fully recyclable reactive dyeing of cotton in green, non-nucleophilic solvents for a sustainable textile industry. *Journal of Cleaner Production, 107*, 550–556. https://doi.org/10.1016/j.jclepro.2015.05.144
Connor, R. (2015). *The United Nations world water development report 2015: Water for a sustainable world* (Vol. 1). UNESCO Publishing.
Da-In Kim, D. (2024). Beyond natural dyes: Embracing sustainable innovations in industrial textile dyeing. *Global Scientific Journals*. https://doi.org/10.26434/chemrxiv-2024-ztldb-v2
Damrath, M., Bhattacharjee, S., & Hoeher, P. A. (2021). Investigation of multiple fluorescent dyes in macroscopic air-based molecular communication. *IEEE Transactions on Molecular, Biological, and Multi-Scale Communications, 7*(2), 78–82. https://doi.org/10.1109/TMBMC.2021.3054877
Darshan, N., & Manonmani, H. K. (2015). Prodigiosin and its potential applications. *Journal of Food Science and Technology, 52*(9), 5393–5407. https://doi.org/10.1007/s13197-015-1740-4
de Oliveira, C. R. S., da Silva Júnior, A. H., Mulinari, J., & Immich, A. P. S. (2021). Textile re-engineering: Eco-responsible solutions for a more sustainable industry. *Sustainable Production and Consumption, 28*, 1232–1248. https://doi.org/10.1016/j.spc.2021.08.001
de Souza Mesquita, L. M., Martins, M., Pisani, L. P., Ventura, S. P. M., & de Rosso, V. V. (2021). Insights on the use of alternative solvents and technologies to recover bio-based food pigments. *Comprehensive Reviews in Food Science and Food Safety, 20*(1), 787–818. https://doi.org/10.1111/1541-4337.12685
Dewi, R. G., Parinderati, R., Hendrawan, I., Dewantoro, M. W. B., & Bayuwega, W. D. (2019, December) Energy Efficiency Monitoring in Textile Industries to Achieve GHG Emissions Reduction Target in Indonesia. In *IOP Conference Series: Earth and Environmental Science, Institute of Physics Publishing*. https://doi.org/10.1088/1755-1315/363/1/012019
L. Dufossé, , Patrick Galaup, Anina Yaron, Shoshana Malis Arad, Philippe Blanc, Kotamballi N. Chidambara Murthy et al., "Microorganisms and microalgae as sources of pigments for food use: A scientific oddity or an industrial reality?," Trends in Food Science and Technology, vol. 16, no. 9, pp. 389–406, 2005, doi: https://doi.org/10.1016/j.tifs.2005.02.006.
Dufossé, L., Fouillaud, M., Caro, Y., Mapari, S. A. S., & Sutthiwong, N. (2014). Filamentous fungi are large-scale producers of pigments and colorants for the food industry. *Current Opinion in Biotechnology, 26*, 56–61. https://doi.org/10.1016/j.copbio.2013.09.007
Dumancas, G., Bello, G., Hughes, J., & Diss, M. (2015, September). Comparison of chemometric algorithms for multicomponent analyses and signal processing: An example from

4-(2-Pyridylazo) resorcinol-metal colored complexes. *Recent Patents on Signal Processing, 4*(2), 106–115. https://doi.org/10.2174/2210686305666150817225812

Dunlap, C. A., Saunders, L. P., Schisler, D. A., Leathers, T. D., Naeem, N., Cohan, F. M., et al. (2016). Bacillus nakamurai sp. Nov., a black-pigment-producing strain. *International Journal of Systematic and Evolutionary Microbiology, 66*(8), 2987–2991. https://doi.org/10.1099/ijsem.0.001135

Durán, M., Ponezi, A. N., Faljoni-Alario, A., Teixeira, M. F. S., Justo, G. Z., & Durán, N. (2012). Potential applications of violacein: A microbial pigment. *Medicinal Chemistry Research, 21*(7), 1524–1532. https://doi.org/10.1007/s00044-011-9654-9

Elbadawi, A. M., & Pearson, J. S. (2003, December). Foam technology in textile finishing. *Textile Progress, 33*(4), 1–31. https://doi.org/10.1080/00405160308688960

Fried, R., Oprea, I., Fleck, K., & Rudroff, F. (2022, January 7). Biogenic colourants in the textile industry—a promising and sustainable alternative to synthetic dyes. *Royal Society of Chemistry*. https://doi.org/10.1039/d1gc02968a

Gobalakrishnan, M., Saravanan, D., & Das, S. (2020). *Sustainable finishing process using natural ingredients* (pp. 129–146). Springer. https://doi.org/10.1007/978-3-030-38545-3_5

Goñi, M. L., Gañán, N. A., Barbosa, S. E., Strumia, M. C., & Martini, R. E. (2017). Supercritical CO_2-assisted impregnation of LDPE/sepiolite nanocomposite films with insecticidal terpene ketones: Impregnation yield, crystallinity and mechanical properties assessment. *Journal of Supercritical Fluids, 130*, 337–346. https://doi.org/10.1016/j.supflu.2017.06.013

Goñi, M. L., Gañán, N. A., Martini, R. E., & Andreatta, A. E. (2018, March). Carvone-loaded LDPE films for active packaging: Effect of supercritical CO_2-assisted impregnation on loading, mechanical and transport properties of the films. *Journal of Supercritical Fluids, 133*, 278–290. https://doi.org/10.1016/j.supflu.2017.10.019

Grisales, C. M., Salazar, L. M., & Garcia, D. P. (2019, February). Treatment of synthetic dye baths by Fenton processes: Evaluation of their environmental footprint through life cycle assessment. *Environmental Science and Pollution Research, 26*(5), 4300–4311. https://doi.org/10.1007/s11356-018-2757-9

Haji, A., & Naebe, M. (2020). Cleaner dyeing of textiles using plasma treatment and natural dyes: A review. *Journal of Cleaner Production, 265*, 121866. https://doi.org/10.1016/j.jclepro.2020.121866

Haque, A. N. M. A., Sultana, N., Sayem, A. S. M., & Smriti, S. A. (2022, September 1). Sustainable adsorbents from plant-derived agricultural wastes for anionic dye removal: A review. *MDPI*. https://doi.org/10.3390/su141711098

Hasanbeigi, A., & Lawrence, E. O. (2013) *Alternative and emerging technologies for an energy-efficient, water-efficient, and low-pollution textile industry.*

Hori, T., & Kongdee, A. (2014). Dyeing of PET/co-PP composite fibers using supercritical carbon dioxide. *Dyes and Pigments, 105*, 163–166. https://doi.org/10.1016/j.dyepig.2014.01.023

Hossain, A. (2018, June). Pollution free dyeing on cotton fabric extracted from Swietenia macrophylla and Musa acuminata as unpolluted dyes and citrus. Limon (L.) as unpolluted mordanting agent. *Trends in Textile Engineering & Fashion Technology, 3*(2). https://doi.org/10.31031/tteft.2018.03.000558

Hossain, M. M., Islam, T., Jalil, M. A., Md. Rakibuzzaman, S., Surid, S. M., Zabed, M. R. I., et al. (2024). Advancements of eco-friendly natural antimicrobial agents and their transformative role in sustainable textiles. *SPE Polymers*. https://doi.org/10.1002/pls2.10135

Hu, Y., Guo, T., Ye, X., Li, Q., Guo, M., Liu, H., et al. (2013). Dye adsorption by resins: Effect of ionic strength on hydrophobic and electrostatic interactions. *Chemical Engineering Journal, 228*, 392–397. https://doi.org/10.1016/j.cej.2013.04.116

Islam, A., Teo, S. H., Taufiq-Yap, Y. H., & Ng, C. H. (2021). Step towards the sustainable toxic dyes and heavy metals removal and recycling from aqueous solution—A comprehensive review. *Resources, Conservation and Recycling, 175*(May), 105849. https://doi.org/10.1016/j.resconrec.2021.105849

Ito, K., Kuwabara, M., Fukunishi, K., & Fujiwara, Y. (1996). Application of clay-cationic dye intercalation to image fixation in thermal dye transfer printing. *Journal of Imaging Science and Technology, 40*(3), 275–280. https://doi.org/10.2352/j.imagingsci.technol.1996.40.3.art00015

Ivanova, E. P., Christen, R., Bizet, C., Clermont, D., Motreff, L., Bouchier, C., et al. (2009). Pseudomonas brassicacearum subsp. neoaurantiaca subsp. nov., orange-pigmented bacteria isolated from soil and the rhizosphere of agricultural plants. *International Journal of Systematic and Evolutionary Microbiology, 59*(10), 2476–2481. https://doi.org/10.1099/ijs.0.009654-0

Iyer, S. N., Behary, N., Nierstrasz, V., Guan, J., & Chen, G. (2019). Study of photoluminescence property on cellulosic fabric using multifunctional biomaterials riboflavin and its derivative Flavin mononucleotide. *Scientific Reports, 9*(1), 1–16. https://doi.org/10.1038/s41598-019-45021-5

Kant, R. (2012). Textile dyeing industry an environmental hazard. *Nature Science (Irvine), 04*(01), 22–26. https://doi.org/10.4236/ns.2012.41004

Karpova, E. E., Kunz, G. I., & Garner, M. B. (2021, July). *Sustainability in the textile and apparel industries*. https://doi.org/10.5040/9781501338700.ch-004

Khan, M. R. (2016). Pigment ink formulation, tests and test methods for pigmented textile inks. *Chemistry and Materials Research, 8*(8), 78–86.

Kim, H. J., Hong, J. P., Kwak, D. S., & Kwon, D. J. (2021, February). Acrylic polymer treatment on nylon fibers improves color strength and durability in thermal transfer printing. *Colloids and Surfaces A: Physicochemical and Engineering Aspects, 615*, 126268. https://doi.org/10.1016/j.colsurfa.2021.126268

Kumar, A., & Choudhury, R. (2018). *Eco-friendly dyes and dyeing*.

Kumar, A., Shankar Vishwakarma, H., Singh, J., & Kumar, M. (2015). Microbial pigments: Production and their applications in various industries. *International Journal of Pharmaceutical, Chemical, and Biological Sciences, 5*(1), 203–212.

Lee, J. H., Kim, Y. S., Choi, T. J., Lee, W. J., & Kim, Y. T. (2004). Paracoccus haeundaensis sp. nov., a gram-negative, halophilic, astaxanthin-producing bacterium. *International Journal of Systematic and Evolutionary Microbiology, 54*(5), 1699–1702. https://doi.org/10.1099/ijs.0.63146-0

Lee, W. K. et al. (2021). Performance Analysis of Ink for Digital Textile Printing Using Natural Indigo, *Textile Coloration and Finishing, Korean Society of Dyers and Finishers, 33*(4), 202–209. https://doi.org/10.5764/TCF.2021.33.4.202.

Lin, Y. B., Wang, X. Y., Fang, H., Ma, Y. N., Tang, J., Tang, M., et al. (2012). Streptomyces shaanxiensis sp. nov., a blue pigment-producing streptomycete from sewage irrigation soil. *International Journal of Systematic and Evolutionary Microbiology, 62*(8), 1725–1730. https://doi.org/10.1099/ijs.0.029959-0

Liu, M., Peng, F., Wang, Y., Zhang, K., Chen, G., & Fang, C. (2009). Kineococcus xinjiangensis sp. nov., isolated from desert sand. *International Journal of Systematic and Evolutionary Microbiology, 59*(5), 1090–1093. https://doi.org/10.1099/ijs.0.004168-0

Liu, Y., Zhang, Y., Yu, Z., Qi, C., Tang, R., Zhao, B., et al. (2020). Microbial dyes: Dyeing of poplar veneer with melanin secreted by Lasiodiplodia theobromae isolated from wood. *Applied Microbiology and Biotechnology, 104*(8), 3367–3377. https://doi.org/10.1007/s00253-020-10478-2

Mabuza, L., Sonnenberg, N., & Marx-Pienaar, N. (2023). Natural versus synthetic dyes: Consumers' understanding of apparel coloration and their willingness to adopt sustainable alternatives. *Resources, Conservation and Recycling Advances, 18*(April). https://doi.org/10.1016/j.rcradv.2023.200146

Mahmud, I., & Kaiser, S. (2020). Recent progress in waterless textile dyeing. *Journal of Textile Science Engineering, 10*(6), 1–3. https://doi.org/10.37421/jtese.2020.10.421

Manikprabhu, D., & Lingappa, K. (2013). γ Actinorhodin a natural and attorney source for synthetic dye to detect acid production of fungi. *Saudi Journal of Biological Sciences, 20*(2), 163–168. https://doi.org/10.1016/j.sjbs.2013.01.004

Mehta, M., Mansi, M., Gaurav, S., & Mehta Mansi, C. (2016). Impact factor: 5.2. *IJAR, 2*(2), 214–218. www.allresearchjournal.com

Melo, M. J. (2023) *History of natural dyes in the ancient Mediterranean civilization*. https://doi.org/10.1002/9781119811749.ch1

Metwally, R. A., El Sikaily, A., El-Sersy, N. A., Ghozlan, H. A., & Sabry, S. A. (2021). Antimicrobial activity of textile fabrics dyed with prodigiosin pigment extracted from marine

Serratia rubidaea RAM_Alex bacteria. *Egyptian Journal of Aquatic Research, 47*(3), 301–305. https://doi.org/10.1016/j.ejar.2021.05.004

Montero, G. A., Smith, C. B., Hendrix, W. A., & Butcher, D. L. (2000). Supercritical fluid technology in textile processing: An overview. *Industrial and Engineering Chemistry Research, 39*(12), 4806–4812. https://doi.org/10.1021/ie0002475

Morales-Oyervides, L., Oliveira, J., Sousa-Gallagher, M., Méndez-Zavala, A., & Montañez, J. C. (2017). Assessment of the dyeing properties of the pigments produced by Talaromyces spp. *Journal of Fungi, 3*(3). https://doi.org/10.3390/jof3030038

Mortensen, A. (2006). Carotenoids and other pigments as natural colorants. *Pure and Applied Chemistry, 78*(8), 1477–1491. https://doi.org/10.1351/pac200678081477

Nawwar, M., Sahu, R. P., Puri, I. K., & Zhitomirsky, I. (2020, March). Functionally decorated carbon nanotube networks for energy storage in supercapacitors. *Frontiers in Energy Research, 8*. https://doi.org/10.3389/fenrg.2020.00046

Nigam, P. S. N., & Pandey, A. (2009). Biotechnology for agro-industrial residues utilisation: Utilisation of agro-residues. In *Biotechnology for agro-industrial residues utilisation: Utilisation of agro-residues* (pp. 1–466). Springer. https://doi.org/10.1007/978-1-4020-9942-7

Niinimäki, K., & Hassi, L. (2011). Emerging design strategies in sustainable production and consumption of textiles and clothing. *Journal of Cleaner Production, 19*(16), 1876–1883. https://doi.org/10.1016/j.jclepro.2011.04.020

Önlü, N. (2017). Sustainability by design in textile products and fashion. *New Design Ideas, 1*(1), 37–49.

Ozturk, E., Cinperi, N. C., & Kitis, M. (2020, May). Improving energy efficiency using the most appropriate techniques in an integrated woolen textile facility. *Journal of Cleaner Production, 254*, 120145. https://doi.org/10.1016/j.jclepro.2020.120145

Peix, A., Berge, O., Rivas, R., Abril, A., & Velázquez, E. (2005). Pseudomonas argentinensis sp. nov., a novel yellow pigment-producing bacterial species, isolated from rhizospheric soil in Córdoba, Argentina. *International Journal of Systematic and Evolutionary Microbiology, 55*(3), 1107–1112. https://doi.org/10.1099/ijs.0.63445-0

Pizzicato, B., Pacifico, S., Cayuela, D., Mijas, G., & Riba-Moliner, M. (2023). Advancements in sustainable natural dyes for textile applications: A review. *Molecules, 28*(16), 1–22. https://doi.org/10.3390/molecules28165954

Pranta, A. D., & Rahaman, M. T. (2024, June). Extraction of eco-friendly natural dyes and biomordants for textile coloration: A critical review. *Nano-Structures and Nano-Objects, 39*, 101243. https://doi.org/10.1016/j.nanoso.2024.101243

Rahman, M. M., Mashud, M., & Rahman, M. M. (Eds.). *Textile science and clothing technology advanced technology in textiles fibre to apparel*. Springer.

Ribul, M., Lanot, A., Tommencioni Pisapia, C., Purnell, P., McQueen-Mason, S. J., & Baurley, S. (2021, December). Mechanical, chemical, biological: Moving towards closed-loop bio-based recycling in a circular economy of sustainable textiles. *Journal of Cleaner Production, 326*, 129325. https://doi.org/10.1016/j.jclepro.2021.129325

Schnitzer, H., Brunner, C., & Gwehenberger, G. (2007). Minimizing greenhouse gas emissions through the application of solar thermal energy in industrial processes. *Journal of Cleaner Production, 15*(13–14), 1271–1286. https://doi.org/10.1016/j.jclepro.2006.07.023

Sengupta, S., & Bhowal, J. (2023, March). Characterization of a blue-green pigment extracted from Pseudomonas aeruginosa and its application in textile and paper dyeing. *Environmental Science and Pollution Research, 30*(11), 30343–30357. https://doi.org/10.1007/s11356-022-24241-9

Shirata, A., Tsukamoto, T., Yasui, H., Hata, T., Hayasaka, S., Kojima, A., et al. (2000). Isolation of bacteria producing bluish-purple pigment and use for dyeing. *Japan Agricultural Research Quarterly, 34*(2), 131–140.

Tiwari, K. (2022). A potential application of microbial pigment: An alternative to synthetic dye & colourants. *International Journal of Current Science Research and Review, 05*(09), 3311–3321. https://doi.org/10.47191/ijcsrr/v5-i9-07

Uğur, Ş. S., & Sariişik, M. (2011). Electrostatic self-assembly dyeing of cotton fabrics. *Coloration Technology, 127*(6), 372–375. https://doi.org/10.1111/j.1478-4408.2011.00328.x

Varadarajan, G., & Venkatachalam, P. (2016). *Sustainable textile dyeing processes*. Springer. https://doi.org/10.1007/s10311-015-0533-3

Venil, C. (2009). An insightful overview on microbial pigment, prodigiosin. *Electronic Journal of Biology, 5*(3), 49–61.

Venil, C. K., Zakaria, Z. A., & Ahmad, W. A. (2013, July). *Bacterial pigments and their applications*. https://doi.org/10.1016/j.procbio.2013.06.006

Vickers, N. J. (2017). Animal communication: When I'm calling you, will you answer too? *Current Biology, 27*(14), R713–R715. https://doi.org/10.1016/j.cub.2017.05.064

Vignali, G., Reid, L. F., Ryding, D., & Henninger, C. E. (n.d.). *Technology-driven sustainability innovation in the fashion supply chain.*

Xiu, Y., Shen, Q., Fan, F., & Wang, C. (2014). Preparation of thermal transfer ink using disperse fluorescent yellow 82 for polyester substrates. *Pigment and Resin Technology, 43*(2), 92–96. https://doi.org/10.1108/PRT-10-2012-0073

Yılmaz, K., Aksu, İ. Ö., Göçken, M., & Demirdelen, T. (2024, September). Sustainable textile manufacturing with revolutionizing textile dyeing: Deep learning-based, for energy efficiency and environmental-impact reduction, pioneering green practices for a sustainable future. *Sustainability (Switzerland), 16*(18). https://doi.org/10.3390/su16188152

Yousaf, M. A., & Aqsa, R. (2023, October). Integrating circular economy, SBTI, digital LCA, and ESG benchmarks for sustainable textile dyeing: A critical review of industrial textile practices. *Global NEST Journal, 25*(7), 39–51. https://doi.org/10.30955/gnj.005145

Zahid, M., Papadopoulou, E. L., Suarato, G., Binas, V. D., Kiriakidis, G., Gounaki, I., et al. (2018). Fabrication of visible light-induced antibacterial and self-cleaning cotton fabrics using manganese doped TiO_2 nanoparticles. *ACS Applied Bio Materials, 1*(4), 1154–1164. https://doi.org/10.1021/acsabm.8b00357

Zaidy, S. S. H., Vacchi, F. I., Umbuzeiro, G. A., & Freeman, H. S. (2019). Approach to waterless dyeing of textile substrates—use of atmospheric plasma. *Industrial and Engineering Chemistry Research, 58*(40), 18478–18487. https://doi.org/10.1021/acs.iecr.9b01260

Zhang, Z., Wang, J., Yao, J., Zhi, L., & Gu, F. (2012). Experimental investigation on electrostatic spray applied in dyeing and finishing system. *Advanced Materials Research, 512–515*, 2221–2225. https://doi.org/10.4028/www.scientific.net/AMR.512-515.2221

Zheng, H., Xu, Y., Zhang, J., Xiong, X., Yan, J., & Zheng, L. (2017, February). An ecofriendly dyeing of wool with supercritical carbon dioxide fluid. *Journal of Cleaner Production, 143*, 269–277. https://doi.org/10.1016/j.jclepro.2016.12.115

Zhou, L., Wang, H., Du, J., Fu, J., & Wang, W. (2018, June). Eco-friendly and durable antibacterial cotton fabrics prepared with polysulfopropylbetaine. *Fibers and Polymers, 19*(6), 1228–1236. https://doi.org/10.1007/s12221-018-8053-y

Zhu, H. H., Guo, J., Yao, Q., Yang, S. Z., Deng, M. R., et al. (2007). Streptomyces vietnamensis sp. nov., a streptomycete with violet-blue diffusible pigment isolated from soil in Vietnam. *International Journal of Systematic and Evolutionary Microbiology, 57*(8), 1770–1774. https://doi.org/10.1099/ijs.0.64774-0

Zhu, H. H., Guo, J., Yao, Q., Yang, S. Z., Deng, M. R., & Li, T. H. (2011). Streptomyces caeruleatus sp. nov., with dark blue diffusible pigment. *International Journal of Systematic and Evolutionary Microbiology, 61*(3), 507–511. https://doi.org/10.1099/ijs.0.017392-0

Zhu, L., Chen, L., Wu, X., & Ding, X. (2018, August). Developing a greenhouse gas management evaluation system for Chinese textile enterprises. *Ecological Indicators, 91*, 470–477. https://doi.org/10.1016/j.ecolind.2018.04.029

Zhu, D., Wan, Z., Zhao, X., Liao, S., Wang, Q., Liu, L., et al. (2022, January). Foaming indigo: An efficient technology for yarn dyeing. *Dyes and Pigments, 197*, 109862. https://doi.org/10.1016/j.dyepig.2021.109862

The manufacturer's authorised representative in the EU is Springer
Nature Customer Service Centre GmbH, Europaplatz 3, 69115 Heidelberg,
Germany. If you have any concerns regarding our products, please
contact ProductSafety@springernature.com

Printed and bound by CPI Group (UK) Ltd, Croydon, CR0 4YY
01/10/2026
02225971-0006